MW00453507

# Typesetting Mathematics with LaTeX

# Typesetting Mathematics with LaTeX

Herbert Voss

UIT
CAMBRIDGE, ENGLAND

Published by
UIT Cambridge Ltd.
PO Box 145
Cambridge
CB4 1GQ
England

Tel: +44 1223 302 041
Web: www.uit.co.uk

ISBN 978-1-906860-17-2

Copyright © 2011 Herbert Voss
All rights reserved.
This book was previously published in German by
Lehmanns Media (**www.lob.de**) in 2009.

The right of Herbert Voss to be identified as the author of
this work has been asserted by him in accordance with the
Copyright, Designs and Patents Act 1988.

The programs and instructions in this book have been
included for their instructional value. Neither the publisher
nor the author offers any warranties or representations in
respect of their fitness for a particular purpose, nor do they
accept accept any liability for any loss or damage arising
from their use.

The publication is designed to provide accurate and
authoritative information in regard to the subject matter
covered. Neither the publisher nor the author makes any
representation, express or implied, with regard to the
accuracy of information contained in this book, nor do they
accept any legal responsibility or liability for any errors or
omissions that may be made. This work is supplied with the
understanding that UIT Cambridge Ltd and its authors are
supplying information, but are not attempting to render
engineering or other professional services. If such services
are required, the assistance of an appropriate professional
should be sought.

Many of the designations used by manufacturers and sellers
to distinguish their products are claimed as trade-marks.
UIT Cambridge Ltd acknowledges trademarks as the
property of their respective owners.

10 9 8 7 6 5 4 3 2 1

# Contents

# Preface

Donald Knuth developed TeX for his math books:

> "Mathematics books and journals do not look as beautiful as they used to. It is not that their mathematical content is unsatisfactory, rather that the old and well-developed traditions of typesetting have become too expensive. Fortunately, it now appears that mathematics itself can be used to solve this problem."
>
> (Donald E. Knuth: *Mathematical Typography*, 1978)

This led to the belief that TeX and LaTeX were only for math publications and were not particularly suited to anything else. In fact, nowadays arts and humanities users employ LaTeX for everything from simple assignments to complex publications.

However, it's still true that many people choose to use LaTeX because it remains one of the best systems in the world for ease and quality of typesetting math. This can be seen from the large number of PDF files with math content on the web that have been created with LaTeX. So far there has been only one book devoted exclusively to typesetting math, George Grätzer's *Math into LaTeX* [10] published in 2000 (and a later edition with the modified title *More Math into LaTeX* [11]). Since 2000, many new LaTeX packages for math typesetting have been developed, so a new book on how to do it seems to be in order.

This book is primarily a practical book; it provides solutions to a very wide range of real-world problems you are likely to meet. So, for example, the book covers fonts, but only to the extent that you will need when writing a LaTeX document; we do cover the internals of the commands used, but only briefly. The book emphasizes "learning by doing", so the examples chapter (??) is extensive.

This book deals mostly with LaTeX rather than TeX, because there are relatively few users of pure TeX. Nevertheless there is a specific chapter for the commands that are defined in TeX which are of course also available to the LaTeX user. First-time users of LaTeX can find introductions to TeX in various publications. For larger projects, more extensive and in-depth literature is required, for example [20].

Explaining a complex subject like LaTeX isn't simple. I am grateful to the following people who have suggested improvements: Hendri Adriaens, Andrea Blomenhofer, Alexander Boronka, Walter Brown, Ci Zhi-jia, Christian Faulhammer, José Luis Gômez Dans, Sebastian Hahn, Azzam Hassam, Martin Hensel, Morten Høgholm, M. Kalidoss, Dan Lasley, Angus Leeming, Tim Love, Dan Luecking, Hendrik Maryns, Heinz Mezera, David Neuway, Joachim Punter, Carl Riehm, Will Robertson, Christoph Rumsmüller, José Carlos Santos, Jens Schwaiger, Uwe Siart, Martin Sievers, Heiko Stamer, Uwe Stöhr, Carsten Thiel, David Weenink, Zou Yuan-Chuan and Michael Zedler. As always, Rolf Niepraschk has given valuable advice. Special thanks go to Monika Hattenbach; she critically checked the original German version and also provided many valuable corrections. And last but not least to Lars Kotthoff; he did the translating of the German version.

All examples in this book can be downloaded as runnable TeX documents from CTAN — `http://mirror.ctan.org/info/examples/Math-E`.

Berlin, August 2010                                                                 Herbert Voß

# Chapter 1

# Introduction

Almost all text processing programmes or typesetting systems use a special mode to handle math. This is primarily because of the special symbols that math requires, although the "eternal construction site", Unicode, may make it easier to access these symbols in the future.

In LaTeXyou need to use math mode to typeset most math symbols because the commands required for them are not available in text mode; for example you need the math mode command \pm for the plus/minus sign in $\pm 12\%$. Apart from symbols, there are other considerations when deciding whether to typeset in text or math mode, as different spacings and character sizes are used. Surprisingly few documents are simple enough to avoid any choices between typesetting in text or math mode — even the minus sign looks very different when set in text mode "-4.321" to when it's set in math mode "$-4.321$".

Getting typeset math and text to look right alongside one another isn't always easy. It can be quite difficult to find a matching math font $y$ for a text font $x$, particularly as typographic conventions for "matching" fonts vary over time. This book does not therefore attempt to explain how to "match" fonts but in Chapter 11 on page 227 you will find suggestions for possible combinations of text and math fonts. The differences between text and math modes, and the many ways of addressing them, complicate any explanation of LaTeX's math typesetting. The following chapters show what you must consider when typesetting math, from simple cases like $y = f(x)$ to relatively complex equations.

$$\iiint\limits_{\mathcal{G}} \left[ u\nabla^2 v - v\nabla^2 u \right] \mathrm{d}^3 V = \oiint\limits_{S} \left( u\frac{\partial v}{\partial n} - v\frac{\partial u}{\partial n} \right) \mathrm{d}^2 A \qquad (1.1)$$

01-00-1

$$\left| \frac{\hat{v}(s) - \hat{v}(t)}{|\tilde{D}u|([t,s[)} - \frac{f(\hat{u}(t) + \frac{\tilde{D}u}{|\tilde{D}u|}(t)|\tilde{D}u|([t,s[)) - f(\hat{u}(t))}{|\tilde{D}u|([t,s[)} \right| \leq K \left| \frac{\hat{u}(s) - \hat{u}(t)}{|\tilde{D}u|([t,s[)} - \frac{\tilde{D}u}{|\tilde{D}u|}(t) \right| \qquad (1.2)$$

01-00-2

$$|I_1| = \left| \int_{\Omega} gRu\,\mathrm{d}\Omega \right|$$

01-00-3

$$\leq C_3 \left[ \int_{\Omega} \left( \int_a^x g(\xi,t)\,\mathrm{d}\xi \right)^2 \mathrm{d}\Omega \right]^{1/2} \times \left[ \int_{\Omega} \left\{ u_x^2 + \frac{1}{k} \left( \int_a^x cu_t\,\mathrm{d}\xi \right)^2 \right\} \mathrm{d}\Omega \right]^{1/2}$$

$$\leq C_4 \left| \left| f \left| \tilde{S}_{a,-}^{-1,0} W_2(\Omega,\Gamma_l) \right| \right| \left| |u| \overset{\circ}{\to} W_2^{\tilde{A}}(\Omega;\Gamma_r,T) \right| \right|.$$

$$|I_2| = \left| \int_0^T \psi(t) \left\{ u(a,t) - \int_{\gamma(t)}^a \frac{\mathrm{d}\theta}{k(\theta,t)} \int_a^\theta c(\xi)u_t(\xi,t)\,\mathrm{d}\xi \right\} \mathrm{d}t \right|$$

$$\leq C_6 \left| \left| f \int_{\Omega} \left| \tilde{S}_{a,-}^{-1,0} W_2(\Omega,\Gamma_l) \right| \right| \left| |u| \overset{\circ}{\to} W_2^{\tilde{A}}(\Omega;\Gamma_r,T) \right| \right|.$$

$$(1.3')$$

# Chapter 2

# Math in inline mode with standard LATEX

## 2.1 Introduction

Inline mode lets you use math elements within a line of ordinary text like this: $f(x) = \int_a^b \frac{\sin x}{x} \mathrm{d}x$. The alternative, displayed mode, places math elements on their own line between lines of text; this is discussed in the next chapter. In both of these chapters we will describe commands that you can use in basic LATEX (i.e. without any additional packages). Most of these standard commands continue to work unchanged when optional packages are added; we will tell you about any exceptions when we come to them. In the following section we cover spacing and character sizing in math mode.

## 2.2 Using inline mode

To create inline elements, you can invoke math mode in three different ways:

```
\(...\)      \begin{math}...\end{math}      $...$
```

The outputs are identical in many cases, for example:

$\sum_{i=1}^{n} i = \frac{1}{2} n \times (n+1)$

$\sum_{i=1}^{n} i = \frac{1}{2} n \times (n+1)$

$\sum_{i=1}^{n} i = \frac{1}{2} n \times (n+1)$

```
\( \sum_{i=1}^{n}i=\frac{1}{2}\,n\times(n+1) \)
```

02-02-1

```
\begin{math}
   \sum_{i=1}^{n}i=\frac{1}{2}\,n\times(n+1)
\end{math}
```

```
$ \sum_{i=1}^{n}i=\frac{1}{2}\,n\times(n+1) $
```

However, there are subtle differences:

1. \(...\) — The \( command opening the environment is fragile (in the TeX sense i.e. it cannot usually be used within other commands such as section headings as it is expanded prematurely before being passed to the higher command (cf. Section 2.2.3 on the next page).
2. \begin{math}...\end{math} — This environment is identical to the first and, likewise, is fragile.
3. $...$ — The $ sequence is usually the best choice as it is robust (i.e. it can be used within other commands as its structure is preserved, not expanded by LaTeX before it is passed on to the higher command). However, within the alltt environment it is regarded as pure text and is not evaluated. So in the alltt environment it is better to use the \(...\) sequence, which does invoke math mode.

There is no formal restriction on the size and content of math elements that you can insert into an ordinary line of text using inline mode. However, large insertions can create layout issues, as here where we place a matrix on a line: $\underline{A} = \begin{bmatrix} a & b & c \\ d & e & f \\ g & h & i \end{bmatrix}$. So for a large element like a matrix it is usually better if you use displayed mode as described in Chapter 3. If you really must place a matrix within a line of ordinary text, the layout will look better if you use the smallmatrix environment from the amsmath package (cf. Section 6.3.5 on page 96): $\underline{A} = \begin{bmatrix} a & b & c \\ d & e & f \\ g & h & i \end{bmatrix}$. Using this environment causes less disruption to the line spacing.

## 2.2.1 Fractions and line height

In inline mode, to avoid increasing the line height, fractions (\frac) are set by default in the font style \scriptstyle (cf. Section 4.9 on page 55). This uses a slightly smaller font size for the fraction, which results in a better vertical spacing: $y = \frac{a}{b+1}$ ($y=\frac{a}{b+1}$). However, this may not be what you want if the numerator or the denominator describe important things that should be highlighted.

There are several ways to change the font size:

1. Use a displayed formula instead of inline mode. This usually defaults to a slightly larger font style for the fraction.
2. Change the font style with the \displaystyle command. This makes the fraction $y = \frac{a}{b+1}$ appear more readable, but increases the line spacing: $y = \dfrac{a}{b+1}$.
3. Use the \dfrac command from the amsmath package (cf. Section 6.7 on page 99). This changes to the larger font style by default.

02-02-2

text $y = \frac{a}{b+1} = \dfrac{a}{b+1}$ text

```
text $ y=\frac{a}{b+1}={\displaystyle\frac{a}{b+1}} $ text
```

Table 4.6 on page 56 shows the math styles recognized by LaTeX.

### 2.2.2  Integration limits and line height

In inline mode, to avoid increasing the line height, integration limits are set by default in the so-called superscript and subscript mode. This makes sense because it is the only way that an expression such as $\int_1^\infty \frac{1}{x^2}\,dx = 1$ can be placed within the normal vertical spacing. However, you can change the placement of the integration limits with the \limits command:

02-02-3

$$\int\limits_1^\infty \frac{1}{x^2}\,\mathrm{d}x = 1$$

```
$ \int\limits_{1}^{\infty}\frac{1}{x^2}\,\mathrm{d}x=1 $
```

The \limits command must follow immediately after a mathematical operator or LaTeX will output an error message.

Changing the placement of the limits again has a negative effect on the line height. This should usually be avoided because it does not improve the typographic appearance of the document. This is especially important if there are multiple limits, as discussed in Sections 4.4 on page 36 and 6.12 on page 108.

### 2.2.3  Math expressions in \part, \chapter, \section etc. headings, e.g. $f(x) = \prod_{i=1}^n \left(i - \frac{1}{2i}\right)$

Any command that is to be passed as an argument to another command, for example to a header (or table of contents or index) command, must be preserved intact, not be expanded by LaTeX before it is passed on to the header command (or in the case of the table of contents, before the information is written to the .toc file.

To insert math elements into a chapter or section heading, you can either use the robust $\dots$ syntax, or else if you need to use a fragile command you can insert the \protect command as described in Section 2.3 on page 11.

As an example, the source code for the heading of this section is:

```
\providecommand*\nxLcs[1]{\texttt{\textbackslash#1}}
\section[Math expressions in headings]% short form for the toc
  {Math expressions in \nxLcs{part},
  \nxLcs{chapter}, \nxLcs{section} etc.\ headings,
e.g.\ $f(x)=\prod_{i=1}^{n}\left(i-\frac{1}{2i}\right)$}
normal text\ldots
```

02-02-4

### 2.2.3   Math expressions in \part, \chapter, \section etc. headings, e.g. $f(x) = \prod_{i=1}^n \left(i - \frac{1}{2i}\right)$

normal text...

We used the optional argument to \section here to give a simple form of the heading text, without any math, to be used in the table of contents.

When you are typesetting headings for sections and chapters etc., remember that the math elements are usually not in bold; the boldface series declaration \bfseries that is used for the text in headings has no effect on the math part. In Section 4.18 on page 65 we show you how you can do this consistently.

Math mode can cause problems with bookmarks in PDF documents. Bookmarks can only contain pure text: math symbols are either ignored or else interpreted as text. Figure 2.1 shows this clearly; the interval limits, subscripts and powers are interpreted as normal text in the bookmarks, which then make no sense.

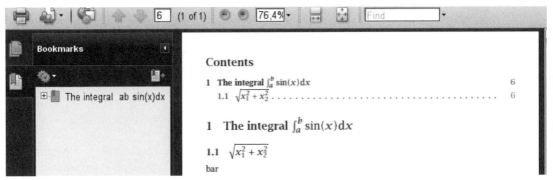

**Figure** 2.1: Headings that contain math elements are treated as text in bookmarks in PDF documents.

You can use the \texorpdfstring command from the hyperref package as a workaround. It requires two arguments, one for the TEX output and one for the bookmarks of the PDF output:

*hyperref*

> \texorpdfstring{*TEX output*}{*PDF output*}

This command makes it easy to specify full math notation to be used in a heading, but give a separate, text-only version for use in the corresponding bookmark.

If we make use of this command for our example from Figure 2.1, the coding and output become:

```
    \usepackage[linktocpage,colorlinks]{hyperref}
\tableofcontents
\section{The \texorpdfstring{integral $ \int_a^b \sin(x)\mathrm{d}x $}{sine integral}}
\subsection{\texorpdfstring{$\sqrt{x_1^2+x_2^2}$}{sqrt(x\_1\^{}2+x\_2\^{}2)}}
bar
```

## Contents

02-02-5

## 1 The integral $\int_a^b \sin(x)\mathrm{d}x$

**1.1** $\sqrt{x_1^2 + x_2^2}$

bar

The corresponding PDF output is shown in Figure 2.2 on the facing page.

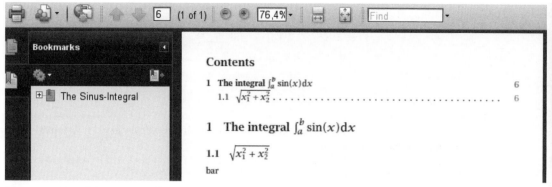

**Figure** 2.2: Use of the \texorpdfstring command allows different outputs for headings and bookmarks in PDF documents.

### 2.2.4  Frames

The \fbox command works whether its argument contains text or math, so you can use it to frame math expressions:

02-02-6

$$\boxed{f(x) = \prod_{i=1}^{n} \left(i - \tfrac{1}{2i}\right)}$$

```
\fbox{$f(x)=\prod_{i=1}^{n}
\left(i-\frac{1}{2i}\right)$}
```

The padding between the contents and the frame (\fboxsep) and the width of the frame line (\fboxrule) are implicit parameters; the default values are 3pt and 0.4pt. You can change these defaults at any time globally in the preamble of the document. You can also temporarily change them locally, leaving the global settings unchanged, by nesting in braces:

```
{\setlength\fboxsep{10pt}% keep local
\fbox{$f(x)=\prod_{i=1}^{n}\left(i-\frac{1}{2i}\right)$}} and
\setlength\fboxrule{1.5pt}
\fbox{$f(x)=\prod_{i=1}^{n}\left(i-\frac{1}{2i}\right)$}
```

02-02-7

$$\boxed{f(x) = \prod_{i=1}^{n} \left(i - \tfrac{1}{2i}\right)} \text{ and } \boxed{\boxed{f(x) = \prod_{i=1}^{n} \left(i - \tfrac{1}{2i}\right)}}$$

The \colorbox and \fcolorbox commands work like \fbox except that you must add colour specifications: only the colour specifications are added:

```
\colorbox{background colour}{contents}
\fcolorbox{frame colour}{background colour}{contents}
```

The following two examples show you how to use these commands and the output from them:

```
\usepackage{xcolor}
{ \setlength\fboxsep{10pt}%
  \colorbox{yellow}{$f(x)=\prod_{i=1}^{n}\left(i-\frac{1}{2i} \right)$} } and
{ \setlength\fboxrule{1.5pt}%
  \colorbox{cyan}{$f(x)=\prod_{i=1}^{n}\left(i-\frac{1}{2i} \right)$} }
```

02-02-8

$$f(x) = \prod_{i=1}^{n}\left(i - \frac{1}{2i}\right) \quad \text{and} \quad f(x) = \prod_{i=1}^{n}\left(i - \frac{1}{2i}\right)$$

```
\usepackage{xcolor}
{ \setlength\fboxsep{10pt}%
  \fcolorbox{cyan}{yellow}{$f(x)=\prod_{i=1}^{n}\left(i-\frac{1}{2i} \right)$} } and
{ \setlength\fboxrule{1.5pt}%
  \fcolorbox{cyan}{gray}{$f(x)=\prod_{i=1}^{n}\left(i-\frac{1}{2i} \right)$} }
```

02-02-9

$$f(x) = \prod_{i=1}^{n}\left(i - \frac{1}{2i}\right) \quad \text{and} \quad f(x) = \prod_{i=1}^{n}\left(i - \frac{1}{2i}\right)$$

### 2.2.5 Line breaks

In math mode LaTeX can insert a line break only if a relation symbol ($=, <, >, \ldots$) or a binary operation symbol ($+, -, \ldots$) occurs, and then only if the symbol appears outside any braces in the math formula. In other words a line break is possible in $a+b+c$, but not in ${a+b+c}$ because the elements are grouped at a deeper level due to the curly braces.

| | |
|---|---|
| Default line-breaking behaviour | $f(x) = a_n x^n + a_{n-1}x^{n-1} + a_{n-2}x^{n-2} + \ldots + a_i x^i +$ $a_2 x^2 + a_1 x^1 + a_0$ |
| The same expression in a so-called LaTeX group {...} | $f(x) = a_n x^n + a_{n-1}x^{n-1} + a_{n-2}x^{n-2} + \ldots + a_i x^i + a_2 x^2 + a_1 x^1$ |
| Without any of the symbols that allow a line break | $f(x) = a_n\left(a_{n-1}\left(a_{n-2}\left(a_{n-3}\left(a_{n-4}(\ldots)\ldots\right)\ldots\right)\ldots\right)\ldots\right)$ |
| Without any of the symbols that allow a line break, but with manual partitioning of the formula | $f(x) \qquad = \qquad a_n\left(a_{n-1}\left(a_{n-2}\left(a_{n-3}\left(a_{n-4}(\ldots)\ldots\right)\ldots\right.\right.\right.$ $\left.\left.\left.\right)\ldots\right)\ldots\right)$ |

The last example above shows that you can force a line break to occur manually; you simply break the formula into two separate, independent parts − $\ldots$ $ $\ldots$ $ and then LaTeX can insert a normal line break between them. One thing to take care over when you want to break a formula is that all sets of parentheses must be both opened and closed within each section of the formula. In the example above we had to add three additional \right. commands to the first section of the formula (the dot makes the bracket invisible in the output cf. Section 4.5 on page 38) and similarly three \left. commands to the second part of the formula. The coding for the example looked like this:

```
$f(x)=a_n\left(a_{n-1}\left(a_{n-2}\left(a_{n-3}\left(a_{n-4}
    \left(\ldots\right)\ldots\right) \ldots \right.\right.\right.$
$\left.\left.\left.\right) \ldots \right) \ldots \right)$
```

If you want to stop LaTeX inserting automatic line breaks globally, for a whole document or within a group, you can set the following values:

| | |
|---|---|
| `\relpenalty=9999` | These values suppress line breaks globally; long equations will |
| `\binoppenalty=9999` | always stick out into the right margin. |

For more discussion on splitting parentheses across line breaks, see Section 4.5.1 on page 41.

### 2.2.6 Horizontal white space

In inline mode, TEX adds a space of length `\mathsurround` before and after every formula. The default value for `\mathsurround` is 0pt. The following three examples illustrate this:

In the first example you can see that in inline mode the default is to have no space between the formula and the surrounding text. In the second example a line of width 20pt is inserted before and after the formula. In the last example the same horizontal distance to the surrounding text is achieved using `\setlength\mathsurround{`*20pt*`}`. (In the examples we have framed the math element, with the frame's padding (`\fboxsep`) set to zero, to show more clearly the difference in extra space due to changing the value of `\mathsurround`).

02-02-10

foo $\boxed{f(x) = \int_1^\infty \frac{1}{x^2}\mathrm{d}x = 1}$ bar

```
foo \fbox{$ f(x)=\int_1^{\infty}
    \frac{1}{x^2}\mathrm{d}x=1 $} bar
```

02-02-11

foo■$\boxed{f(x) = \int_1^\infty \frac{1}{x^2}\mathrm{d}x = 1}$■bar

```
foo\rule{20pt}{1.5ex}\fbox{%
    $ f(x)=\int_1^{\infty}
    \frac{1}{x^2} \mathrm{d}x=1 $}%
    \rule{20pt}{1.5ex}bar
```

02-02-12

foo$\boxed{\quad f(x) = \int_1^\infty \frac{1}{x^2}\mathrm{d}x = 1 \quad}$bar

```
\setlength\mathsurround{20pt}

foo\fbox{$ f(x)=\int_1^{\infty}
    \frac{1}{x^2}\mathrm{d}x=1 $}bar
```

## 2.3 TEXnicalities

In Section 2.2.3 on page 7 we mentioned that some commands are fragile (in TEX terminology: they incur premature expansion) and said that this problem can usually be averted by prefixing a `\protect`. Let's look at some examples of the effects of this fragility.

To make the examples concrete, we'll use the example file `protect.tex`, which contains the following code:

```
\documentclass{article}
\begin{document}
\tableofcontents
\section{The integral \( \int \sin(x)\mathrm{d}x \) }
foo
\end{document}
```

The first LaTeX or PDFLaTeX run completes without errors:

```
This is pdfTeX, Version 3.1415926-1.40.10 (TeX Live 2009)
entering extended mode
(./protect.tex
LaTeX2e <2009/09/24>
Babel <v3.81> and hyphenation patterns for english, usenglishmax, dumylang, noh
yphenation, german-x-2009-06-19, ngerman-x-2009-06-19, ancientgreek, ibycus, ar
abic, basque, bulgarian, catalan, pinyin, coptic, croatian, czech, danish, dutc
h, esperanto, estonian, farsi, finnish, french, galician, german, ngerman, mono
greek, greek, hungarian, icelandic, indonesian, interlingua, irish, italian, ku
rmanji, latin, latvian, lithuanian, mongolian, mongolian2a, bokmal, nynorsk, po
lish, portuguese, romanian, russian, sanskrit, serbian, slovak, slovenian, span
ish, swedish, turkish, ukenglish, ukrainian, uppersorbian, welsh, loaded.
(/usr/local/texlive/2009/texmf-dist/tex/latex/base/article.cls
Document Class: article 2007/10/19 v1.4h Standard LaTeX document class
(/usr/local/texlive/2009/texmf-dist/tex/latex/base/size10.clo))
No file protect.aux.
No file protect.toc.
[1] (./protect.aux) )
Output written on protect.dvi (1 page, 484 bytes).
Transcript written on protect.log.
```

If we look at the table-of-contents file `protect.toc` that was created during this first LATEX run, we see that the entire mathematical part of the heading has been expanded:[1]

```
\contentsline{section}{\numberline{1}The integral \relax $ \intop \nolimits
\mathop {\mathgroup \symoperators sin}\nolimits (x)\mathrm {d}x \relax
\GenericError { }{LaTeX Error: Bad math environment delimiter}{See the
LaTeX manual or LaTeX Companion for explanation.}{Your command was ignored.
\MessageBreak
Type I <command> <return> to replace it with another
command,\MessageBreak or <return> to continue without it.} }{1}
```

What has happened is that when the `\(` is written to the `.toc` file, it is correctly expanded to `\relax $`, but it is not executed. Therefore TEX stays in text mode and does not switch to math mode. When TEX reaches the closing `\)` the system is not in math mode, so TEX gives the error message, which halts any further processing.

During further LATEX runs, this error messages becomes visible to the user:

```
This is pdfTeXk, Version 3.1415926-1.40.9 (Web2C 7.5.7)
 %&-line parsing enabled.
entering extended mode
(./protect.tex
LaTeX2e <2005/12/01>
[ ... ]
Document Class: article 2005/09/16 v1.4f Standard LaTeX document class
(/usr/local/texlive/2008/texmf-dist/tex/latex/base/size10.clo)) (./protect.aux)
(./protect.toc

! LaTeX Error: Bad math environment delimiter.

See the LaTeX manual or LaTeX Companion for explanation.
Type  H <return>  for immediate help.
 ...
```

---

[1]The actual output in the `.toc` file appears on a single line.

```
1.1 ...k or <return> to continue without it.\} \}\{1\}
```

However, if a \protect is placed in front of both the \( and the \), the .toc file for the table of contents produced by the first LaTeX run now has the following contents[2] with the math mode commands preserved:

```
\contentsline {section}{\numberline {1}The integral \( \intop \nolimits
  \mathop {\mathgroup \symoperators sin}\nolimits (x)\@mathrm {d}x \) }{1}
```

We mentioned earlier in Section 2.2 on page 5 that you can avoid the problems described above by using the $ ... $ syntax; the basic TEX $ command is robust by definition because it is an active character directly assigned to the category "math shift" and therefore cannot be expanded any further:

```
\catcode'\$=3 % dollar sign is math shift
```

In contrast to that, the other two syntaxes for invoking math mode for inline math elements are defined in the main LaTeX file latex.ltx:

```
\def\({\relax\ifmmode\@badmath\else$\fi}
\def\){\relax\ifmmode\ifinner$\else\@badmath\fi\else \@badmath\fi}
\let\math=\(
\let\endmath=\)
```

---

[2]The actual output in the .toc file is again on a single line.

## 2.4 Examples

Table 2.1: Fractions, series, and products to approximate $\pi$.

| *author* | *year* | *expression* |
|---|---|---|
| Brahe | 1570 | $\pi \approx \dfrac{88}{\sqrt{785}}$ |
| Vieta | 1592 | $\dfrac{2}{\pi} = \sqrt{\dfrac{1}{2}} \times \sqrt{\dfrac{1}{2} + \dfrac{1}{2}\sqrt{\dfrac{1}{2}}} \times \sqrt{\dfrac{1}{2} + \dfrac{1}{2}\sqrt{\dfrac{1}{2} + \dfrac{1}{2}\sqrt{\dfrac{1}{2}}}} \times \cdots$ |
| Wallis | | $\dfrac{\pi}{2} = \dfrac{2 \times 2 \times 4 \times 4 \times 6 \times 6 \ldots}{1 \times 3 \times 3 \times 5 \times 5 \times 7 \ldots}$ |
| Brouncker | 1660 | $\dfrac{\pi}{4} = 1 + \cfrac{1^2}{2 + \cfrac{3^2}{2 + \cfrac{5^2}{2 + \cfrac{7^2}{2 + \cdots}}}}$ |
| Newton | 1666 | $\pi = \dfrac{3\sqrt{3}}{4} + 24 \times \left( \dfrac{1}{12} - \dfrac{1}{5 \times 2^5} - \dfrac{1}{28 \times 2^7} - \dfrac{1}{72 \times 2^9} - \cdots \right)$ |
| Gregory/Leibniz | 1673 | $\dfrac{\pi}{4} = 1 - \dfrac{1}{3} + \dfrac{1}{5} - \dfrac{1}{7} + \dfrac{1}{9} - \cdots$ |
| Gregory/Sharp | 1700 | $\dfrac{\pi}{6} = \dfrac{1}{\sqrt{3}} \left( 1 - \dfrac{1}{3 \times 3^1} + \dfrac{1}{5 \times 3^2} - \dfrac{1}{7 \times 3^3} + \cdots \right)$ |
| Euler | 1750 | $\dfrac{\pi^2}{6} = \dfrac{1}{1^2} + \dfrac{1^2}{2} + \dfrac{1^2}{3} + \dfrac{1^2}{4} + \dfrac{1^2}{5} + \cdots$ |
| | | $\dfrac{\pi^2}{8} = \dfrac{1}{1^2} + \dfrac{1}{3^2} + \dfrac{1}{5^2} + \dfrac{1}{7^2} + \dfrac{1}{9^2} + \cdots$ |
| | | $\dfrac{\pi^2}{12} = \dfrac{1}{1^2} - \dfrac{1}{2^2} + \dfrac{1}{3^2} - \dfrac{1}{4^2} + \dfrac{1}{5^2} - \cdots$ |
| Gauß | 1830 | $\pi = 48 \arctan \dfrac{1}{18} + 32 \arctan \dfrac{1}{57} - 20 \arctan \dfrac{1}{239}$ |
| Ramanujan | 1910 | $\dfrac{1}{\pi} = \dfrac{\sqrt{8}}{9801} \sum_{n=0}^{\infty} \dfrac{(4n!)(1103 + 26390n)}{(n!)^4 \times 396^{4n}}$ |

# Chapter 3

# Math in display mode with standard LaTeX

## 3.1 Introduction

Display mode, or displaymath mode, lets you place math elements in their own paragraph between lines of text like this:

$$f(x) = \int_a^b \frac{\sin x}{x} \mathrm{d}x$$

A displayed formula is offset in its own paragraph as opposed to inline mode where it is part of the current line of text (cf. Chapter 2 on page 5). In this chapter we will tell you about the standard LaTeX environments for display mode, and point out differences to the way you use commands in inline mode. However, bear in mind that standard LaTeX environments for display mode are now rarely used as the amsmath package, described in Chapter 6 on page 77, provides much better functionality. Nevertheless the traditional environments are still favoured by users who want to avoid loading the amsmath package.

## 3.2 Environments for display mode

### 3.2.1 Single-line environments

Use the \[ ...\] sequence to display a *single-line*, unnumbered formula.

> \[ ... *formula* ... \]

$$f(x) = \prod_{i=1}^{n} \left( i - \frac{1}{2i} \right)$$

```
\[
   f(x)=\prod_{i=1}^{n}\left(i-\frac{1}{2i} \right)
\]
```
03-02-1

Alternatively you could use the `displaymath` environment:

> ```
> \begin{displaymath}
> ...formula...
> \end{displaymath}
> ```

This is identical to the \[ ...\] sequence but, as its syntax is longer, it is rarely used in practice.

A normal line of text before the `displaymath` environment.

$$f(x) = \prod_{i=1}^{n} \left( i - \frac{1}{2i} \right)$$

A normal line of text after the `displaymath` environment.

```
A normal line of text before the
\texttt{displaymath} environment.
%
\begin{displaymath}
   f(x)=\prod_{i=1}^{n}\left(i-\frac{1}{2i}\right)
\end{displaymath}
%
A normal line of text after the
\texttt{displaymath} environment.
```
03-02-2

*Do not use $$...$$*  Avoid using the $$ ... $$ command from TeX, because in LaTeX it modifies the vertical spacing of a displayed formula.[1] Here is how it is defined in LaTeX:

```
\def\[{%
   \relax\ifmmode
      \@badmath
   \else
      \ifvmode
         \nointerlineskip
         \makebox[.6\linewidth]%
      \fi
      $$%%$$ BRACE MATCH HACK
   \fi
}
\def\]{%
   \relax\ifmmode
      \ifinner
         \@badmath
      \else
         $$%%$$ BRACE MATCH HACK
      \fi
```

---

[1] The TeX syntax should always be avoided as it has several disadvantages.

```
    \else
        \@badmath
    \fi
    \ignorespaces
}
```

Use the `equation` environment to display a single-line, numbered formula:

```
\begin{equation}
...formula...
\end{equation}
```

The equation number is right-justified by default.

<table>
<tr><td>03-02-3</td><td>

A normal line of text before the equation environment.

$$f(x) = \prod_{i=1}^{n} \left( i - \frac{1}{2i} \right) \qquad (3.1)$$

A normal line of text after the equation environment.

</td><td>

```
A normal line of text before the
\texttt{equation} environment.
%
\begin{equation}
  f(x)=\prod_{i=1}^{n}\left(i-\frac{1}{2i}\right)
\end{equation}
%
A normal line of text after the
\texttt{equation} environment.
```

</td></tr>
</table>

If you load the `amsmath` package, you can also use the starred version of the `equation`  *equation\** environment. Using `equation*` suppresses the output of the equation number. Standard LATEX, however, does not have a starred version.

<table>
<tr><td>03-02-4</td><td>

A normal line of text before the equation environment.

$$f(x) = \prod_{i=1}^{n} \left( i - \frac{1}{2i} \right)$$

A normal line of text after the equation environment.

</td><td>

```
\usepackage{amsmath}

A normal line of text before the
\texttt{equation} environment.
%
\begin{equation*}
  f(x)=\prod_{i=1}^{n}\left(i-\frac{1}{2i}\right)
\end{equation*}
%
A normal line of text after the
\texttt{equation} environment.
```

</td></tr>
</table>

### 3.2.2 TEXnicalities...

LATEX defines the `equation` environment differently to the short form \[ ... \]:

```
\def\equation{$$\refstepcounter{equation}}
\def\endequation{\eqno \hbox{\@eqnnum}$$\@ignoretrue}
```

Let's look at the different behaviour in the output of the two environments:

```
\begin{minipage}{0.49\linewidth}
The line above the equation.
```

```
\[ f(x)=\prod_{i=1}^n\left(i-\frac{1}{2i}\right) \]
```

```
The line below the equation.
\end{minipage}\hfill\begin{minipage}{0.49\linewidth}
The line above the equation.
```

```
\begin{equation} f(x)=\prod_{i=1}^{n}\left(i-\frac{1}{2i}\right) \end{equation}
```

```
The line below the equation.
\end{minipage}
```

The line above the equation.    The line above the equation.    03-02-5

$$f(x) = \prod_{i=1}^{n} \left(i - \frac{1}{2i}\right)$$    $$f(x) = \prod_{i=1}^{n} \left(i - \frac{1}{2i}\right) \tag{3.2}$$

The line below the equation.    The line below the equation.

You can see that the equation environment (on the right-hand side) has uneven vertical spacing above and below the equation. This is caused by the blank lines before and after the formula in the coding. The blank lines in the coding have no effect on the \[ ... \] sequence, however, as its LaTeX definition automatically corrects the spacing with the instruction \nointerlineskip.

If we remove the blank lines from the coding, both environments produce the same output:

```
\begin{minipage}{0.49\linewidth}
The line above the equation.
%
\[ f(x)=\prod_{i=1}^n\left(i-\frac{1}{2i}\right) \]
%
The line below the equation.
\end{minipage}\hfill\begin{minipage}{0.49\linewidth}
The line above the equation.
%
\begin{equation} f(x)=\prod_{i=1}^{n}\left(i-\frac{1}{2i}\right) \end{equation}
%
The line below the equation.
\end{minipage}
```

The line above the equation.    The line above the equation.    03-02-6

$$f(x) = \prod_{i=1}^{n} \left(i - \frac{1}{2i}\right)$$    $$f(x) = \prod_{i=1}^{n} \left(i - \frac{1}{2i}\right) \tag{3.3}$$

The line below the equation.    The line below the equation.

When you use the equation environment, make sure you do not have blank lines in the coding before and after displayed formulae to prevent uneven vertical spacing.

### 3.2.3 Multiline environments

In standard LaTeX the environment available for multiline formulae is the eqnarray environment. Its structure is defined as a matrix, consisting of three columns and an arbitrary number of rows. However, this environment has many deficiencies and you should avoid using it. [18]  *Do not use eqnarray!* Alternative and better environments are provided by the amsmath package, as we will discuss later in Chapter 6 on page 77. We are mentioning the eqnarray environment here only for completeness and to help you avoid the problems that it can cause.

The syntax of this environment is:

```
\begin{eqnarray}
...&...&...\\
...&...&...\\
...\\
...&...&...
\end{eqnarray}
```

In principle it is an array environment with column alignment rcl, which stands for "right — centred — left" (cf. Section 4.1 on page 32). If you needed to change this alignment, you would need to change the definition of the environment.[2] In addition math elements in the middle column are not typeset in the normal math style, but in \scriptstyle, which uses a slightly smaller font size (cf. Section 4.9 on page 55). You can see this in the example below, and it is worth bearing in mind when considering using eqnarray:

```
\begin{eqnarray}
\textrm{right-aligned} & \textrm{centred} & \textrm{left-aligned}\label{eq:eqnarray}\\
\frac{1}{\sqrt{n}} & \frac{\sqrt{n}}{n} & \frac{n}{n\sqrt{n}}
\end{eqnarray}
```

03-02-7

$$\text{right-aligned} \quad \text{centred} \quad \text{left-aligned} \tag{3.4}$$

$$\frac{1}{\sqrt{n}} \quad \frac{\sqrt{n}}{n} \quad \frac{n}{n\sqrt{n}} \tag{3.5}$$

As you can see, each line of the eqnarray environment has its own equation number. (We will show you how you can change the format of equation numbers in Section 3.4 on page 21.) Because of the varying font size, only use the eqnarray environment for a math element of the form $a = b$, i.e. where you can put a math expression in the first column of the matrix, a comparison or binary symbol in the central column, and then another math expression in the final column:

03-02-8

$$y = d \tag{3.6}$$
$$y = cx + d \tag{3.7}$$
$$y = bx^2 + cx + d \tag{3.8}$$
$$y = ax^3 + bx^2 + cx + d \tag{3.9}$$

Equations 3.6 and 3.9 have been furnished with a label such that they can be referenced in the text.

```
\begin{eqnarray}
y & = & d        \label{eq:2}\\
y & = & cx+d\\
y & = & bx^{2}+cx+d\\
y & = & ax^{3}+bx^{2}+cx+d\label{eq:5}
\end{eqnarray}
Equations~\ref{eq:2} and~\ref{eq:5} have
been furnished with a label such that
they can be referenced in the text.
```

---

[2]The definition can be found in the file $TEXMF/tex/latex/base/latex.ltx

You can suppress equation numbers by using the starred version eqnarray∗:

$$y = d$$
$$y = cx + d$$
$$y = bx^2 + cx + d$$
$$y = ax^3 + bx^2 + cx + d$$

As Equation 3.10 shows …

$$y = x^4 \qquad (3.10)$$

```
\begin{eqnarray*}
  y & = & d     \label{eq:3}\\
  y & = & cx+d              \\
  y & = & bx^{2}+cx+d       \\
  y & = & ax^{3}+bx^{2}+cx+d
\end{eqnarray*}
As Equation~\ref{eq:3} shows \ldots
\begin{eqnarray}
  y & = & x^4
\end{eqnarray}
```

03-02-9

If you just want to suppress numbering for individual lines, insert the command \nonumber **before** the end of the line:

$$y = d \qquad (3.11)$$
$$y = cx + d$$
$$y = bx^2 + cx + d$$
$$y = ax^3 + bx^2 + cx + d \qquad (3.12)$$

```
\begin{eqnarray}
  y & = & d                    \\
  y & = & cx+d        \nonumber \\
  y & = & bx^{2}+cx+d\nonumber \\
  y & = & ax^{3}+bx^{2}+cx+d
\end{eqnarray}
```

03-02-10

These examples illustrate the unappealing default layout of the eqnarray environment, with too much space before and after the equals sign. You can change the horizontal spacing manually by modifying the length \arraycolsep. The default value is 5.0pt.

$$f(x) = \int \frac{\sin x}{x}\,dx$$

```
\begin{eqnarray*}
f(x) & = & \int\frac{\sin x}{x}\,\mathrm{d}x
\end{eqnarray*}
```

03-02-11

$$f(x) = \int \frac{\sin x}{x}\,dx$$

```
\setlength\arraycolsep{1.4pt}
\begin{eqnarray*}
f(x) & = & \int\frac{\sin x}{x}\,\mathrm{d}x
\end{eqnarray*}
```

## 3.3 Short forms for macros

The commands described above are defined in the base file latex.ltx[3] as follows:

```
\def\displaymath{\[}
\def\enddisplaymath{\]\@ignoretrue}
\@definecounter{equation}
\def\equation{$$\refstepcounter{equation}}
\def\endequation{\eqno \hbox{\@eqnnum}$$\@ignoretrue}
```

Similarly, you can define custom short forms if you want to use shorter names for the environments or custom extensions. The LaTeX-internal command \@ifstar is also useful; it reads a starred argument if one is present and branches accordingly.

---

[3]$TEXMF/tex/latex/base/latex.ltx

In the following example, we will define \be and \ee as shorter commands for beginning and ending the equation environment, and we will incorporate the \@ifstar command to suppress equation numbers when required:

03-03-1

$$f(x) = \int \frac{\sin x}{x}\, \mathrm{d}x \qquad (3.13)$$

$$f(x) = \int \frac{\sin x}{x}\, \mathrm{d}x$$

```
\makeatletter
\newcommand\be{\@ifstar{\[}{\equation}}
\newcommand\ee{\@ifstar{\]}{\endequation}}
\makeatother

\be  f(x)=\int\frac{\sin x}{x}\,\mathrm{d}x \ee
\be* f(x)=\int\frac{\sin x}{x}\,\mathrm{d}x \ee*
```

Be careful of abbreviating environments, however, as it can often lead to problems when the user does not know exactly how environments are defined in LaTeX.

## 3.4  Equation numbers

Usually an equation number can be created for every line of an equation. This may be done or not done automatically, depending on the environment used, or individual lines may be excluded from the numbering with the macro \nonumber.

The following conditions apply for the numbering:

- If the document class is of type article, the numbering is done consecutively across all parts and sections. The counter can only be reset manually; we will show how to do this in the following sections.
- If the document class is of type book, the numbering is done chapter by chapter, i.e. the counter is reset when a new chapter starts. The number of the chapter is usually put in front of the number of the equation to be able to distinguish between them.
- In standard LaTeX there is a problem with the numbering of lines if the equation is too long. If this happens the number of the equation is output over the top of the actual equation. To correct this, the user has to manually intervene or use the amsmath package, which corrects the problem by placing the number below the equation (cf. Equation 06-11-2 on page 104).

We will discuss the numbering of "subequations" in Section 6.11.2 on page 107.

### 3.4.1  Changing the style of equation numbers

The \theequation command holds the style that is used to typeset the equation numbers and it is defined in the respective document class, for example in scrbook[4] as:

```
\renewcommand*\theequation{\thechapter.\arabic{equation}}
```

You can redefine the character style at any point in the document; it will not affect equations before the redefinition. In the following example we redefine the style from the Arabic number set to the Alpha character set. Then the numbering of what would have been equation "3.15" becomes "3-O" (O being the 15th character in the Alpha character set):

---

[4]http://www.ctan.org/tex-archive/macros/latex/contrib/koma-script/

$$f(x) = \int \frac{\sin x}{x} \, dx \qquad (3.14)$$

$$f(x) = \int \frac{\sin x}{x} \, dx \qquad (3\text{-}O)$$

```
\newcommand*\diff{\mathop{}\!\mathrm{d}}
```
03-04-1
```
\begin{equation}
f(x)= \int\frac{\sin x}{x}\diff x \end{equation}
\renewcommand*\theequation{%
    \thesection-\Alph{equation}}
\begin{equation}
f(x)= \int\frac{\sin x}{x}\diff x
\end{equation}
```

If you want to change the outward appearance rather than the numbering style, for example square brackets instead of parentheses, you need to redefine the \@eqnnum command. This has to be embedded in \makeatletter...\makeatother because of the @ sign in the command name. The normal definition is:

```
\def\@eqnnum{{\normalfont \normalcolor (\theequation)}}
```

Let's now redefine the command to output square brackets for the equation numbers:

$$f(x) = \int \frac{\sin x}{x} \, dx \qquad [3.16]$$

```
\makeatletter
\renewcommand*\@eqnnum{{%
    \normalfont\normalcolor[\theequation]}}
\makeatother
\newcommand*\diff{\mathop{}\!\mathrm{d}}

\begin{equation}
f(x)= \int\frac{\sin x}{x}\diff x
\end{equation}
```
03-04-2

amsmath    The above redefinition has no effect when the amsmath package is loaded, as a different command is responsible for the outward appearance; this is discussed in Section 6.11 on page 103.

### 3.4.2 Resetting counters

In a document class of type book, when a new chapter starts the equation counter is reset by default (though the chapter counter continues). This is independent of any redefinition of the equation numbering style shown in the previous section. The following Example 03-04-3 shows the default reset of the counter in action:

```
\newcommand*\diff{\mathop{}\!\mathrm{d}}
\chapter{foo}
The equations are counted chapter by chapter.
\begin{equation} f(x)= \int\frac{\sin x}{x}\diff x \end{equation}
\begin{equation} f(x)= \int\frac{1}{x^2}\diff x    \end{equation}
\chapter{bar}
The counting of the \texttt{equation}s is now reset because of the new chapter
and starts again.
\begin{equation} f(x)= \int\frac{\sin x}{x}\diff x \end{equation}
\begin{equation} f(x)= \int\frac{1}{x^2}\diff x    \end{equation}
```

03-04-3

**Chapter 1**

**foo**

The equations are counted chapter by chapter.

$$f(x) = \int \frac{\sin x}{x}\,dx \qquad (1.1)$$

$$f(x) = \int \frac{1}{x^2}\,dx \qquad (1.2)$$

**Chapter 2**

**bar**

The counting of the equations is now reset because of the new chapter and starts again.

$$f(x) = \int \frac{\sin x}{x}\,dx \qquad (2.1)$$

$$f(x) = \int \frac{1}{x^2}\,dx \qquad (2.2)$$

The chngcntr package lets you remove the reset switch. This is usually done by an according entry in the preamble of the document, but it can also be done at any position within the document. Here is an example:

```
\usepackage{chngcntr}  \counterwithout{equation}{chapter}
\newcommand*\diff{\mathop{}\!\mathrm{d}}
\chapter{foo}
\begin{equation} f(x)= \int\frac{\sin x}{x}\diff x \end{equation}
\begin{equation} f(x)= \int\frac{1}{x^2}\diff x \end{equation}
\chapter{bar}
The counting of the \texttt{equation}s is now not reset by the new chapter.
\begin{equation} f(x)= \int\frac{\sin x}{x}\diff x \end{equation}
\begin{equation} f(x)= \int\frac{1}{x^2}\diff x \end{equation}
```

03-04-4

**Chapter 1**

**foo**

$$f(x) = \int \frac{\sin x}{x}\,dx \qquad (1)$$

$$f(x) = \int \frac{1}{x^2}\,dx \qquad (2)$$

**Chapter 2**

**bar**

The counting of the equations is now not reset by the new chapter.

$$f(x) = \int \frac{\sin x}{x}\,dx \qquad (3)$$

$$f(x) = \int \frac{1}{x^2}\,dx \qquad (4)$$

The same effect can also be achieved with the `remreset` package. Remember when using this package for redefining the counter with the macro `\@removefromreset` that you have to embed this reset command in a `\makeatletter`...`\makeatother` sequence though.

### 3.4.3 Left-aligned equation numbers

Equation numbers are right-aligned by default for almost all document classes. The document class option `leqno` changes this globally for the whole document:

Normal text on the line, which is wrapped at some point.

$$(3.17) \qquad f(x) = x^2 + x - 1$$

$$(3.18) \qquad g(x) = \frac{x^2}{x+1}$$

$$(3.19) \qquad h(x) = \cosh(x-1)$$

Again normal text on the line.

```
\documentclass[leqno]{article}

Normal text on the line, which
is wrapped at some point.
%
\begin{equation}f(x)=x^2 +x-1
\end{equation}
\begin{equation}g(x)=\frac{x^2}{x+1}
\end{equation}
\begin{equation}h(x)=\mathrm{cosh}(x-1)
\end{equation}
%
Again normal text on the line.
```

03-04-5

If the `leqno` option is not available, you can achieve the same effect with the package `leqno`.[5] However, only use this in exceptional cases; in principle this package is obsolete — it only loads the file `leqno.clo` anyway. The extension `.clo` stands for "class option".

The `\@eqnnum` command that we mentioned in Section 3.4.1 on page 21 can also be redefined to let you set some individual equations to have left-aligned equation numbers, among otherwise right-aligned ones, by locally overriding the default right-align setting:

```
\documentclass{article}
\makeatletter
\newcommand*\Left{\begingroup\renewcommand\@eqnnum{\hb@xt@.01\p@{}%
    \rlap{\normalfont\normalcolor\hskip-\displaywidth(\theequation)}}}
\makeatother
Normal text in a line, which does not have any additional meaning.
%
\begin{equation} f(x)=x^2 +x-1          \end{equation}\Left
\begin{equation} g(x)=\frac{x^2}{x+1}    \end{equation}
\[      g^{\prime}(x)=\frac{2x(x+1)-x^2}{(x+1)^2}  \] \endgroup
\begin{equation} h(x)=\mathrm{cosh}(x-1) \end{equation}
%
Normal text in a line, which does not have any additional meaning.
```

---

[5]`http://www.ctan.org/tex-archive/macros/latex/unpacked/leqno.sty`

03-04-6

Normal text in a line, which does not have any additional meaning.

$$f(x) = x^2 + x - 1 \tag{3.20}$$

$$\tag{3.21} g(x) = \frac{x^2}{x+1}$$

$$g'(x) = \frac{2x(x+1) - x^2}{(x+1)^2}$$

$$h(x) = \cosh(x-1) \tag{3.22}$$

Normal text in a line, which does not have any additional meaning.

This option only works if the `amsmath` package is *not* loaded; amsmath redefines the  *amsmatl* `\@eqnnum` command itself so that your own manual redefinition has no effect (see Section 6.11.1 on page 106).

## 3.5 Labels

You can give any numbered equation or equation line an external label in addition to the internal number assigned to it by the counter. These labels can then be referred to in the text with a reference:

03-05-1

$$
\begin{array}{rcll}
y & = & d & (3.23) \\
y & = & cx + d & (3.24) \\
y & = & bx^2 + cx + d & (3.25) \\
y & = & ax^3 + bx^2 + cx + d & (3.26)
\end{array}
$$

In the coding we gave equations 3.23 and 3.26 a label so that they could be referred to in the text.

```
\begin{eqnarray}
y & = & d        \label{eq:2}\\
y & = & cx+d\\
y & = & bx^{2}+cx+d\\
y & = & ax^{3}+bx^{2}+cx+d\label{eq:5}
\end{eqnarray}
In the coding we gave equations~\ref{eq:2}
and~\ref{eq:5} a label so that they could
be referred to in the text.
```

Remember that:
- You can choose any label name provided that it does not contain spaces or any of the active TeX special characters. Altogether there are 10 characters that have a special meaning in LaTeX so they cannot be used arbitrarily in the text: { } # & _ % $ \ ^ ~. When you want to use them in the text they have to be masked, usually by preceding them with a backslash, as in \{ to output {. Alternatively, the commands from the \textcomp package can be used, for example \textasciicircum to output ^.
- When referred to in the text, the labels themselves are not output; TeX replaces them with the corresponding equation numbers.

Beware of trying to use labels in the `eqnarray*` environment where the equations are not numbered:

$$\begin{aligned} y &= d \\ y &= cx + d \\ y &= bx^2 + cx + d \\ y &= ax^3 + bx^2 + cx + d \end{aligned}$$

As Equation 3.27 shows ...

$$y = x^4 \qquad (3.27)$$

03-05-2

```
\begin{eqnarray*}
  y & = & d    \label{eq:3}\\
  y & = & cx+d              \\
  y & = & bx^{2}+cx+d       \\
  y & = & ax^{3}+bx^{2}+cx+d
\end{eqnarray*}
As Equation~\ref{eq:3} shows \ldots
\begin{eqnarray}
  y & = & x^4
\end{eqnarray}
```

In the coding we labelled the first equation, and then tried to refer to it in the text that followed with \ref{eq:3} as before. However, with the equation numbers suppressed, and therefore no internal counter associated with the equations, LaTeX ends up referencing the next numbered equation and the output makes no sense. LaTeX simply refers internally to the current value of the equation counter at the time when the label occurs and does not check whether that number actually applies to the labelled equation.

If instead of referring to the equations by their number you want to specify a symbol, for example (∗), you can do this by redefining \theequation for standard LaTeX.

```
\begin{equation}
  f(x)=\int_1^{\infty}\frac{1}{x^2}\mathrm{d}x=1
\end{equation}\renewcommand*\theequation{*}
%
\begin{equation}\label{eq:*}
  f(x)=\int_1^{\infty}\frac{1}{x^2}\mathrm{d}x=1
\end{equation}
%
When we want to refer to the equation labelled with a star we use
\verb+\ref{eq:*}+ in the coding and this results in
\ref{eq:*} --- the star is output correctly in the text.
```

$$f(x) = \int_1^\infty \frac{1}{x^2} \mathrm{d}x = 1 \qquad (3.28)$$

03-05-3

$$f(x) = \int_1^\infty \frac{1}{x^2} \mathrm{d}x = 1 \qquad (*)$$

When we want to refer to the equation labelled with a star we use \ref{eq:*} in the coding and this results in ∗ — the star is output correctly in the text.

*amsmath*    A much better method for labelling equations is provided in the amsmath package by the \tag command, which is described in Section 6.11.1 on page 106. The package also provides the \eqref command to reference such tags.

## 3.6 Frames

Displayed formulae can be framed in a similar way to math elements in inline mode by using the \fbox command (see Section 2.2.4 on page 9). However, within the \fbox command displayed formulae have to be put inside a \parbox or minipage because \fbox expects horizontal material:

```
Similar to inline mode, displayed formulae can be framed, but have to be \ldots
(This text is to illustrate the line width.)
```

```
\fbox{\parbox{\linewidth}{%
\begin{equation}
  f(x)=\int_1^{\infty}\frac{1}{x^2}\,\mathrm{d}x=1
\end{equation}}}
```

03-06-1

Similar to inline mode, displayed formulae can be framed, but have to be ... (This text is to illustrate the line width.)

$$f(x) = \int_1^\infty \frac{1}{x^2}\,\mathrm{d}x = 1 \tag{3.30}$$

The width of a displayed equation with an equation number is the same as the width of a whole line of text (or wider for long formulae); the \fbox command then allows space for padding on each side (\fboxsep), and for the width of the \fboxrule, resulting in a frame that is wider than the lines of text. It aligns with the left side, but the right side hangs outside the right-hand text margin.

The calc package corrects this by specifying the correct \parbox width through \linewidth-2\fboxsep-2\fboxrule.

```
\usepackage{calc}
Similar to inline mode, displayed formulae can be framed, but have to be \ldots
(This text is to illustrate the line width.)
```

```
\fbox{\parbox{\linewidth-2\fboxsep-2\fboxrule}{%
\begin{equation} f(x)=\int_1^{\infty}\frac{1}{x^2}\,\mathrm{d}x=1 \end{equation}}}
```

03-06-2

Similar to inline mode, displayed formulae can be framed, but have to be ... (This text is to illustrate the line width.)

$$f(x) = \int_1^\infty \frac{1}{x^2}\,\mathrm{d}x = 1 \tag{3.31}$$

By default the equation number is placed inside the frame, but you can change this, either by using additional packages (e. g. mathtool) or else by defining a new command; the trick is to place a numbered "phantom equation" alongside an unnumbered framed equation. It is there only to generate the equation number and is otherwise empty. The example \myMathBox command has four arguments altogether, two optional and two required:

**#1:** Optional argument for the frame colour; defaults to black.
**#2:** Optional argument for the background colour; defaults to white. This argument is only evaluated if the first optional argument is given as well. If only one optional argument is given, it is always assumed to be the frame colour.
**#3:** Mathematical object (or can be pure text).
**#4:** Label; required, but can be empty.

Let's look at four identical equations. The first has been typeset without a frame for comparison to the other three, which all use the \myMathBox command; the second uses one optional argument, so has a red frame; the third uses both optional arguments, to colour its frame and background; the fourth has no optional arguments, so has the default frame and background colours.

```
\usepackage{color,calc}
\makeatletter
\def\myMathBox{\@ifnextchar[{\my@MBoxi}{\my@MBoxi[black]}}
\def\my@MBoxi[#1]{\@ifnextchar[{\my@MBoxii[#1]}{\my@MBoxii[#1][white]}}
\def\my@MBoxii[#1][#2]#3#4{%
  \par\noindent\fcolorbox{#1}{#2}{%
      \parbox{\linewidth-1.5\labelwidth-2\fboxrule-2\fboxsep}{#3}%
   \parbox{1.5\labelwidth}{\begin{eqnarray}\label{#4}\end{eqnarray}}\par}
\makeatother
Similar to inline mode, displayed formulae can be framed, but have to be \ldots
(This text is to illustrate the line width.)
%
\begin{equation}
  f(x)=x^2 +x
\end{equation}
%
\myMathBox[red]{ \[f(x)=x^2 +x \]}{eq:frame3}
\myMathBox[red][yellow]{ \[f(x)=x^2 +x\] }{eq:frame4}
\myMathBox{ \[f(x)=x^2 +x\] }{eq:frame5}
```

Similar to inline mode, displayed formulae can be framed, but have to be . . . (This text is to illustrate the line width.)

03-06-3

$$f(x) = x^2 + x \tag{3.32}$$

$$\boxed{\quad f(x) = x^2 + x \quad} \tag{3.33}$$

$$\boxed{\quad f(x) = x^2 + x \quad} \tag{3.34}$$

$$\boxed{\quad f(x) = x^2 + x \quad} \tag{3.35}$$

The first equation is not in line horizontally with the other three equations because this equation is centred with respect to the whole width of the line, whereas the framed ones are centred within their frames.

Again, the amsmath package offers better solutions for framing displayed formulae (cf. $\mathcal{AMS}$ Section 6.16 on page 118). Another possibility is the empheq package by Morten Høgholm, which is described in detail in Section 9.12 on page 176.

Unnumbered equations cause a problem as the width of the equation is not initially known. The varwidth package by Donald Arseneau provides an elegant solution to this problem. Its varwidth environment expects the maximum width of the box as a parameter and then calculates the actual width internally.

```
\usepackage{varwidth}
\usepackage{calc}
Similar to inline mode, displayed formulae can be framed, but have to be \ldots
(This text is to illustrate the line width.)
\begin{center}
\fbox{\begin{varwidth}{\linewidth-2\fboxsep-2\fboxrule}
\[ f(x)=\int\limits_1^{\infty}\frac{1}{x^2}\,\mathrm{d}x=1 \]
\end{varwidth}}

\fbox{\begin{varwidth}{\linewidth-2\fboxsep-2\fboxrule}
\[ f(x)=\int\limits_1^{\infty}\frac{1}{x^2}\,\mathrm{d}x=1 \]
\vspace{\belowdisplayshortskip}
\end{varwidth}}
\end{center}
```

03-06-4  Similar to inline mode, displayed formulae can be framed, but have to be ... (This text is to illustrate the line width.)

$$\boxed{\quad f(x) = \int\limits_{1}^{\infty} \frac{1}{x^2}\,\mathrm{d}x = 1 \quad}$$

$$\boxed{\quad f(x) = \int\limits_{1}^{\infty} \frac{1}{x^2}\,\mathrm{d}x = 1 \quad}$$

Note that we enclosed the \fbox commands in a center environment; this ensured that the equations would be centred on the line, because as we mentioned earlier the \fbox command does not centre equations automatically across the whole line. Furthermore you can see in the first frame that the vertical spacing is not the same above and below the equation. In the second *amsmath* frame we corrected this manually using \vspace{\belowdisplayshortskip}; the amsmath package corrects the vertical spacing automatically.

## 3.7 Examples

The conservation principles for mass, torque, and energy can be given in differential or integral form: 03-07-1

**Differential form**

$$\frac{\partial \varrho}{\partial t} + \operatorname{div}(\varrho \vec{v}) = 0$$

$$\varrho \frac{\partial \vec{v}}{\partial t} + (\varrho \vec{v} \times \nabla) \vec{v} = \vec{f}_0 + \operatorname{div} \mathsf{T} = \vec{f}_0 - \operatorname{grad} p + \operatorname{div} \mathsf{T}' \tag{3.36}$$

$$\varrho T \frac{\mathrm{d}s}{\mathrm{d}t} = \varrho \frac{\mathrm{d}e}{\mathrm{d}t} - \frac{p}{\varrho} \frac{\mathrm{d}\varrho}{\mathrm{d}t} = -\operatorname{div} \vec{q} + \mathsf{T}' : \mathsf{D}$$

**Integral form**

$$\frac{\partial}{\partial t} \iiint \varrho \, \mathrm{d}^3 V + \oiint \varrho(\vec{v} \times \vec{v} ecn) \, \mathrm{d}^2 A = 0 \tag{3.37}$$

$$\frac{\partial}{\partial t} \iiint \varrho \vec{v} \, \mathrm{d}^3 V + \oiint \varrho \vec{v} (\vec{v} \times \vec{n}) \, \mathrm{d}^2 A = \iiint f_0 \, \mathrm{d}^3 V + \oiint \vec{n} \times \mathsf{T} \, \mathrm{d}^2 A \tag{3.38}$$

$$\frac{\partial}{\partial t} \iiint \left(\frac{1}{2}v^2 + e\right) \varrho \, \mathrm{d}^3 V + \oiint \left(\frac{1}{2}v^2 + e\right) \varrho \, (\vec{v} \times \vec{n}) \, \mathrm{d}^2 A = \tag{3.39}$$

$$- \oiint (\vec{q} \times \vec{v} ecn) \, \mathrm{d}^2 A + \iiint \left(\vec{v} \times \vec{f}_0\right) \mathrm{d}^3 V + \oiint (\vec{v} \times \vec{n} \, \mathsf{T}) \, \mathrm{d}^2 A.$$

The $\nabla$ operator in Cartesian coordinates:

$$\vec{\nabla} = \frac{\partial}{\partial x} \vec{e}_x + \frac{\partial}{\partial y} \vec{e}_y + \frac{\partial}{\partial z} \vec{e}_z \;,\; \operatorname{grad} f = \vec{\nabla} f = \frac{\partial f}{\partial x} \vec{e}_x + \frac{\partial f}{\partial y} \vec{e}_y + \frac{\partial f}{\partial z} \vec{e}_z$$

$$\operatorname{div} \vec{a} = \vec{\nabla} \times \vec{a} = \frac{\partial a_x}{\partial x} + \frac{\partial a_y}{\partial y} + \frac{\partial a_z}{\partial z} \;,\; \nabla^2 f = \frac{\partial^2 f}{\partial x^2} + \frac{\partial^2 f}{\partial y^2} + \frac{\partial^2 f}{\partial z^2}$$

$$\operatorname{rot} \vec{a} = \vec{\nabla} \times \vec{a} = \left(\frac{\partial a_z}{\partial y} - \frac{\partial a_y}{\partial z}\right) \vec{e}_x + \left(\frac{\partial a_x}{\partial z} - \frac{\partial a_z}{\partial x}\right) \vec{e}_y + \left(\frac{\partial a_y}{\partial x} - \frac{\partial a_x}{\partial y}\right) \vec{e}_z$$

# Math elements from standard LaTeX

The elements in this chapter can be used in inline mode as well as display mode. All the points discussed about inline mode in Chapter 2 on page 5 apply.

## 4.1 `array` environment

This environment is in principle the same as `eqnarray` (cf. Section **??** on page **??**), except that it allows a variable number of columns and that the whole formula only gets **one** equation number. The `array` environment does not activate math mode itself, so it must be used within another math command or environment. Its syntax is in fact the same as the `tabular` environment; TeX does not distinguish between the two internally:

> \begin{array} [position] {*column definition*}
> ...&...&...&...\\ [*line feed*]
> ...&...&...&...\\ \hline
> ...
> ...&...&...&...\\ \cline{*from—to*}
> ...&...&...&...
> \end{array}

Optional arguments are highlighted above. The default position parameter is c (centred) if nothing else is specified. We do not discuss column definitions in this book, but there is further explanation in [29]. You can also insert optional line feeds and lines; the latter do not have the same significance as in a table.

The following three examples are constructed to show the layout options achievable with the `array` environment; they do not really make math sense:

$$
\begin{array}{@{} rcl @{}}
\mathrm{right} & \mathrm{centre} & \mathrm{left} \\[10pt]
a & b & c \\ \hline
1 & 2 & 3 \\ \cline{2-3}
A & \multicolumn{2}{|c|}{BC}
\end{array}
\tag{4.1}
$$

04-01-1

```
\begin{equation}
\begin{array}{@{} rcl @{}}
\mathrm{right} & \mathrm{centre}
   & \mathrm{left} \\[10pt]
a & b & c                      \\\hline
1 & 2 & 3                      \\\cline{2-3}
A & \multicolumn{2}{|c|}{B C} \\\cline{2-3}
\end{array}
\end{equation}
```

To get a better horizontal spacing around the "'='" character the space between two columns is set to a smaller value.

```
\setlength\arraycolsep{1.4pt}
\[
  z=f(x,y) \rightarrow
  \begin{array}[b]{|rl|} y &=f(x)\\ y &= 3x^2      \end{array} \rightarrow
  \begin{array}[c]{|rl|} x &=f(u)\\ x &= \sqrt{u} \end{array} \rightarrow
  \begin{array}[t]{|rl|} v &=f(w)\\ v &= u^2-w^3  \end{array} \rightarrow
  f(w)
\]
```

04-01-2

$$
z = f(x,y) \rightarrow
\begin{vmatrix} y = f(x) \\ y = 3x^2 \end{vmatrix}
\rightarrow
\begin{vmatrix} x = f(u) \\ x = \sqrt{u} \end{vmatrix}
\rightarrow
\begin{vmatrix} v = f(w) \\ v = u^2 - w^3 \end{vmatrix}
\rightarrow f(w)
$$

```
\begin{equation}
\left.
\begin{array}{r@{\hspace{.75em}}ccrr}
  \textrm{a}) & y & = & d            & \textrm{(constant)}   \\[1pt]
  \textrm{b}) & y & = & cx+d         & \textrm{(linear)}     \\
  \textrm{c}) & y & = & bx^2+cx+d    & \textrm{(quadratic)}\\
  \textrm{d}) & y & = & ax^3+bx^2+cx+d & \textrm{(cubic)}
\end{array}
\right\} \textrm{polynomials}
\end{equation}
```

04-01-3

$$
\left.
\begin{array}{r c r l}
\text{a)} & y & = & d & \text{(constant)} \\
\text{b)} & y & = & cx+d & \text{(linear)} \\
\text{c)} & y & = & bx^2+cx+d & \text{(quadratic)} \\
\text{d)} & y & = & ax^3+bx^2+cx+d & \text{(cubic)}
\end{array}
\right\} \text{polynomials} \tag{4.2}
$$

The individual cells behave in the same way as in a normal `tabular` environment. As with the `eqnarray` environment, the distance between columns in the array can look too large. The same solution works: change the `\arraycolsep`. This is defined in `latex.ltx`, but its value is set in the respective document class; in the one used here it is set to 5.0pt. If instead we set `\arraycolsep` (locally) to a value similar to other environments for displayed formulae, the layout is much more appealing:

```
\setlength\arraycolsep{1.4pt}
\begin{equation}
\left.
\begin{array}{r@{\hspace{.75em}}ccrr}
  \textrm{a}) & y & = & d            & \textrm{(constant)}   \\[1pt]
  \textrm{b}) & y & = & cx+d         & \textrm{(linear)}     \\
  \textrm{c}) & y & = & bx^2+cx+d    & \textrm{(quadratic)}\\
  \textrm{d}) & y & = & ax^3+bx^2+cx+d & \textrm{(cubic)}
\end{array}
\right\} \textrm{polynomials}
\end{equation}
```

04-01-4

$$
\left.
\begin{array}{r l}
\text{a)} & y = d \quad \text{(constant)} \\
\text{b)} & y = cx+d \quad \text{(linear)} \\
\text{c)} & y = bx^2+cx+d \quad \text{(quadratic)} \\
\text{d)} & y = ax^3+bx^2+cx+d \quad \text{(cubic)}
\end{array}
\right\} \text{polynomials} \tag{4.3}
$$

If this is to be done globally, change the value in the preamble; if it only needs to apply locally, set it within a group with `\begingroup...\endgroup`.

Arrays are a particularly good way of typesetting case differentiations. The amsmath package has a specific `cases` environment, which is based on the `array` environment (cf. Section ?? on page ??). If you don't want to load this package, you can achieve the same effect yourself with the `array` environment:

```
\begin{equation}
  x=\left\{ \begin{array}{cl}
```

```
    0 & \textrm{if A=\ldots}\\
    1 & \textrm{if B=\ldots}\\
    x & \textrm{This text can be as long as you want it to be,
                    but it will not be wrapped\ldots}
\end{array}\right.
\end{equation}
```

$$
x = \left\{ \begin{array}{ll} 0 & \text{if A=\ldots} \\ 1 & \text{if B=\ldots} \\ x & \text{This text can be as long as you want it to be, but it will not be wrapped\ldots} \end{array} \right.
$$ (4.4)

04-01-5

Use a \parbox for the text if it is so long that it will need to wrap.

```
\begin{equation}
x = \left\{
  \begin{array}{ l p{0.7\textwidth} }
    0 & if A=\ldots\tabularnewline
    1 & if B=\ldots\tabularnewline
    x & \parbox[t]{0.75\columnwidth}{This text can be as long as you want it
    to be, and now it will even wrap onto several lines because of the
    \texttt{\textbackslash parbox} used. Isn't that nice. After all it doesn't
    look good if the text protrudes into the margin like it did before.}
  \end{array}
  \right.
\end{equation}
```

$$
x = \left\{ \begin{array}{ll} 0 & \text{if A=\ldots} \\ 1 & \text{if B=\ldots} \\ x & \text{This text can be as long as you want it to be, and now it will even wrap onto several lines because of the \parbox used. Isn't that nice. After all it doesn't look good if the text protrudes into the margin like it did before.} \end{array} \right.
$$ (4.5)

04-01-6

*delarray*  For more choice over which delimiters are used with an array, for example the type of braces, use the delarray package (cf. Section 9.10 on page 175).

## 4.2 Matrix

TeX has three alternative commands for typesetting a matrix: \bordermatrix, \matrix, and \pmatrix. These same commands are also available in standard LaTeX; the only thing to note
*\cr* ⚷  is that TeX expects \cr (carriage return) as a line ending. These commands work differently, however, when the amsmath package is loaded as it redefines them (cf. Section 6.3.5 on page 96). We are only covering the commands in standard LaTeX for completeness; the definitions in the amsmath package provide much better support for typesetting matrices.
*amsmath*  Unlike the array environment, with matrix commands the number of columns does not have to be defined beforehand; TeX determines it from the number of column delimiters:

<table>
<tr><td>04-02-1</td></tr>
</table>

$$\begin{matrix} A & B & C \\ d & e & f \\ 1 & 2 & 3 \end{matrix} \qquad \begin{pmatrix} A & B & C \\ d & e & f \\ 1 & 2 & 3 \end{pmatrix}$$

$$\begin{matrix} & 0 & 1 & 2 \\ 0 & \begin{pmatrix} A & B & C \\ d & e & f \\ 1 & 2 & 3 \end{pmatrix} \\ 1 \\ 2 \end{matrix}$$

```
$\matrix{ A & B & C \cr  d & e & f \cr
   1 & 2 & 3 }$ \qquad
$\pmatrix{ A & B & C \cr  d & e & f \cr
   1 & 2 & 3 }$ \\[5pt]
$\bordermatrix{
     & 0 & 1 & 2 \cr
   0 & A & B & C \cr
   1 & d & e & f \cr
   2 & 1 & 2 & 3 }$
```

When using \bordermatrix, more flexibility is available if you use the command's extended definition. This lets you align the outer column and row on the base and right of the matrix (by using the starred version of the command) or change the delimiters (by using an optional argument, which defaults to normal parentheses). To use this, the redefinition of the command must be put into the preamble of the document. It is not given here for space reasons, but you can see it in the source code of the example.

```
$\bordermatrix{ &  1 &  2 \cr 1 & x1 & x2 \cr 2 & x3 & x4 \cr 3 & x5 & x6 }$
$\bordermatrix[{[]}]{ &  1 &  2 \cr 1 & x1 & x2 \cr 2 & x3 & x4 \cr 3 & x5 & x6 }$
$\bordermatrix[\{\}]{ &  1 &  2 \cr 1 & x1 & x2 \cr 2 & x3 & x4 \cr 3 & x5 & x6 }$
$\bordermatrix*{ x1 & x2 & 1 \cr x3 & x4 & 2 \cr x5 & x6 & 3 \cr 1 &  2 }$

$\bordermatrix*[{[]}]{ x1 & x2 & 1 \cr x3 & x4 & 2 \cr x5 & x6 & 3 \cr 1 &  2 }$
$\bordermatrix*[\{\}]{ x1 & x2 & 1 \cr x3 & x4 & 2 \cr x5 & x6 & 3 \cr 1 &  2 }$
```

<table>
<tr><td>04-02-2</td></tr>
</table>

$$\begin{matrix} & 1 & 2 \\ 1 & \begin{pmatrix} x1 & x2 \\ x3 & x4 \\ x5 & x6 \end{pmatrix} \\ 2 \\ 3 \end{matrix} \quad \begin{matrix} & 1 & 2 \\ 1 & \begin{bmatrix} x1 & x2 \\ x3 & x4 \\ x5 & x6 \end{bmatrix} \\ 2 \\ 3 \end{matrix} \quad \begin{matrix} & 1 & 2 \\ 1 & \begin{Bmatrix} x1 & x2 \\ x3 & x4 \\ x5 & x6 \end{Bmatrix} \\ 2 \\ 3 \end{matrix} \quad \begin{matrix} \begin{pmatrix} x1 & x2 \\ x3 & x4 \\ x5 & x6 \end{pmatrix} & \begin{matrix} 1 \\ 2 \\ 3 \end{matrix} \\ 1 \quad 2 \end{matrix}$$

$$\begin{matrix} \begin{bmatrix} x1 & x2 \\ x3 & x4 \\ x5 & x6 \end{bmatrix} & \begin{matrix} 1 \\ 2 \\ 3 \end{matrix} \\ 1 \quad 2 \end{matrix} \quad \begin{matrix} \begin{Bmatrix} x1 & x2 \\ x3 & x4 \\ x5 & x6 \end{Bmatrix} & \begin{matrix} 1 \\ 2 \\ 3 \end{matrix} \\ 1 \quad 2 \end{matrix}$$

## 4.3 Roots

The \sqrt command typesets square roots (or any other root by using the command's optional argument):

> \sqrt [root exponent] {radicand}

<table>
<tr><td>04-03-1</td></tr>
</table>

$$\sqrt{x} \qquad \sqrt[3]{x} \qquad \sqrt[\pi]{\pi}$$

```
\large
$\sqrt{x} \quad  \sqrt[3]{x} \quad \sqrt[\pi]{\pi}$
```

If there are multiple root expressions after each other, the individual roots may have different heights, as in the following example:

$$\sqrt{a}\;\sqrt{T}\;\sqrt{2\alpha k_{B_1}T^i} \qquad (4.9)$$

```
\begin{equation}
\sqrt{a}\,\sqrt{T}\,\sqrt{2\alpha k_{B_1}T^i}
\end{equation}
```

`04-03-2`

This can be fixed with \vphantom, the vertical version of the \phantom command. This allows enough space for typesetting whatever expression is in its argument without actually printing it:

| \phantom{*expression*} | Reserves the space *expression* would take. |
|---|---|
| \hphantom{*expression*} | Reserves only the horizontal space. |
| \vphantom{*expression*} | Reserves only the vertical space. |

When typesetting multiple roots, use \vphantom commands within each \sqrt command to ensure that each one contains the highest and lowest elements. Then each root will be the same height. Here is a corrected version of the previous example:

$$\sqrt{a}\;\sqrt{T}\;\sqrt{2\alpha k_{B_1}T^i} \qquad (4.9)$$

```
\begin{equation}
\sqrt{a\vphantom{k_{B_1}T^{\scriptscriptstyle i}}}\,
\sqrt{T\vphantom{k_{B_1}T^{\scriptscriptstyle i}}}\,
\sqrt{2\alpha k_{B_1}T^i}
\end{equation}
```

`04-03-3`

The \smash command offers another possibility to optimise typesetting root expressions. The last one of the following series of equations shows its use in standard LaTeX; the result is rather bizarre here though. The amsmath package redefines this command so it can be used *amsmath* more effectively; this is shown in Section 6.8 on page 101.

$$a\sqrt{x}+b\sqrt{y}+c\sqrt{z}$$
$$a\sqrt{x}+b\sqrt{y}+c\sqrt{z}$$
$$a\sqrt{x}+b\sqrt{y}+c\sqrt{z}$$
$$a\sqrt{x}+b\sqrt{y}+c\sqrt{z}$$

```
$a\sqrt{x}+b\sqrt{y}+c\sqrt{z}$\\
$a\sqrt{\vphantom{y}x}+b\sqrt{\vphantom{x} y}
  +c\sqrt{\vphantom{y}z}$\\
$a\sqrt{\vphantom{yl}x}+b\sqrt{\vphantom{l} y}
  +c\sqrt{\vphantom{yl}z}$\\
$a\sqrt{x}+b\sqrt{\smash{y}}+c\sqrt{z}$
```

`04-03-4`

Roots are displayed with different symbols depending on their size; they are all listed in Table 11.5 on page 247. This simple demonstration makes this clearer:

$$\sqrt{1+\sqrt{1+\sqrt{1+\sqrt{1+\sqrt{1+\sqrt{1+x}}}}}}$$

```
\[
  \sqrt{1+ \sqrt{1+ \sqrt{1+ \sqrt{1
    + \sqrt{1 + \sqrt{1+x}}}}}}
\]
```

`04-03-5`

## 4.4 Super-/subscript and limits

There are several alternatives for typesetting indices (subscript):

$$a_{min} \leftrightarrow a_{max}\; a_{\min} \leftrightarrow a_{\max}$$
$$a_{\min} \leftrightarrow a_{\max};\; a_{\min} \leftrightarrow a_{\max}$$

```
$a_{min}\leftrightarrow a_{max}$
$a_{\mbox{min}}\leftrightarrow a_{\mbox{max}}$\\[3pt]
$a_{\mathrm{min}}\leftrightarrow a_{\mathrm{max}}$;
$ a_{\min}\leftrightarrow a_{\max}$
```

`04-04-1`

The same alternatives work for exponents. The last example uses the two predefined math operators \max and \min (cf. Section 4.13 on page 62). Multiple indices, or stacked indices, are fairly common in math for determining the sum or the product over two variables. (For general information about limits see Section 2.2.2 on page 7). The TEX command \atop can also be used in LATEX to stack limits. The syntax is:

{ above \atop below }

This command can be nested arbitrarily; the number of stacked symbols is theoretically unlimited — $\{a\atop\{b\atop c...\}\}$ etc. Example 04-04-2 shows this:

04-04-2

$$\sum_{\substack{1\le i\le p\\1\le j\le q\\1\le k\le r}} a_{ij}b_{jk}c_{ki} \qquad (4.6)$$

```
\begin{equation}
\sum_{1\le i\le p\atop {%
    {1\le j\le q\atop 1\le k\le r}}%
}a_{ij}b_{jk}c_{ki}
\end{equation}
```

Other ways to typeset multiple limits include using an array environment, but this can look strange due to the big column distance (cf. Section 4.1). The amsmath package provides additional useful commands such as \substack (cf. Section 6.12) and \underset and \overset *amsmath* (cf. Section 6.18). Layout problems can arise when indices are particularly wide. Example 04-04-3 illustrates this, using frames to show the layout distortion more clearly:

04-04-3

$$\boxed{\sum_{\substack{1\le i\le \sqrt{p^{-3}}\\1\le j\le q\\1\le k\le r}}}\boxed{a_{ij}b_{jk}c_{ki}} \qquad (4.7)$$

```
\begin{equation}
\fboxsep=0pt\fboxrule=0.1pt
\fbox{$\sum_{{1\le i\le \sqrt{p^{-3}}}\atop{
    {1\le j\le q\atop 1\le k\le r}}}$}
\fbox{$a_{ij}b_{jk}c_{ki}$}
\end{equation}
```

The amsmath package provides an elegant solution for this problem, described in Section 6.12 on page 108. If not using this package, the best option is to use the \makebox command.

To use it for limits, assign its optional width argument to 0pt; then the indices will be centred and will not affect the horizontal space to the next math expression. However, there is a new problem; \makebox does not know about the current math style. The first equation in Example 04-04-4 shows what can happen; the second one manually changes the font size to the correct font style.

```
\begin{equation}
\sum_{\makebox[0pt]{${{1\le i\le \sqrt{p^{-3}}}\atop{%
    {1\le j\le q\atop 1\le k\le r}}}$}}a_{ij}b_{jk}c_{ki}
\end{equation}
\begin{equation}
\sum_{\makebox[0pt]{${{\scriptscriptstyle 1\le i\le \sqrt{p^{-3}}}\atop{%
    {1\le j\le q\atop 1\le k\le r}}}$}}a_{ij}b_{jk}c_{ki}
\end{equation}
```

$$\sum_{\substack{1\le i\le \sqrt{p^{-3}} \\ 1\le j\le q \\ 1\le k\le r}} a_{ij}b_{jk}c_{ki}$$ (4.8)

04-04-4

$$\sum_{\substack{1\le i\le \sqrt{p^{-3}} \\ 1\le j\le q \\ 1\le k\le r}} a_{ij}b_{jk}c_{ki}$$ (4.9)

## 4.5 Brackets, braces, and parentheses

Typesetting parentheses at different levels and of different types is one of the standard problems in math typesetting. In LaTeX, the term parenthesis covers all of the following: ( ) [ ] \ / { } | ‖ ⌊ ⌋ ⌈ ⌉ ⟨ ⟩ ↑ ⇑ ↓ ⇓ ↕ ⇕

Parentheses symbols are usually used within a combination of \left and \right commands; the type of parenthesis symbol is the argument of the respective command. The advantage of typesetting parentheses within these commands is that LaTeX automatically tries to adjust the size to keep the expression in parentheses smaller than the parentheses themselves. This also happens with nested parentheses.[1] Every \left requires a corresponding \right; however, if you don't want the corresponding parenthesis to be typeset, specify a dot as the respective command's argument. For example to achieve a case differentiation, which only requires a left bracket, the coding is:

$$\text{sgn}(x) = \begin{cases} -1 & \text{for } x < 1 \\ 0 & \text{for } x = 0 \\ 1 & \text{for } x > 1 \end{cases}$$

```
\[ \mathrm{sgn}(x)=\left\{
   \begin{array}{rl}
      -1 & \textrm{for } x<1 \\
       0 & \textrm{for } x=0 \\
       1 & \textrm{for } x>1
   \end{array}  \right. \]
```

04-05-1

*curly braces*

The curly brace can only be assigned to the \left and \right commands if it is masked by the backslash (escaped). Sometimes you might want to specify a fixed size for the parentheses, either because LaTeX is not typesetting them appropriately or else because you want larger parentheses. Four commands are available for this:

| | |
|---|---|
| \big⟨parenthesis⟩ | \Big⟨parenthesis⟩ |
| \bigg⟨parenthesis⟩ | \Bigg⟨parenthesis⟩ |

The following five examples show all the parenthesis symbols that are defined as delimiters.[2] They can all be resized arbitrarily with the four commands listed above. (Example 04-05-2 with the "\left−\right" combinations is only given for comparison; the size of the symbols in this case is usually determined by the size of the enclosed environment.)

```
$\left(\right)~\left[\,\right]~\left<\right>~\left/\right|\left\backslash\right.~
\left\lmoustache\,\right\rmoustache~\left\arrowvert\right\vert~
\left\Vert\,\right\Arrowvert~\left\uparrow\right\downarrow~\left\updownarrow\right.~
\left\Uparrow\right\Downarrow~\left\Updownarrow\right.~\left\langle\right\rangle~
```

---

[1] Cf. Section 4.5.1 on page 41.
[2] Cf. Section 4.5.2 on page 43

```
\left\lbrace\right\rbrace~\left\lceil\,\right\rceil~\left\lfloor\,\right\rfloor~
\left\lgroup\right\rgroup~\left\bracevert\right.$
```

04-05-2

$$( ) \; [ \; ] \; \langle \; \rangle \; / | \backslash \; \int \Big\lbrace \; \| \; \|\| \; \uparrow\downarrow \; \updownarrow \; \Uparrow\Downarrow \; \Updownarrow \; \langle \; \rangle \; \{ \; \} \; \lceil \; \rceil \; \lfloor \; \rfloor \; \left[ \; \right] \; |$$

```
$\big(\big)~\big[\,\big]~\big<\big>~\big/\big|\big\backslash~
\big\lmoustache\,\big\rmoustache~\big\arrowvert\big\vert~
\big\Vert\,\big\Arrowvert~\big\uparrow\big\downarrow~\big\updownarrow~
\big\Uparrow\big\Downarrow~\big\Updownarrow~\big\langle\big\rangle~
\big\lbrace\big\rbrace~\big\lceil\,\big\rceil~\big\lfloor\,\big\rfloor~
\big\lgroup\big\rgroup~\big\bracevert$
%\end{LTXexample}
```

04-05-3

$$\Big( \Big) \; \Big[ \; \Big] \; \Big\langle \; \Big\rangle \; / | \backslash \; \int \Big\lbrace \; | \; \| \; \|\| \; \uparrow\downarrow \; \updownarrow \; \Uparrow\Downarrow \; \Updownarrow \; \langle \; \rangle \; \{ \; \} \; \lceil \; \rceil \; \lfloor \; \rfloor \; ( \; ) \; |$$

```
$\Big(\Big)~\Big[\,\Big]~\Big<\Big>~\Big/\Big|\Big\backslash~
\Big\lmoustache\,\Big\rmoustache~\Big\arrowvert\Big\vert~
\Big\Vert\,\Big\Arrowvert~\Big\uparrow\Big\downarrow~\Big\updownarrow~
\Big\Uparrow\Big\Downarrow~\Big\Updownarrow~\Big\langle\Big\rangle~
\Big\lbrace\Big\rbrace~\Big\lceil\,\Big\rceil~\Big\lfloor\,\Big\rfloor~
\Big\lgroup\Big\rgroup~\Big\bracevert$
```

04-05-4

$$\bigg( \bigg) \; \bigg[ \; \bigg] \; \bigg\langle \; \bigg\rangle \; / | \backslash \; \int \bigg\lbrace \; | \; \| \; \|\| \; \uparrow\downarrow \; \updownarrow \; \Uparrow\Downarrow \; \Updownarrow \; \langle \; \rangle \; \{ \; \} \; \lceil \; \rceil \; \lfloor \; \rfloor \; ( \; ) \; |$$

```
$\bigg(\bigg)~\bigg[\,\bigg]~\bigg<\bigg>~\bigg/\bigg|\bigg\backslash~
\bigg\lmoustache\,\bigg\rmoustache~\bigg\arrowvert\bigg\vert~
\bigg\Vert\,\bigg\Arrowvert~\bigg\uparrow\bigg\downarrow~\bigg\updownarrow~
\bigg\Uparrow\bigg\Downarrow~\bigg\Updownarrow~\bigg\langle\bigg\rangle~
\bigg\lbrace\bigg\rbrace~\bigg\lceil\,\bigg\rceil~\bigg\lfloor\,\bigg\rfloor~
\bigg\lgroup\bigg\rgroup~\bigg\bracevert$
```

04-05-5

$$\Bigg( \Bigg) \; \Bigg[ \; \Bigg] \; \Bigg\langle \; \Bigg\rangle \; / | \backslash \; \int \Bigg\lbrace \; | \; \| \; \|\| \; \uparrow\downarrow \; \updownarrow \; \Uparrow\Downarrow \; \Updownarrow \; \langle \; \rangle \; \{ \; \} \; \lceil \; \rceil \; \lfloor \; \rfloor \; ( \; ) \; |$$

```
$\Bigg(\Bigg)~\Bigg[\,\Bigg]~\Bigg<\Bigg>~\Bigg/\Bigg|\Bigg\backslash~
\Bigg\lmoustache\,\Bigg\rmoustache~\Bigg\arrowvert\Bigg\vert~\Bigg\Vert\,
    \Bigg\Arrowvert~
\Bigg\uparrow\Bigg\downarrow~\Bigg\updownarrow~\Bigg\Uparrow\Bigg\Downarrow~
\Bigg\Updownarrow~\Bigg\langle\Bigg\rangle~\Bigg\lbrace\Bigg\rbrace~
\Bigg\lceil\,\Bigg\rceil~\Bigg\lfloor\,\Bigg\rfloor~\Bigg\lgroup\Bigg\rgroup~
\Bigg\bracevert$
```

04-05-6

$$\Bigg( \Bigg) \; \Bigg[ \; \Bigg] \; \Bigg\langle \; \Bigg\rangle \; / | \backslash \; \int \Bigg\lbrace \; | \; \| \; \|\| \; \uparrow\downarrow \; \updownarrow \; \Uparrow\Downarrow \; \Updownarrow \; \langle \; \rangle \; \{ \; \} \; \lceil \; \rceil \; \lfloor \; \rfloor \; ( \; ) \; |$$

The above commands do not distinguish between left and right parenthesis, and do not need to be used in pairs; this gives you the flexibility to typeset expressions such as:

$$\bigg) \times \frac{a}{b} \times \bigg( \tag{4.11}$$

```
\begin{equation}
    \bigg)\times \frac{a}{b} \times\bigg(
\end{equation}
```

04-05-7

In this case the horizontal spaces between the math expression and the parentheses are a bit too large. To reduce the horizontal spacing use the left/right variants of the commands:

| | |
|---|---|
| \bigl⟨*parenthesis*⟩ | \bigr⟨*parenthesis*⟩ |
| \Bigl⟨*parenthesis*⟩ | \Bigr⟨*parenthesis*⟩ |
| \biggl⟨*parenthesis*⟩ | \biggr⟨*parenthesis*⟩ |
| \Biggl⟨*parenthesis*⟩ | \Biggr⟨*parenthesis*⟩ |

LATEX interprets the \biggl and \biggr commands as Mathopen symbols, inserting a smaller horizontal space between them and the math expression by default:

$$\bigg) \times \frac{a}{b} \times \bigg( \tag{4.12}$$

```
\begin{equation}
    \biggl)\times \frac{a}{b} \times\biggr(
\end{equation}
```

04-05-8

On the other hand the "m" variant, which stands for "middle", increases the horizontal spacing around the delimiter. It is again available for all the commands:

| | |
|---|---|
| \bigm⟨*parenthesis*⟩ | \Bigm⟨*parenthesis*⟩ |
| \biggm⟨*parenthesis*⟩ | \Biggm⟨*parenthesis*⟩ |

The following example shows the effect of using the "m" version on the spacing around the central vertical delimiter; the first expression uses the normal version of \bigg, while the second expression uses the "m" version:

$$\left(\frac{1}{3} \middle| \frac{3}{4}\right)$$
$$\left(\frac{1}{3} \middle| \frac{3}{4}\right)$$

```
\everymath{\displaystyle}

$\bigg(\frac{1}{3}\bigg|\frac{3}{4}\bigg)$\\
$\bigg(\frac{1}{3}\biggm|\frac{3}{4}\bigg)$
```

04-05-9

**Table 4.1**: Examples for the different "big" commands

| symbol | code | example | code |
|---|---|---|---|
| ( ) | () | $3\left(a^2 + b^{c^2}\right)$ | 3\Big( a^2+b^{c^2}\Big) |
| [ ] | [] | $3\left[a^2 + b^{c^2}\right]$ | 3\Big[ a^2+b^{c^2}\Big] |
| / \ | /\backslash | $3\big/a^2 + b^{c^2}\big\backslash$ | 3\Big/<br>a^2+b^{c^2}\Big\backslash |
| { } | \{\} | $3\left\{a^2 + b^{c^2}\right\}$ | 3\Big\{ a^2+b^{c^2}\Big\} |

continued...

... continued

| symbol | code | example | code |
|--------|------|---------|------|
| \| ‖ | \|\Vert | $3\big\lvert a^2+b^{c^2}\big\rVert$ | `3\Big|a^2+b^{c^2}\Big\Vert` |
| ⌊ ⌋ | \lfloor \rfloor | $3\lfloor a^2+b^{c^2}\rfloor$ | `3\Big\lfloor a^2+b^{c^2}` `\Big\rfloor` |
| ⌈ ⌉ | \lceil\rceil | $3\lceil a^2+b^{c^2}\rceil$ | `3\Big\lceil a^2+b^{c^2}` `\Big\rceil` |
| ⟨ ⟩ | \langle\rangle | $3\langle a^2+b^{c^2}\rangle$ | `3\Big\langle` `a^2+b^{c^2}\Big\rangle` |
| ↑ ⇑ | \uparrow \Uparrow | $3\lvert a^2+b^{c^2}\Uparrow$ | `3\Big\uparrow` `a^2+b^{c^2}\Big\Uparrow` |
| ↓ ⇓ | \downarrow \Downarrow | $3\lvert a^2+b^{c^2}\Downarrow$ | `3\Big\downarrow a^2+b^{c^2}` `\Big\Downarrow` |
| ↕ ⇕ | \updownarrow \Updownarrow | $3\lvert a^2+b^{c^2}\Updownarrow$ | `3\Big\updownarrow a^2+b^{c^2}` `\Big\Updownarrow` |

## 4.5.1  Examples

**Parentheses across line breaks**

The following equation is too wide to fit on one line; it protrudes into the margins and the equation number is forced to appear underneath it.[3]

```
\everymath{\displaystyle}
\begin{equation}
\frac{1}{2}\,\Delta(f_{ij}\,f^{ij})=2\times\left(\sum_{i<j}\chi_{ij}
    \left(\sigma_{i}-\sigma_{j}\right)^{2}+f^{ij}\nabla_{j}\nabla_{i}
    (\Delta f)+\nabla_{k}f_{ij}\nabla^{k}f^{ij}+f^{ij}f^{k}
    \left[2\nabla_{i}R_{jk}-\nabla_{k}R_{ij}\right]\right)
\end{equation}
```

04-05-10

$$\frac{1}{2}\Delta(f_{ij}\,f^{ij}) = 2\times\left(\sum_{i<j}\chi_{ij}\left(\sigma_i-\sigma_j\right)^2 + f^{ij}\nabla_j\nabla_i(\Delta f) + \nabla_k f_{ij}\nabla^k f^{ij} + f^{ij}f^k\left[2\nabla_i R_{jk} - \nabla_k R_{ij}\right]\right)$$

(4.13)

Using an `array` environment (cf. Section 4.1 on page 32) is one way to split an equation across two lines, by splitting the `\left[ ... \right]` coding into two parts: `\left[ ... \right.` and `\left. ... \right]`.

```
\setlength\arraycolsep{2pt}
\begin{equation}
\begin{array}{rcl}
```

---

[3]Depending on the document class it may also be the case that the equation number appears in the normal place and therefore overwrites the equation itself.

```
\displaystyle % to achieve equal sizes
\frac{1}{2}\,\Delta(f_{ij}\,f^{ij}) & = & 2\times\left(\sum\limits_{i<j}\chi_{ij}
    \left(\sigma_{i}-\sigma_{j}\right)^{2}+f^{ij}\nabla_{j}\nabla_{i}
    (\Delta f) \right. \\[8pt]
    & & {}+\left.\nabla_{k}f_{ij}\nabla^{k}f^{ij}+f^{ij}f^{k}
    \left[2\nabla_{i}R_{jk}-\nabla_{k}R_{ij}\right]\right)
\end{array}
\end{equation}
```

$$\frac{1}{2}\Delta(f_{ij}\,f^{ij}) = 2\times\left(\sum_{i<j}\chi_{ij}\left(\sigma_i-\sigma_j\right)^2 + f^{ij}\nabla_j\nabla_i(\Delta f)\right.$$
$$\left. + \nabla_k f_{ij}\nabla^k f^{ij} + f^{ij}f^k\left[2\nabla_i R_{jk}-\nabla_k R_{ij}\right]\right) \tag{4.14}$$

04-05-11

However, the height of each row of the `array` is determined only by the contents of that row's cells. The second line does not know about the height of the first line, so the final parenthesis is not the same size as the corresponding opening one. Use the \Bigg command to solve this problem:

```
\setlength\arraycolsep{2pt}
\begin{equation}
\begin{array}{rcl}
\displaystyle % same sizes
\frac{1}{2}\,\Delta(f_{ij}\,f^{ij}) & = & 2\times\Bigg(\sum\limits_{i<j}\chi_{ij}
    \left(\sigma_{i}-\sigma_{j}\right)^{2}+f^{ij}\nabla_{j}\nabla_{i}
    (\Delta f) \\[8pt]
    & & {}+\nabla_{k}f_{ij}\nabla^{k}f^{ij}+f^{ij}f^{k}
    \left[2\nabla_{i}R_{jk}-\nabla_{k}R_{ij}\right]\Bigg)
\end{array}
\end{equation}
```

$$\frac{1}{2}\Delta(f_{ij}\,f^{ij}) = 2\times\Bigg(\sum_{i<j}\chi_{ij}\left(\sigma_i-\sigma_j\right)^2 + f^{ij}\nabla_j\nabla_i(\Delta f)$$
$$+ \nabla_k f_{ij}\nabla^k f^{ij} + f^{ij}f^k\left[2\nabla_i R_{jk}-\nabla_k R_{ij}\right]\Bigg) \tag{4.15}$$

04-05-12

Another way to split an equation across two lines is to break it into smaller parts and treat it as a set of equations, using a multiline environment. Using the amsmath package is the best *amsmath* way of doing this (cf. Section 6.3.2 on page 89), but here is an example using the eqnarray environment from standard LᴬTEX, as discussed in Section ?? on page ??. Note that equation numbers are suppressed until the last line:

```
\setlength\arraycolsep{2pt}
\begin{eqnarray}
  B(r,\phi,\lambda) & = & \,\frac{\mu}{r}
    \Bigg[\sum_{n=2}^{\infty}\Bigg(\left(\frac{R_e}{r}\right)^n J_nP_n(s\phi)\nonumber\\
    & & {}+\sum_{m=1}^n \left( \frac{R_e}{r} \right) ^n
    (C_{nm}\cos m\lambda+S_{nm}\sin m\lambda)P_{nm}(s\phi) \Bigg)\Bigg]
\end{eqnarray}
```

04-05-13

$$B(r,\phi,\lambda) = \frac{\mu}{r}\left[\sum_{n=2}^{\infty}\left(\left(\frac{R_e}{r}\right)^n J_n P_n(s\phi)\right.\right.$$

$$\left.\left.+ \sum_{m=1}^{n}\left(\frac{R_e}{r}\right)^n (C_{nm}\cos m\lambda + S_{nm}\sin m\lambda)P_{nm}(s\phi)\right)\right] \tag{4.16}$$

### "Middle bar"

The \middle command works similarly to the \left and \right commands, though can appear on its own. However, it does require the amsmath package (cf. Chapter 6).

 *amsmath*

```
\usepackage{mathrsfs}
\begin{equation}
\mathcal{W} := \left\{ \bigcup^\infty_{j=0} W_j \; \middle| \;
  W_j \subset \tilde P_{*j} \textrm{ open},\; W_j \subset W_{j+1}\right\}
\end{equation}
```

04-05-14

$$\mathcal{W} := \left\{\bigcup_{j=0}^{\infty} W_j \;\middle|\; W_j \subset \tilde{P}_{*j} \text{ open}, \; W_j \subset W_{j+1}\right\} \tag{4.17}$$

For further examples of the use of vertical bars, see Section 9.7 on page 172 about the braket package.

### 4.5.2  Defining delimiters

The syntax of the \DeclareMathDelimiter command for defining delimiters is:

\DeclareMathDelimiter{*name/symbol*}{*type*}{*font1*}{*No.*}{*font2*}{*No.*}

*Font1* applies to the small symbols and *Font2* to the large ones. If no small symbol is available, the symbol from the "largesymbols" is stated twice; this is what causes the size problems discussed earlier in this section. The delimiters (or parenthese) in LaTeX are defined in the file fontmath.ltx.[4]

```
\DeclareMathDelimiter{(}{\mathopen} {operators}{"28}{largesymbols}{"00}
\DeclareMathDelimiter{)}{\mathclose}{operators}{"29}{largesymbols}{"01}
\DeclareMathDelimiter{[}{\mathopen} {operators}{"5B}{largesymbols}{"02}
\DeclareMathDelimiter{]}{\mathclose}{operators}{"5D}{largesymbols}{"03}
\DeclareMathDelimiter{<}{\mathopen}{symbols}{"68}{largesymbols}{"0A}
\DeclareMathDelimiter{>}{\mathclose}{symbols}{"69}{largesymbols}{"0B}
\DeclareMathDelimiter{/}{\mathord}{operators}{"2F}{largesymbols}{"0E}
\DeclareMathDelimiter{|}{\mathord}{symbols}{"6A}{largesymbols}{"0C}
\expandafter\DeclareMathDelimiter\@backslashchar{\mathord}{symbols}{"6E}{largesymbols}{"0F}
\DeclareMathDelimiter{\lmoustache}{\mathopen}{largesymbols}{"7A}{largesymbols}{"40}
\DeclareMathDelimiter{\rmoustache}{\mathclose}{largesymbols}{"7B}{largesymbols}{"41}
\DeclareMathDelimiter{\arrowvert}{\mathord}{symbols}{"6A}{largesymbols}{"3C}
\DeclareMathDelimiter{\Arrowvert}{\mathord}{symbols}{"6B}{largesymbols}{"3D}
\DeclareMathDelimiter{\Vert}{\mathord}{symbols}{"6B}{largesymbols}{"0D}
```

---

[4]The file can be found in the directory $TEXMF/tex/latex/base/.

```
\let\|=\Vert
\DeclareMathDelimiter{\vert}{\mathord}{symbols}{"6A}{largesymbols}{"0C}
\DeclareMathDelimiter{\uparrow}{\mathrel}{symbols}{"22}{largesymbols}{"78}
\DeclareMathDelimiter{\downarrow}{\mathrel}{symbols}{"23}{largesymbols}{"79}
\DeclareMathDelimiter{\updownarrow}{\mathrel}{symbols}{"6C}{largesymbols}{"3F}
\DeclareMathDelimiter{\Uparrow}{\mathrel}{symbols}{"2A}{largesymbols}{"7E}
\DeclareMathDelimiter{\Downarrow}{\mathrel}{symbols}{"2B}{largesymbols}{"7F}
\DeclareMathDelimiter{\Updownarrow}{\mathrel}{symbols}{"6D}{largesymbols}{"77}
\DeclareMathDelimiter{\backslash}{\mathord}{symbols}{"6E}{largesymbols}{"0F}
\DeclareMathDelimiter{\rangle}{\mathclose}{symbols}{"69}{largesymbols}{"0B}
\DeclareMathDelimiter{\langle}{\mathopen}{symbols}{"68}{largesymbols}{"0A}
\DeclareMathDelimiter{\rbrace}{\mathclose}{symbols}{"67}{largesymbols}{"09}
\DeclareMathDelimiter{\lbrace}{\mathopen}{symbols}{"66}{largesymbols}{"08}
\DeclareMathDelimiter{\rceil}{\mathclose}{symbols}{"65}{largesymbols}{"07}
\DeclareMathDelimiter{\lceil}{\mathopen}{symbols}{"64}{largesymbols}{"06}
\DeclareMathDelimiter{\rfloor}{\mathclose}{symbols}{"63}{largesymbols}{"05}
\DeclareMathDelimiter{\lfloor}{\mathopen}{symbols}{"62}{largesymbols}{"04}
\DeclareMathDelimiter{\lgroup}{\mathopen}{largesymbols}{"3A}{largesymbols}{"3A}
\DeclareMathDelimiter{\rgroup}{\mathclose}{largesymbols}{"3B}{largesymbols}{"3B}
\DeclareMathDelimiter{\bracevert}{\mathord}{largesymbols}{"3E}{largesymbols}{"3E}
```

You can define a custom symbol in the same way. Let's define \Norm to be the symbol with the number $42_{16}$ (decimal 66) from the symbol set cmex10:[5]

```
\DeclareMathDelimiter{\Norm}{\mathopen}{largesymbols}{"42}{largesymbols}{"42}
```

The symbol is a short but thick vertical line:

|               `\fontencoding{OMX}\fontfamily{cmex}\selectfont\char"42` | 04-05-15 |

If this symbol is typeset in text " ", it looks very strange as it has depth but no height. In math mode, however, parentheses are centred vertically to the base line so this is not a problem. After defining \Norm, we can then use it in the usual manner:

| $\|*BLA*\|$ | `\DeclareMathDelimiter{\Norm}{\mathopen}%` | 04-05-16 |
| | `  {largesymbols}{"42}{largesymbols}{"42}` | |
| $\left\|\frac{*BLA*}{*BLUB*}\right\|$ | `$\left\Norm *BLA* \right\Norm$\\[10pt]` | |
| | `%` | |
| | `$\left\Norm \frac{*BLA*}{*BLUB*} \right\Norm$\\[10pt]` | |
| $\|\big\|\bigg\|\Big\|\Bigg\|$ | `%` | |
| | `$\Norm \big\Norm \bigg\Norm \Big\Norm \Bigg\Norm$` | |

### 4.5.3 Problems with nested parentheses

In the following example all the parentheses are the same height:

$$\int_\gamma F'(z)dz = \int_\alpha^\beta F'(\gamma(t)) \times \gamma'(t)dt$$

| `\[ \int_\gamma F^\prime(z)\mathrm{d}z=` | 04-05-17 |
| `   \int_\alpha^\beta F'\left(\gamma(t)` | |
| `   \right)\times\gamma^\prime(t)\mathrm{d}t \]` | |

---

[5]This is only available with the font encoding OMX, cf. Figure 11.5 on page 247.

However, nested parantheses look better if the inner parentheses are smaller than the outer ones. You can help TEX to determine what height the parentheses should be by changing the value of either the length \delimitershortfall or the TEX counter \delimiterfactor. They are set to the following values:

> \delimitershortfall=5pt
> \delimiterfactor=901

The relative size of parentheses to a given formula is given by the value of \delimiterfactor/1000. This may be too short by up to \delimitershortfall. These values are only taken into account by TEX at the end of the formula, but it is best to set them immediately before the formula, and within a group if they need to apply locally.

If in our example, we set \delimitershortfall to −2pt, the parentheses may be shorter by the negative length −2pt, i.e. they may be 2pt longer:

04-05-18

$$\int_\gamma F'(z)\,\mathrm{d}z = \int_\alpha^\beta F'\left(\gamma(t)\right) \times \gamma'(t)\,\mathrm{d}t$$

```
\newcommand*\diff{\mathop{}\!\mathrm{d}}

{\setlength\delimitershortfall{-2pt}% keep local
\[\int_\gamma F^\prime(z)\diff z=
   \int_\alpha^\beta F'\left(\gamma(t)\right)
   \times\gamma^\prime(t)\diff t\]}
```

The same result can be achieved by setting the counter \delimiterfactor to 1002.

04-05-19

$$\int_\gamma F'(z)\,\mathrm{d}z = \int_\alpha^\beta F'\left(\gamma(t)\right) \times \gamma'(t)\,\mathrm{d}t$$

```
\newcommand*\diff{\mathop{}\!\mathrm{d}}

{\delimiterfactor=1002  % keep it local
\[\int_\gamma F^\prime(z)\diff z
   =\int_\alpha^\beta F'\left(\gamma(t)\right)
   \times\gamma^\prime(t)\diff t\]}
```

The effect of changing the value of either parameter is demonstrated clearly in the following expression, though in practice you would never need to typeset parentheses like this. The size of the various nested parentheses in the Example 04-05-20 is the default output, but is adjusted in the Example 04-05-21 by changing \delimitershortfall and in the Example 04-05-22 on the next page by changing \delimiterfactor:

```
\[ \left(\left(\left(\left(\left(\left(\left(\left(\left(\left(
   \left(\left(\left(\left( A \right)\right)\right)\right)\right)\right)
   \right)\right)\right)\right)\right)\right)\right)\right) \]
```

04-05-20

$$((((((((((((((A))))))))))))))$$

```
{ \setlength\delimitershortfall{-1pt} % keep local
\[ \left(\left(\left(\left(\left(\left(\left(\left(\left(\left(
   \left(\left(\left(\left( A \right)\right)\right)\right)\right)\right)
   \right)\right)\right)\right)\right)\right)\right)\right) \] }
```

04-05-21

```
{\delimiterfactor=1002 % keep local
\[ \left(\left(\left(\left(\left(\left(\left(\left(\left(
   \left(\left(\left(\left( A \right)\right)\right)\right)\right)\right)
   \right)\right)\right)\right)\right)\right)\right)\right) \]}
```

04-05-22

## 4.6 Text in math mode

Normal text should be typeset in upright form; only math expressions should be in italics: $math$ text $math$ (cf. Table 4.5 on page 50). We have already mentioned that letters and numbers in math mode are taken from a different character set than the ones in normal text. While in math mode, there are several ways to change to normal text mode:

| | |
|---|---|
| \mathrm | The characters are taken from the math character set, but typeset upright. |
| \textrm | The characters are taken from the text character set, which are upright by default. |
| \mbox | The characters are taken from the text character set, but are typeset in the \textstyle math style (cf. Section 4.9 on page 55); beware when using \mbox for exponents or indices. |

```
$math - \mathrm{upright} - \textrm{upright} - \mbox{upright}$
\par\Large
$A^{\mathrm{text}}_{\mathrm{text}}$ \quad $A^{\textrm{text}}_{\textrm{text}}$\quad
$A^{\mbox{text}}_{\mbox{text}}$     \quad $A^{\textnormal{text}}_{\textnormal{text}}$
```

$$math – \text{upright} – \text{upright} – \text{upright}$$
$$A^{\text{text}}_{\text{text}} \qquad A^{\text{text}}_{\text{text}} \qquad A^{\text{text}}_{\text{text}} \qquad A^{\text{text}}_{\text{text}}$$

04-06-1

$\mathrm$  Only use \mathrm to insert text within math when you are using the roman character set for normal text; otherwise the inserted text will not match the rest. In math mode there is no line break to wrap text, so put longer pieces of text in a \parbox or minipage:

```
\usepackage{ragged2e}
\begin{equation}
a+b+c+d+ef = g+h+i+j+k\qquad\textrm{\parbox[t]{.3\linewidth}{\RaggedRight%
  This is a very long description of a formula.}}
\end{equation}
```

04-06-2

$$a + b + c + d + ef = g + h + i + j + k \qquad \text{This is a very long description of a formula.} \qquad (4.18)$$

The amsmath package offers a better command for typesetting text in math mode (cf. Section ?? on page ??).

*amsmath*

## 4.7 Font commands

TeX commands for changing font styles (for example, for bold or italic) are still supported in LaTeX. However, they are all pretty much obsolete now, and should not be used as they can lead to irritating results; we are only mentioning them for the sake of completeness. All of them are switches, i.e. they do not expect any arguments and apply as long as they are not changed: \bf, \cal, \it, \rm, \tt, \normalfont. Example 04-07-1 shows these old commands and their effects:

04-07-1

**test** $\mathcal{TEST}$ test `test` *test test*

```
$\bf test\ \cal TEST\ \rm test\
\tt test\ \it test\ \normalfont test$
```

LaTeX's new font commands are different; they are not switches but commands requiring an argument. Every change of font is only local. Table 4.2 shows a summary of the new commands \mathrm, \mathit, \mathcal, \mathsf, \mathtt, and \mathbf.

Table 4.2: LaTeX font commands in math mode without loading additional packages (\mathcal only for capital letters)

04-07-2

| command | test |
|---------|------|
| default | $ABCDEFGHIJKLMNOPQRSTUVWXYZ$ |
|         | $abcdefghijklmnopqrstuvwxyz$ |
| \mathrm | ABCDEFGHIJKLMNOPQRSTUVWXYZ |
|         | abcdefghijklmnopqrstuvwxyz |
| \mathit | *ABCDEFGHIJKLMNOPQRSTUVWXYZ* |
|         | *abcdefghijklmnopqrstuvwxyz* |
| \mathcal | $\mathcal{ABCDEFGHIJKLMNOPQRSTUVWXYZ}$ |
| \mathsf | ABCDEFGHIJKLMNOPQRSTUVWXYZ |
|         | abcdefghijklmnopqrstuvwxyz |
| \mathtt | ABCDEFGHIJKLMNOPQRSTUVWXYZ |
|         | abcdefghijklmnopqrstuvwxyz |
| \mathbf | **ABCDEFGHIJKLMNOPQRSTUVWXYZ** |
|         | **abcdefghijklmnopqrstuvwxyz** |

## 4.8 Whitespace

TeX and LaTeX define three special math lengths, with the following values, in the file `fontmath.ltx`:[6]

| | |
|---|---|
| `\thinmuskip` | Whitespace between normal characters and operators; set to 3mu. |
| `\medmuskip` | Whitespace between normal characters and operators in display mode and for the math style `\textstyle`; set to 4mu plus 2mu minus 4mu (cf. Section 8.1 on page 149). |
| `\thickmuskip` | Whitespace between normal characters and relations in display mode and for the math style `\textstyle`; set to 5mu plus 5mu. |

Here mu stands for **math unit**. It is defined in relation to the width of a capital letter M (1em), and is therefore a dynamic length, varying with the font size, here 0.55557pt.

$$1\text{mu} = \frac{1}{18}\,\text{em} = 0.55557\text{pt} \Leftrightarrow 1\text{pt} = 2.182\text{mu}$$

All three lengths can be adjusted to act as elastic (or glue), and are used by TeX to give different spacings between math characters, especially symbols and operators. The effects of changing these lengths are shown in Table 4.3; each one is reduced in turn to 0pt and the result is framed to emphasize the difference:

Table 4.3: The effect of changing the math lengths

| | |
|---|---|
| default | $\boxed{f(x) = \sqrt[3]{x^2} + 3x_0 \times \sin^2 x}$ |
| `\thinmuskip=0mu` | $\boxed{f(x) = \sqrt[3]{x^2} + 3x_0 \times \sin^2\!x}$ |
| `\medmuskip=0mu` | $\boxed{f(x) = \sqrt[3]{x^2}{+}3x_0{\times}\sin^2 x}$ |
| `\thickmuskip=0mu` | $\boxed{f(x){=}\sqrt[3]{x^2} + 3x_0 \times \sin^2 x}$ |
| all set to 0mu | $\boxed{f(x){=}\sqrt[3]{x^2}{+}3x_0{\times}\sin^2\!x}$ |

You can see that:

- `\thinmuskip=0mu` reduces the space between sin and $x$
- `\medmuskip=0mu` reduces the space to the left and right of + and ×
- `\thickmuskip=0mu` reduces the space to the left and right of =

A complete summary of the additional horizontal space for the different math "atoms" is *math atom* shown in Table 4.4 on the next page. (An atom is an expression that cannot be decomposed any further into the math base units like operator, relation, etc.)

### 4.8.1 Horizontal space in math mode

LaTeX additionally defines the short forms `\!`, `\>`, and `\;` for spaces in math mode:

```
\def\!{\mskip-\thinmuskip}
\def\>{\mskip\medmuskip}
\def\;{\mskip\thickmuskip}
```

---

[6]The file can be found at `$TEXMF/tex/latex/base`, notes on the values in [13].

**Table** 4.4: Overview of the additional whitespace to be inserted between two atoms in math mode (according to [13])

| | | ord | op | bin | rel | open | close | punct | inner |
|---|---|---|---|---|---|---|---|---|---|
| | | | | | *right side* | | | | |
| | ord | 0 | 1 | (2) | (3) | 0 | 0 | 0 | (1) |
| | op | 1 | 1 | – | (3) | 0 | 0 | 0 | (1) |
| | bin | (2) | (2) | – | – | (2) | – | – | (2) |
| *left* | rel | (3) | (3) | – | 0 | (3) | 0 | 0 | (3) |
| *side* | open | 0 | 0 | – | 0 | 0 | 0 | 0 | 0 |
| | close | 0 | 1 | (2) | (3) | 0 | 0 | 0 | (1) |
| | punct | (1) | (1) | – | (1) | (1) | (1) | (1) | (1) |
| | inner | (1) | 1 | (2) | (3) | (1) | 0 | (1) | (1) |

| | | | |
|---|---|---|---|
| 0 | no additional whitespace | ord | normal character |
| 1 | \thinmuskip | op | large operator |
| 2 | \medmuskip | bin | binary operation |
| 3 | \thickmuskip | rel | relation |
| () | only for \displaystyle and \textstyle | open | opening |
| – | impossible combinations | close | closing |
| | | punct | punctuation |
| | | inner | a delimited subformula |

04-08-1

$ab\ ab\ a\ b\ a\ b$        `$ a b\ a\!b\ a\>b\ a\;b $`

\mskip expects math mode, so you will get an error message if these commands are used in text mode in standard LaTeX. However, the amsmath package redefines the commands, so they can be used in text mode.

Small spaces always play an important role in typesetting math because there are many different combinations of operators, link characters, and other special symbols. Spaces of different widths are often required to be inserted between them. For example, a derivative in an expression like $L\frac{\mathrm{d}i}{\mathrm{d}t}$ usually looks better when typeset with a small space between it and the preceding text (here the $L$): $L\,\frac{\mathrm{d}i}{\mathrm{d}t}$ (`$L\:\frac{\mathrm{d}i}{\mathrm{d}t}$`, cf. Section 4.20.4 on page 70). Table 4.5 on the next page shows a list of the commands that can be used in math mode. The respective "visible" space inserted is illustrated by the distance between the framed characters [a] and [b].

In some cases the short forms can cause problems when used with other packages, where they are redefined for other purposes, so the long forms should be used instead.

## 4.8.2 Problems

The use of the \phantom command and the corresponding \hphantom and \vphantom in math mode can cause problems; math objects lose their meaning for TeX if they occur within another command. Take a look at the following example:

Table 4.5: Space in math mode, the long forms \medspace, \thickspace, \negmedspace, and \negthickspace are only available with amsmath.

| positive space | | negative space | |
|---|---|---|---|
| $ab$ | `a b` | | |
| $a b$ | `a b` | | |
| $a\ b$ | `a  b` | | |
| $a\mbox{\textvisiblespace}b$ | `a ␣ b` | | |
| $a\,b$ ($a\thinspace b$) | `a b` | $a\! b$ | `ab` |
| $a\: b$ ($a\medspace b$) | `a b` | $a\negmedspace b$ | `ab` |
| $a\; b$ ($a\thickspace b$ | `a b` | $a\negthickspace b$ | `ab` |
| $a\quad b$ | `a   b` | | |
| $a\qquad b$ | `a     b` | | |
| $a\hspace{0.5cm}b$ | `a    b` | $a\hspace{-0.5cm}b$ | `ba` |
| $a\kern0.5cm b$ | `a    b` | $a\kern-0.5cm b$ | `ba` |
| $a\mkern31.04mu b$ | `a    b` | $a\mkern-31.04mu b$ | `ba` |
| $a\hphantom{xx}b$ | `a   b` | | |
| $axxb$ | `a xx b` | | |

$$a \rightarrow b \qquad (4.19)$$

$$a \quad b \qquad (4.20)$$

```
\begin{equation}a \rightarrow b\end{equation}
\begin{equation}
a \hphantom{\rightarrow} b
\end{equation}
```

04-08-2

The second line has different horizontal spacing to the first line; this is because the \rightarrow occurs as an argument within a \hphantom command, so is no longer treated by TeX as a special math symbol requiring the insertion of extra horizontal space on either side. Thus the space reserved by the \hphantom command is just the width of the arrow itself, treating the \rightarrow command as a normal text character. This can be adjusted either by manually inserting appropriate-sized spaces to those that TeX would have used, or by using the \mathrel command. Both are illustrated in this example:

$$a \rightarrow b \qquad (4.21)$$

$$a \quad b \qquad (4.22)$$

$$a \quad b \qquad (4.23)$$

```
\begin{equation}a \rightarrow b\end{equation}
\begin{equation}
a \mkern\thickmuskip\hphantom{\rightarrow}
 \mkern\thickmuskip b
\end{equation}
\begin{equation}
a \mathrel{\hphantom{\rightarrow}} b
\end{equation}
```

04-08-3

Depending on the type of symbol, the additional whitespace may have to be replaced by the length \medmuskip or \thinmuskip. More information on the type of symbols can be found in the file `fontmath.ltx`[7] or in the `amssymb` package[8]. Alternatively, you can output the definition into the log file during the TeX run with `\show\rightarrow`; this leads to the following entry in the log file:[9]

```
> \rightarrow=\mathchar"3221.
1.20 \show\rightarrow
```

3221 is the code of the type. The first digit corresponds to the type of symbol:

| | |
|---|---|
| 0 : ordinary | 1 : large operator | 2 : binary operation |
| 3 : relation | 4 : opening | 5 : closing |
| 6 : punctuation | 7 : variable family | |

Another problem is the grouping[10] of parts of the math expression. Compare for example:

04-08-4

$$50 \times 10^{12}$$
$$50{\times}10^{12}$$

```
$ 50\times10^{12}  $\\
$ 50{\times}10^{12}$
```

In the second example `\times` was enclosed in parentheses, which caused no additional horizontal space to be inserted. In fact this version is the more appealing layout here; for frequent use the package `numprint` can be used.[11]

### 4.8.3  Point or comma

The decimal separator used in Europe is different to that in the UK or USA; most modern software automatically adapts to a European or Anglo-American locale. Switching in LaTeX is a bit more difficult because the type of the character also has to be changed — the space around the thousands separator is different from the space around the decimal separator. In the following example, the first option is the default (Anglo-American) version, and the second option shows what happens if the characters are just swapped:

04-08-5

| | |
|---|---|
| $1,234,567.89$ | default |
| $1.234.567,89$ | wrong spaces |
| $1.234.567,89$ | German standard |
| $1.234.567,89$ | correct spaces |
| $1.234.567,89$ | correct spaces |

```
$\begin{array}{@{}l l@{}}
1,234,567.89       & \textrm{default}         \\[3pt]
1.234.567,89       & \textrm{wrong spaces}\\[3pt]
1{.}234{.}567{,}89 & \textrm{German standard}\\[3pt]
1\mathpunct{.}234\mathpunct{.}567{,}89
  & \textrm{correct spaces}\\[3pt]
1\mathpunct{.}234\mathpunct{.}567\mathord{,}89
  & \textrm{correct spaces}
\end{array}$
```

The original definition in `fontmath.ltx`[12] is:

```
\DeclareMathSymbol{,}{\mathpunct}{letters}{"3B}
\DeclareMathSymbol{.}{\mathord}{letters}{"3A}
```

---

[7]`$TEXMF/tex/latex/base`
[8]`$TEXMF/tex/latex/amsfonts`
[9]The compilation run stops there and waits for input such that the user can see the output there already.
[10]Through {...} or `\bgroup`...`\egroup` or `\begingroup`...`\endgroup`
[11]`ftp://ftp.dante.de/tex-archive/macros/latex/contrib/numprint/`
[12]`$TEXMF/tex/latex/base/`

Therefore by definition, the comma is of type \mathpunct but the dot is of type \mathord. This is not as strange as it might appear; the type only refers to the difference in horizontal space that TeX should insert.

A document-wide change can be achieved by swapping the definitions in the preamble. Individual changes can be made by bracketing (i.e./grouping) the comma { , } as in the third and fourth options in Example 04-08-5. The requirement for insertion of space is lost in the sense that the contents of the curly braces are no longer recognized as a special math symbol within a larger expression. Anything set in curly braces is treated as a separate formula, i.e. there is no space automatically inserted before and after. The same behaviour can be achieved with \mathord{,} (last option in Example 04-08-5).[13]

### 4.8.4 Vertical space

#### Displayed formulae

There are four different lengths that together control the vertical space options between the text and a displayed formula. All of them are dynamic spaces, defined as follows in latex.ltx (cf. Section 8.1 on page 149):

```
\abovedisplayskip=12pt plus 3pt minus 9pt
\abovedisplayshortskip=0pt plus 3pt
\belowdisplayskip=12pt plus 3pt minus 9pt
\belowdisplayshortskip=7pt plus 3pt minus 4pt
```

TeX looks at where the preceding line of text ends horizontally in relation to the beginning of the formula. If it ends before the start of the following formula, short skips are used. Otherwise the normal skips are used. With the standard definitions, both options should appear to have the same spacing. To illustrate the fact that different skips are used, the following example sets short skips to 0pt and normal ones to 20pt without glue.

---

Ends before the formula.
$$f(x) = \int \frac{\sin x}{x} \, dx \tag{4.24}$$
Now ends not *before*, but *after* the following formula.

$$f(x) = \int \frac{\sin x}{x} \, dx \tag{4.25}$$

Ends before again.
And on the next line there is a normal line of text as usual.

---

As \abovedisplayshortskip was set to 0pt, no space was inserted before the first formula (Equation 4.24). For the following formula (Equation 4.24), however, 20pt spaces were inserted. Note that it is the length of the preceding text line that determines whether short or normal skips are used both before *and* after a formula; the length of the line of text that follows the formula has no impact.

---

[13]To swap comma and point for the whole document the icomma package can be used — CTAN://macros/latex/contrib/was/.

The following result is achieved with the skips set to the standard lengths. It clearly shows the advantage of having dynamic lengths ("plus" − "minus" values, cf. Section 8.1 on page 149):

Ends before the formula.
$$f(x) = \int \frac{\sin x}{x}\,\mathrm{d}x \tag{4.26}$$
Now ends not *before*, but *after* the following formula.
$$f(x) = \int \frac{\sin x}{x}\,\mathrm{d}x \tag{4.27}$$
Ends before again.
And on the next line there is a normal line of text as usual.

## Within display environments

\\[*length*]   This method, used in text mode to increase the spacing between lines of text, works the same for displayed formulae with possible line breaks (see Section 6.2 on page 80).

04-08-6

$$f(x) = \frac{1}{x^2} \tag{4.28}$$
$$f'(x) = -\frac{2}{x^3} \tag{4.29}$$

$$\int f(x)\,\mathrm{d}x = -\frac{1}{x} \tag{4.30}$$

```
\begin{eqnarray}
f(x)                    &=&\hphantom{-}\frac{1}{x^2}\\
f^\prime(x)             &=&-\frac{2}{x^3}        \\[20pt]
\int f(x)\,\mathrm{d}x&=&-\frac{1}{x}
\end{eqnarray}
```

(Note the use of \hphantom in Example 04-08-6 to reserve space for a negative sign in the first equation in order to achieve a matching horizontal alignment to the other equations.)

\jot   The space between the individual lines of an eqnarray environment can be influenced globally through the length \jot. It will be inserted by LaTeX after every line as additional line feed and is determined by:

```
\newdimen\jot
\jot=3pt
```

The following examples show the same formulae three times, with the default value, with \jot=0pt, and with \jot=10pt.

```
\usepackage{tabularx}
\begin{tabularx}{\linewidth}{@{}XXX@{}}
\begin{eqnarray*}
y & = & d\\ y & = & c\frac{1}{x}+d\\ y & = & b\frac{1}{x^{2}}+cx+d
\end{eqnarray*}
&
\setlength\jot{0pt}
\begin{eqnarray*}
y & = & d\\ y & = & c\frac{1}{x}+d\\ y & = & b\frac{1}{x^{2}}+cx+d
```

```
\end{eqnarray*}
&
\setlength\jot{10pt}
\begin{eqnarray*}
y & = & d\\ y & = & c\frac{1}{x}+d\\ y & = & b\frac{1}{x^{2}}+cx+d
\end{eqnarray*}
\end{tabularx}
```

$$y = d$$
$$y = c\frac{1}{x}+d$$
$$y = b\frac{1}{x^2}+cx+d$$

$$y = d$$
$$y = c\frac{1}{x}+d$$
$$y = b\frac{1}{x^2}+cx+d$$

$$y = d$$
$$y = c\frac{1}{x}+d$$
$$y = b\frac{1}{x^2}+cx+d$$

04-08-7

If using \jot frequently, it makes sense to define a new environment to be able to specify the value for \jot as a parameter. For example:

$$y = d \tag{4.31}$$
$$y = c\frac{1}{x}+d \tag{4.32}$$
$$y = b\frac{1}{x^2}+cx+d \tag{4.33}$$

```
\newenvironment{mathspace}[1][3pt]{%
  \setlength{\jot}{#1}\ignorespaces\eqnarray
}{\endeqnarray\ignorespacesafterend}

\begin{mathspace}[0pt]
  y & = & d\\
  y & = & c\,\frac{1}{x}+d\\
  y & = & b\,\frac{1}{x^{2}}+cx+d
\end{mathspace}
```

04-08-8

**\arraystretch**   In contrast to \jot, \arraystretch is a command and not a length. Its value is the factor used to multiply the current line spacing. Therefore the default value is 1:

```
\renewcommand\arraystretch{1}
```

As the name suggests, the command works for array environments, either explicit, or implicit as in the matrix environment. Tables also use this value. To change this value you must use \renewcommand; the change is global for the current group and all groups below. The effect is similar to setting \jot to a constant value. In the following example we first use the default value, and then set the value to 2:

$$y = d$$
$$y = c\frac{1}{x}+d$$
$$y = b\frac{1}{x^2}+cx+d$$

$$y = d$$
$$y = c\frac{1}{x}+d$$
$$y = b\frac{1}{x^2}+cx+d$$

```
$\begin{array}{rcl}
  y & = & d\\
  y & = & c\frac{1}{x}+d\\
  y & = & b\frac{1}{x^{2}}+cx+d
\end{array}$
{\renewcommand\arraystretch{2}% keep local
$\begin{array}{rcl}
  y & = & d\\
  y & = & c\frac{1}{x}+d\\
  y & = & b\frac{1}{x^{2}}+cx+d
\end{array}$}
```

04-08-9

**\vspace**   The \vspace command, frequently used in text mode, can be used in the same way in math mode if it is given as the argument to a \noalign command. \noalign is a TEX primitive and is designed to insert vertical material between two lines in a table or array. It is usually used to insert horizontal lines, but can also be "abused" to create vertical spacing.

04-08-10

$$\left(\begin{array}{*6{c}}
0 & 1 & 1 & 0 & 0 & 1 \\
1 & 0 & 0 & 1 & 1 & 0 \\ \hline
0 & 1 & 1 & 0 & \frac{1}{\sqrt{2}} & 1 \\ \hline
1 & 0 & 1 & 0 & 1 & 0 \\
0 & 1 & 0 & 1 & 0 & 1
\end{array}\right)$$

```
\[ \left(\begin{array}{*6{c}}
0 & 1 & 1 & 0 & 0 & 1 \\
1 & 0 & 0 & 1 & 1 & 0 \\\hline
\noalign{\vspace{12pt}}
0 & 1 & 1 & 0 & \frac{1}{\sqrt{2}} & 1\\
\noalign{\vspace{12pt}}\hline
1 & 0 & 1 & 0 & 1 & 0 \\
0 & 1 & 0 & 1 & 0 & 1 \\
\end{array}\right) \]
```

If the \vspace command is used in math mode without a \noalign command, TEX would abort the processing with this error message:

```
! Misplaced \noalign.
\hline ->\noalign
                 {\ifnum 0=`}\fi \let \hskip \vskip \let \vrule \hrule \let ...
1.8 \vspace{12pt}\hline
```

**The setspace package**   The setspace package by Geoffrey Tobin is actually meant for the typesetting of normal text, but it also affects formulae and can therefore be used for math mode as well. If using it, then it is best to define a new environment to be able to change the vertical line spacing effectively and keep the changes local.

04-08-11

$$\begin{vmatrix} a & = & b \\ 1 & = & y \\ X & = & z \end{vmatrix}$$

$$\text{Text}\begin{vmatrix} a & = & b \\ 1 & = & y \\ X & = & z \end{vmatrix}\text{Text}$$

```
\usepackage{setspace}
\newenvironment{Array}[2][1]
    {\setstretch{#1}\array{#2}}
    {\endarray}

\[ \begin{Array}[0.75]{|ccc|}
a &=& b\\ 1 &=& y\\ X &=& z
\end{Array} \]

Text $\begin{Array}[2]{|ccc|}
a &=& b\\ 1 &=& y\\ X &=& z
\end{Array}$ Text
```

## 4.9 Math font styles

LATEX knows about the four math font styles listed in Table 4.6 on the following page, which are also called $D$, $S$, $SS$ and $T$. There are four additional styles to improve the typesetting of exponents and indices: $D'$, $S'$, $SS'$ and $T'$.

$$x^2 \to D^S \qquad x_1 \to D_S \qquad x_1^2 \to D'^S_{S'} \qquad x^{2^a} \to D^{S^{SS}} \qquad x_{a_1} \to D_{S_{SS}} \qquad x_{a_1}^{2^a} \to D'^{S^{SS}}_{S_{SS'}}$$

The default font style of TEX depends on the math environment that is in use. An expression in inline mode (cf. Chapter 2 on page 5) is typeset in \textstyle. It does not have a set font

Table 4.6: Summary of the math font styles and their effect for inline and display mode.

| style | inline | display |
|---|---|---|
| default | $f(t) = \frac{T}{2\pi} \int \frac{1}{\sin\frac{\omega}{t}} \, \mathrm{d}t$ | $f(t) = \frac{T}{2\pi} \int \frac{1}{\sin\frac{\omega}{t}} \, \mathrm{d}t$ |
| \displaystyle | $f(t) = \frac{T}{2\pi} \int \frac{1}{\sin\frac{\omega}{t}} \, \mathrm{d}t$ | $f(t) = \frac{T}{2\pi} \int \frac{1}{\sin\frac{\omega}{t}} \, \mathrm{d}t$ |
| \scriptstyle | $f(t) = \frac{T}{2\pi} \int \frac{1}{\sin\frac{\omega}{t}} \, \mathrm{d}t$ | $f(t)=\frac{T}{2\pi}\int\frac{1}{\sin\frac{\omega}{t}}\,\mathrm{d}t$ |
| \scriptscriptstyle | $f(t)=\frac{T}{2\pi}\int\frac{1}{\sin\frac{\omega}{t}}\,\mathrm{d}t$ | $f(t)=\frac{T}{2\pi}\int\frac{1}{\sin\frac{\omega}{t}}\,\mathrm{d}t$ |
| \textstyle | $f(t) = \frac{T}{2\pi} \int \frac{1}{\sin\frac{\omega}{t}} \, \mathrm{d}t$ | $f(t) = \frac{T}{2\pi} \int \frac{1}{\sin\frac{\omega}{t}} \, \mathrm{d}t$ |

size. Normal characters are typeset in the same size as the surrounding text, but all fractions are typeset in a slightly smaller font size, as can be seen here:

normal text $\frac{a}{b}$ normal text

```
normal text $\frac{a}{b}$ normal text
```

04-09-1

For displayed formulae (display mode, cf. Chapter 3 on page 15), the style \displaystyle is used. This has the same font size as for the surrounding text for normal characters and for the main fractions. Within a line, \displaystyle can be suboptimal because of the larger line spacing that therefore occurs:

```
Within a line this can be suboptimal because of the larger line
spacing that is created to make space for the fraction --- $\displaystyle\frac{a}{b}$.
Within a line this can be suboptimal because of the larger line
spacing that is created to make space for the fraction. It just does not look good
and it is much better to use the \texttt{\textbackslash textstyle} variant.
```

Within a line this can be suboptimal because of the larger line spacing that is created to make space for the fraction $-\frac{a}{b}$. Within a line this can be suboptimal because of the larger line spacing that is created to make space for the fraction. It just does not look good and it is much better to use the \textstyle variant.

04-09-2

There will no doubt be many occasions when you want to overwrite the default style. Apart from these two main styles, you will find \scriptstyle useful for indices and exponents, and \scriptscriptstyle useful for stacked indices or exponents and compound fractions. All four styles are matched with one another by LaTeX or the used document class; changes should be done with care. Table 4.6 shows a summary of the styles and their respective meaning for inline mode and display mode.

Independent of these styles, the size of the math parts is adapted to the font size of the surrounding text. Commands to change the font size are switches; they do not take any arguments and only take effect when used in text mode. So in the following example the first fraction is huge like the text before it, while the second fraction is normal, unaffected by the \huge command enclosed within the $...$ syntax:

```
\huge\verb+\huge+ $\frac{a}{b}$\normalsize normal $\huge\frac{a}{b}$ normal
```

04-09-3

$$\text{\huge \Large\huge} \quad \frac{a}{b}\text{normal } \tfrac{a}{b} \text{ normal}$$

Any change to inter-line spacing affects the whole paragraph, even if the font size was only changed for a part of a formula. In such cases it makes more sense to work with a new environment that changes the font size for the text, but keeps the old value for \baselineskip, which controls the line spacing. The following example illustrates this:

If you want to set a formula in very small font, you must change the font size to a very small value using the \tiny command outside the display environment, and end the paragraph after the formula. Before the next paragraph reset the font size with the \normalsize command. This paragraph is more dense than the others; the line spacing was reduced because of the \tiny command. LaTeX decides at the end of the paragraph which vertical spacing will be used.

```
If you want to set a formula in very small font, do not use \verb=\tiny= in this way.
The interline spacing of the lines in this paragraph will be incorrect. See the next
example for correct spacing.
\tiny
\begin{equation}
\int_1^2\,\frac{1}{x^2}\,\mathrm{d}x=0.5
\end{equation}
\normalsize
If you want to set a formula in very small font, do not use \verb=\tiny= in this way.
The spacing of the lines in the above paragraph is incorrect. See the next example.
```

04-09-4

If you want to set a formula in very small font, do not use \tiny in this way. The interline spacing of the lines in this paragraph will be incorrect. See the next example for correct spacing.

$$\int_1^2 \frac{1}{x^2}\,dx = 0.5 \tag{4.34}$$

If you want to set a formula in very small font, do not use \tiny in this way. The spacing of the lines in the above paragraph is incorrect. See the next example.

Defining a new environment for typesetting a "small" formula will make LaTeX always assume the normal font size for the inter-line spacing. For example:

```
\makeatletter
\newenvironment{smallequation}[1]{%
   \skip@=\baselineskip#1\baselineskip=\skip@\equation
}{\endequation \ignorespacesafterend}
\makeatother
If you want to set a formula in very small font, you must change
the font size to a very small value using the \texttt{\textbackslash tiny}
```

```
command outside the display environment, and end the paragraph after the
formula\ldots
\begin{smallequation}{\tiny}
    \int_1^2\,\frac{1}{x^2}\,\mathrm{d}x=0.5
\end{smallequation}
```

If you want to set a formula in very small font, you must change the font size to a very small value using the \tiny command outside the display environment, and end the paragraph after the formula...

$$\int_1^2 \frac{1}{x^2} \, dx = 0.5 \tag{4.35}$$

04-09-5

If the formula needs to be very large, different problems arise. They are dealt with in Section 9.15 on page 180.

## 4.10 Dots

Dots are useful to indicate a continuous structure, especially in connection with matrices. Table 4.7 shows the point symbols available for standard LaTeX. More are provided by the amsmath package (cf. Table 6.6 on page 120). It is not always easy to see the difference between individual series of dots.

Table 4.7: Dots in math mode for standard LaTeX

| \cdots | $\cdots$ | \ddots | $\ddots$ | \ldots | $\ldots$ | \vdots | $\vdots$ |
|---|---|---|---|---|---|---|---|

To align the \ddots in exactly the opposite direction, use the \reflectbox command from the graphicx package:

$\ddots\ \reflectbox{\ddots}$

```
\usepackage{graphicx}

$\ddots$ \reflectbox{$\ddots$}
```

04-10-1

\reflectbox works in text mode; therefore its argument has to be enclosed in $...$ even when it is used in math mode.

As an example, the determinant of $n$th degree of the matrix $\mathbf{A} = (a_{ik})$ is given as follows:

$$\Delta = \| a_{ik} \| = \begin{vmatrix} a_{11} & a_{12} & a_{13} & \cdots & a_{1n} \\ a_{21} & a_{22} & a_{23} & \cdots & a_{2n} \\ \vdots & \vdots & \vdots & \ddots & \vdots \\ a_{n1} & a_{n2} & a_{n3} & \cdots & a_{nn} \end{vmatrix}$$

```
\usepackage{graphicx}

\[
\Delta = \| a_{ik} \| = \left|
\begin{array}{ccccc}
a_{11} &a_{12} &a_{13} &\cdots &a_{1n}\\
a_{21} &a_{22} &a_{23} &\cdots &a_{2n}\\
\vdots &\vdots &\vdots &\ddots &\vdots\\
a_{n1} &a_{n2} &a_{n3} &\cdots &a_{nn}
\end{array} \right|
\]
```

04-10-2

## 4.11 Accents

Accents are a topic in themselves in every character set or typesetting programme. LaTeX is no different; there is an almost unmanageable faculty for different symbols with accents. Here we will give only the ones from standard LaTeX. Table 4.8 shows a summary of the accents that can be placed above or below a character. The amssymb package makes it easy to define more. Also see Section 9.1 on page 167, where we discuss the accents package.

Table 4.8: Accents in math mode; the ones marked with a ˙ need the amsmath package.

| name | example | name | example |
|---:|:---|---:|:---|
| \acute | $\acute{a}$ | \bar | $\bar{a}$ |
| \breve | $\breve{a}$ | \check | $\check{a}$ |
| \dddot˙ | $\dddot{a}$ | \ddot | $\ddot{a}$ |
| \dot | $\dot{a}$ | \grave | $\grave{a}$ |
| \hat | $\hat{a}$ | \mathring | $\mathring{a}$ |
| \overbrace | $\overbrace{a}$ | \overleftarrow | $\overleftarrow{a}$ |
| \overleftrightarrow˙ | $\overleftrightarrow{a}$ | \overline | $\overline{a}$ |
| \overrightarrow | $\overrightarrow{a}$ | \tilde | $\tilde{a}$ |
| \underbar | $\underline{a}$ | \underbrace | $\underbrace{a}$ |
| \underleftarrow˙ | $\underleftarrow{a}$ | \underleftrightarrow˙ | $\underleftrightarrow{a}$ |
| \underline | $\underline{a}$ | \underrightarrow˙ | $\underrightarrow{a}$ |
| \vec | $\vec{a}$ | \widehat | $\widehat{a}$ |
| \widetilde | $\widetilde{a}$ | | |

Accents above the letters i and j are often problematic; it is difficult to place a second accent sensibly if a dot is already present and two stacked accents look rather strange. Alternatively, use the \imath and \jmath commands as a base for accents. The possibilities for placing accents are almost unlimited, for example to use them to cross out characters with a horizontal line.[14]

04-11-1

$\vec{ı}\ \ddot{ı}\ a\ a$

```
$\vec{\imath}\ \ddot{\imath}\ \mathaccent`-a\
\mathaccent\mathcode`-a$
```

### 4.11.1 Over- and underbrace

The \overbrace and \underbrace commands are used relatively frequently in typesetting math.

```
\underbrace{argument}
\overbrace{argument}
```

They can also be nested if you want to typeset material both above and below the argument:

---

[14]A better solution for crossing out whole words can be found in Section 9.8 on page 173.

$$f(x) = \overbrace{x^2 + 2x + 1}^{\text{factorise}}\underbrace{\phantom{x^2+2x+1}}_{(x+1)^2}$$

```
$\begin{array}{r@{\kern1.5pt}c@{\kern1.5pt}c}
     &   & \textrm{factorise}\\
f(x) & = & \underbrace{\overbrace{x^{2}+2x+1}}\\
     &   & \quad\left(x+1\right)^{2}
\end{array}$
```

04-11-2

In contrast there are no similar \underbracket or \overbracket commands for LaTeX, but they can be defined as follows:

```
\makeatletter
\def\underbracket{%
  \@ifnextchar[{\@underbracket}{\@underbracket[\@bracketheight]}}
\def\@underbracket[#1]{%
  \@ifnextchar[{\@under@bracket[#1]}{\@under@bracket[#1][0.4em]}}
\def\@under@bracket[#1][#2]#3{%\message {Underbracket: #1,#2,#3}
  \mathop{\vtop{\m@th\ialign{##\crcr $\hfil \displaystyle {#3}\hfil $%
  \crcr\noalign{\kern 3\p@ \nointerlineskip }\upbracketfill {#1}{#2}
  \crcr\noalign{\kern 3\p@ }}}}\limits}
\def\upbracketfill#1#2{$\m@th \setbox \z@ \hbox {$\braceld$}
  \edef\@bracketheight{\the\ht\z@}\bracketend{#1}{#2}
  \leaders \vrule \@height #1 \@depth \z@ \hfill
  \leaders \vrule \@height #1 \@depth \z@ \hfill \bracketend{#1}{#2}$}
\def\bracketend#1#2{\vrule height #2 width #1\relax}
\makeatother
```

You will only understand these definitions if you have knowledge of TeX programming. However, once these definitions are made, you can typeset for example the following:

```
\setlength\arraycolsep{0pt}
$\begin{array}{@{}ccccc@{}}
\textrm{hate \LaTeX\ }
  & \underbracket[0.5pt]{1\rightarrow2\rightarrow3\rightarrow4}
  & \underbracket[0.75pt][0.75em]{\rightarrow5\rightarrow6\rightarrow7}
  & \underbracket[1pt][1em]{\rightarrow8\rightarrow9\rightarrow10}
  & \textrm{~love \LaTeX}\\
  & \textrm{little} & \textrm{medium} & \textrm{lots}
\end{array}$
```

$$\text{hate \LaTeX\ } \underbracket{1 \to 2 \to 3 \to 4}_{\text{little}} \underbracket{\to 5 \to 6 \to 7}_{\text{medium}} \underbracket{\to 8 \to 9 \to 10}_{\text{lots}} \text{ love \LaTeX}$$

04-11-3

In the above definition, \underbracket has two optional parameters, as also does \overbracket, which is defined implicitly further down in Example 04-11-5 on the facing page.

| \underbracket [line width] [symbol height] {*argument*} |
|---|
| \overbracket [line width] [symbol height] {*argument*} |

line width      Given by the value with a valid unit. Default 1pt.
symbol height   Given by the value with a valid unit. Default 1em.

If neither of the optional parameters are given, the default ones, which were set similarly to \underbrace, are assumed. If only one optional parameter is given, it is interpreted as the line width. The same applies to \overbracket. The following summary shows the possible combinations:

| | |
|---|---|
| \underbracket{...} | The default values for line width (1pt) and symbol height (1em) are assumed. |
| \underbracket [...] {...} | The line width is given by the optional parameter. |
| \underbracket [...] [...] {...} | The line width is given by the first optional parameter and the symbol height by the second. |

04-11-4

*foo bar foo bar foo bar*

`$\underbracket{foo~bar}$ $\underbracket[.1pt]{foo~bar}$`
`$\underbracket[2pt][1em] {foo~bar}$`

The definition of \overbracket is almost identical to that of \underbracket so we will not print it here, but it can be found in the non-visible part of the preamble of the next example:

```
\setlength\arraycolsep{0pt}
$\begin{array}{@{}ccccc@{}}
\textrm{hate \LaTeX\ }
  & \overbracket[0.5pt]{1\rightarrow2\rightarrow3\rightarrow4}
  & \overbracket[0.75pt][0.75em]{\rightarrow5\rightarrow6\rightarrow7}
  & \overbracket[1pt][1em]{\rightarrow8\rightarrow9\rightarrow10}
  & \textrm{\ love \LaTeX}\\
  & \textrm{little} & \textrm{medium} & \textrm{lots}
\end{array}$
```

04-11-5

hate LATEX $1 \rightarrow 2 \rightarrow 3 \rightarrow 4 \rightarrow 5 \rightarrow 6 \rightarrow 7 \rightarrow 8 \rightarrow 9 \rightarrow 10$ love LATEX

little        medium        lots

### 4.11.2  Vectors

Vectors are in essence a special case of an accent. LATEX defines the \vec command, which creates an arrow the width of a single character:

04-11-6

$\vec{a}\ \vec{ab}\ \vec{A}\ \vec{A_1^n}\ \vec{\imath}$

`$\vec{a}$ $\vec{ab}$ $\vec{A}$`
`$\vec{A_1^n}$ $\vec{\imath}$`

Since vector arrows are often required for several characters this command is not really good enough. It is better to load the esvect package, which avoids this problem. Examples are given in Section 9.13 on page 178.

## 4.12  Exponents and indices

The two active characters _ and ^ may only be used in math mode; this is a common trap. A_B in text mode cannot be written as A_B in the source code, but only as A\_B. Alternatively you can write them in math mode, though this will typeset the letters in italics.

The order of exponent and index is not significant, i.e. $a_1^2$ and $a^2_1$ each result in $a_1^2$. The only important point to remember is to bracket together an index or exponent of more than one character. Individual characters may also be bracketed, but this is not necessary. Correct bracketing is vital with compound exponents or indices in order to get the correct result. For example:

```
\[ 7a_{11}^{13}+a_1x+a_0 \quad x^{2}+y^2 = r^2 \quad a_{i-1}+a_{i+1}<a_i \]
%
\[ ((x^2)^3)^4 = {({{\left({\left(x^2\right)}^3\right)}^4 \]
```

$$7a_{11}^{13} + a_1 x + a_0 \quad x^2 + y^2 = r^2 \quad a_{i-1} + a_{i+1} < a_i$$

04-12-1

$$((x^2)^3)^4 = ((x^2)^3)^4 = \left(\left(x^2\right)^3\right)^4$$

## 4.13 Operators

An operator may be a symbol, or a name that is then typeset in upright letters. The assignment of this category is arbitrary. Nevertheless the European and the American standards are very similar. The definition of a symbol or word as an operator also affects the space to be inserted before and after it. We have already mentioned the different spacings in Section 4.8 on page 48.

Table 4.9 shows a summary of the operator symbols defined in fontmath.ltx[15] and Table 4.10 on the facing page shows the operator names defined in latex.ltx[16].

Table 4.9: Summary of the operators defined in fontmath.ltx

| | | | | |
|---|---|---|---|---|
| \coprod ∐ | \bigvee ⋁ | \bigwedge ⋀ | \biguplus ⊎ | \bigcap ⋂ |
| \bigcup ⋃ | \intop ∫ | \int ∫ | \prod ∏ | \sum Σ |
| \bigotimes ⊗ | \bigoplus ⊕ | \bigodot ⊙ | \ointop ∮ | \oint ∮ |
| \bigsqcup ⊔ | \smallint ∫ | | | |

The two symbols \oint (∮) and \ointop (∮) and the two symbols \int (∫) and \intop (∫) appear to be identical. However, if we use them in an equation with limits, as in the following example, you can see the difference; \ointop puts the limits above and below the integral symbol whereas \oint puts them as super- and subscripts. The same applies for \int and \intop.

```
\newcommand*\diff{\mathop{}\!\mathrm{d}}
\[ \oint_{U(r)}H\diff r=\ointop_{U(r)}H\diff r=\frac{I}{2\pi r_0}\oint_{U(r)}
   \diff r=\frac{I}{2\pi r_0}\oint_{U(r)}\diff r=I \]
```

$$\oint_{U(r)} H\, dr = \oint_{U(r)} H\, dr = \frac{I}{2\pi r_0} \oint_{U(r)} dr = \frac{I}{2\pi r_0} \oint_{U(r)} dr = I$$

04-13-1

---

[15]$TEXMF/tex/latex/base
[16]ditto

Table 4.10: Summary of the operators defined in `latex.ltx` (`ltmath.dtx`)

| | | | |
|---|---|---|---|
| \log log | \lg lg | \ln ln | \lim lim |
| \limsup lim sup | \liminf lim inf | \sin sin | \arcsin arcsin |
| \sinh sinh | \cos cos | \arccos arccos | \cosh cosh |
| \tan tan | \arctan arctan | \tanh tanh | \cot cot |
| \coth coth | \sec sec | \csc csc | \max max |
| \min min | \sup sup | \inf inf | \arg arg |
| \ker ker | \dim dim | \hom hom | \det det |
| \exp exp | \Pr Pr | \gcd gcd | \deg deg |
| \bmod mod | \pmoda (mod $a$) | | |

You can define new text operators by using the `\mathop` command. In the following example we define two new operators, `\foo` and `\baz`: [17]

04-13-2

$$\mathrm{foo}_1^2 = \mathrm{baz}_1^{\,2} = \mathrm{foo}_1^{\,2}$$

```
\makeatletter
\newcommand\foo{\mathop{\operator@font foo}\nolimits}
\newcommand\baz{\mathop{\operator@font baz}}
\makeatother

\[ \foo_1^2 = \baz_1^2 = \foo\limits_1^2 \]
```

`\foo` was defined as an operator with `\nolimits`; therefore the index and exponent are typeset as super- and subscript. The definition of `\baz` did not contain that instruction; therefore the exponent and index are placed above and below in display mode. This layout can be enforced for `\foo` as well by using `\limits` in the coding.

Further operators are defined in the `amsmath` package; they are described in Section 6.13 on page 113.                                                        *amsmath*

## 4.14  Greek letters

Greek letters are a common cause of confusion in LaTeX because they are usually not all contained within a single math font, i.e. there are not all of the small and the capital letters in normal, bold and italic versions. For example, the `amsmath` package simulates a bold version by superimposing characters with light kerning; this is called "poor man's bold".

Furthermore there are usually no bold and upright lowercase Greek letters. In Section 6.17 on page 119 we will show you how to achieve bold lowercase letters with commands.       *amsmath*

Capital letters only have their own name if the character does not correspond to a normal capital letter. For example, the Greek uppercase letter for α is identical to A so does not have its own name. Table 4.11 only shows those capital letters that have their own name. Sometimes there is a choice of characters for a lowercase letter; the alternative option has `var` prefixed to its command name.

---

[17] `\bar` was not used here because there already is a command of that name with a different meaning.

Table 4.11: The Greek letters (empty cells designate the normal character).

| name | character | uppercase | character | \mathbf | \mathit |
|------|-----------|-----------|-----------|---------|---------|
| \alpha | α | | | | |
| \beta | β | | | | |
| \gamma | γ | \Gamma | Γ | **Γ** | *Γ* |
| \delta | δ | \Delta | Δ | **Δ** | *Δ* |
| \epsilon | ϵ | | | | |
| \varepsilon | ε | | | | |
| \zeta | ζ | | | | |
| \eta | η | | | | |
| \theta | θ | \Theta | Θ | **Θ** | *Θ* |
| \vartheta | ϑ | | | | |
| \iota | ι | | | | |
| \kappa | κ | | | | |
| \delta | λ | \Lambda | Λ | **Λ** | *Λ* |
| \mu | μ | | | | |
| \nu | ν | | | | |
| \xi | ξ | \Xi | Ξ | **Ξ** | *Ξ* |
| \pi | π | \Pi | Π | **Π** | *Π* |
| \varpi | ϖ | | | | |
| \rho | ρ | | | | |
| \varrho | ϱ | | | | |
| \sigma | σ | \Sigma | Σ | **Σ** | *Σ* |
| \varsigma | ς | | | | |
| \tau | τ | | | | |
| \upsilon | υ | \Upsilon | Υ | **Υ** | *Υ* |
| \phi | ϕ | \Phi | Φ | **Φ** | *Φ* |
| \varphi | φ | | | | |
| \chi | χ | | | | |
| \psi | ψ | \Psi | Ψ | **Ψ** | *Ψ* |
| \omega | ω | \Omega | Ω | **Ω** | *Ω* |

For bold and upright lowercase Greek letters, use the upgreek packet by Walter Schmidt or the bm packet by David Carlisle and Frank Mittelbach. Examples can be found in Section 7.12 and Section 9.6 on page 171.

## 4.15 Page break

As a rule a page break is not possible in a displayed formula. In standard LaTeX, the only time that you might want one to insert one would be in an eqnarray environment; it is impossible in an array environment as this is treated as a single unit because of its single equation number. The amsmath package provides two commands — similar to \pagebreak — for page breaks *amsmath* within displayed formulae (cf. Section 6.5 on page 97).

## 4.16  Stacked symbols

You may sometimes need to create a stacked symbol if LaTeX or one of the many packages does not already offer an appropriate one; for example "$\stackrel{\triangle}{=}$". The syntax for the \stackrel command is:

\stackrel{*above*}{*base*}

Frequently used symbols should be defined in the preamble of the document so you can then just use a short form, e.g. \eqdef for the symbol above. If \ensuremath is used with the definition, the new command will set its argument *always* in math mode, which could avoid any problems with the short form. This can be seen in the following example, where the first symbol is created in text mode.

04-16-1

eqdef: $\vec{x} \stackrel{\mathrm{def}}{=} (x_1, \ldots, x_n)$

```
\newcommand{\eqdef}{%
   \ensuremath{%
      \mathbin{\stackrel{\mathrm{def}}{=}}}}

eqdef: $\vec{x}\eqdef\left(x_{1},\ldots,x_{n}\right)$
```

The font size for the stacked symbol is slightly smaller; in this case it corresponds to the math style \scriptsize.

## 4.17  Binomials

\choose is primarily used for binomials. It is now obsolete, but we are mentioning it for completeness. It is similar to \atop (cf. page 37).

above \choose below

The parentheses are not required, but they can be useful to distinguish the arguments of the command from any surrounding text.

04-17-1

$$\binom{m+1}{n} = \binom{m}{n} + \binom{m}{k-1}$$

```
\[
   {m+1\choose n}={m\choose n}+{m\choose k-1}
\]
```

The amsmath package has better ways to typeset binomials (cf. Section 6.7 on page 99). *amsmath*

## 4.18  Boldmath — bold math font

If a complete formula should be typeset in bold font, it would be tedious to have to make all the individual elements bold with the \mathbf command. However, you also can't use text mode's \bfseries command as this has no effect in math mode:

04-18-1

text **text** $f(x) = a^2 + b^2 x$ text

```
text {\bfseries text $f(x)=a^2+b^2x$} text
```

This is because \bfseries switches to bold text font, but the bold math characters come from a different font. Instead, use the \boldmath command for math mode; it switches to the bold math font but does not affect the normal text. It must be inserted **outside** math mode, i.e. before the formula, however:

text text $f(x) = a^2 + b^2 x$ text

text \boldmath text $f(x)=a^2+b^2x$ text

04-18-2

\boldmath works as a switch, so remains valid until normal mode is enabled again with an \unboldmath command (which again needs to be inserted while in text mode).

A $f(x) = a^2 + b^2 x y = f(x)$ B

A \boldmath$f(x)=a^2+b^2x$\unboldmath $y=f(x)$ B

04-18-3

If either are used within math mode, they have no effect:

A $f(x) = a^2 + b^2 x \; y = f(x)$ B

A $\boldmath f(x)=a^2+b^2x$
\boldmath$\unboldmath y=f(x)$ B

04-18-4

An alternative to \boldmath is \mathversion. This command allows a global setting for math mode. Possible parameters are "normal" and "bold". This command, too, **must** be used in text mode.

```
\usepackage{tabularx}
\begin{tabularx}{\linewidth}{@{} XX @{}}
\[ \sum_{\makebox[0pt]{${\scriptscriptstyle 1\le i\le p\atop {%
        {1\le j\le q\atop 1\le k\le r}}}$}}a_{ij}b_{jk}c_{ki} \]
&
\mathversion{bold}
\[ \sum_{\makebox[0pt]{${\scriptscriptstyle 1\le i\le p\atop {%
        {1\le j\le q\atop 1\le k\le r}}}$}}a_{ij}b_{jk}c_{ki} \]
\end{tabularx}
```

$$\sum_{\substack{1 \le i \le p \\ 1 \le j \le q \\ 1 \le k \le r}} a_{ij} b_{jk} c_{ki}$$

$$\sum_{\substack{1 \le i \le p \\ 1 \le j \le q \\ 1 \le k \le r}} a_{ij} b_{jk} c_{ki}$$

04-18-5

| | |
|---|---|
| \boldmath | Switches to bold math font. **Must** be used outside math mode. |
| \unboldmath | Switches to normal math font. Also **must** be used outside math mode. |
| \mathversion{ ... } | Switches to *normal* or *bold* depending on the parameter. Also **must** be used outside math mode. |

If you only want to set part of a formula in bold font, one option is to use the \mathbf command. However, this always typesets its argument upright and not slanted, which may not be what you want:

$$\sum_{\substack{1 \le i \le p \\ 1 \le j \le q \\ 1 \le k \le r}} a_{ij} \mathbf{b_{jk}} c_{ki}$$

```
\[ \sum_{%
    \makebox[0pt]{${{\scriptscriptstyle 1\le i\le p\atop {%
        {1\le j\le q\atop 1\le k\le r}}}}$%
    $}}a_{ij}\mathbf{b_{jk}}c_{ki}
\]
```

04-18-6

Even the indices are typeset in upright font here. This can be prevented by reverting to text mode locally or by using the bm package, as shown in Section 9.6 on page 171.

Local reversal into text mode is possible with a \mbox command. To do this for Example 04-18-6, open an \mbox command, insert a \boldmath command, switch to math mode with the $ ... $ syntax and insert within it the expression \mathbf{b_{jk}}. Then insert an \unboldmath command and close the box. This works because \mbox switches into text mode, thereby allowing the \boldmath and \unboldmath commands to switch the font type.

04-18-7

$$\sum_{\substack{1\le i\le p\\1\le j\le q\\1\le k\le r}} a_{ij}\boldsymbol{b}_{jk}c_{ki}$$

```
\[ \sum_{%
    \makebox[0pt]{${\scriptscriptstyle 1\le i\le p\atop {%
       {1\le j\le q\atop 1\le k\le r}}}%
$}}a_{ij}\mbox{\boldmath$b_{jk}$\unboldmath}c_{ki} \]
```

However, when using \mbox, TEX does not know anything about the size of the font outside the box. It simply assumes that \displaystyle is the active font. If you are using an \mbox to make just an exponent or index bold, it will apply the wrong font size:

04-18-8

$$\sum_{\substack{1\le i\le p\\1\le j\le q\\1\le k\le r}} a_{ij}\boldsymbol{b}_{jk}c_{ki}$$

```
\[
\sum_{\makebox[0pt]{${\scriptscriptstyle 1\le i\le p\atop {%
       {1\le j\le q\atop 1\le k\le r}}}%
$}}a_{ij}b_{\mbox{\boldmath$jk$\unboldmath}}c_{ki}
\]
```

This can be fixed with a small trick — simply don't insert the _ character to switch to the index until you are **inside** the \mbox — \mbox{\boldmath$_{jk}$\unboldmath}. The result is then correct:

04-18-9

$$\sum_{\substack{1\le i\le p\\1\le j\le q\\1\le k\le r}} a_{ij}\boldsymbol{b}_{jk}c_{ki}$$

```
\[
\sum_{\makebox[0pt]{${\scriptscriptstyle 1\le i\le p\atop {%
       {1\le j\le q\atop 1\le k\le r}}}%
$}}a_{ij}b_{\mbox{\boldmath$_{jk}$\unboldmath}}c_{ki}
\]
```

Usually titles of chapters, sections, etc. are typeset in bold font, as are items in the description environment. However, we have already mentioned that any elements in math mode in titles are not affected by the default bold setting. Math elements have to be treated separately if bold font should be used for both math and text. Otherwise, the wrong result will occur:

04-18-10

# 1   Function $f(x) = x^2$

**This is** $y = f(x)$ only a demonstration.

**And** $z = f(x, y)$ another demonstration.

```
\section{Function $f(x)=x^2$}
\begin{description}
\item[This is $y=f(x)$] only a demonstration.
\item[And $z=f(x,y)$] another demonstration.
\end{description}
```

To apply a bold font for the math part in titles and the description environment, you can use the solutions presented above or you can redefine the corresponding environments as here:

```
\let\itemOld\item\makeatletter
\renewcommand\item[1][]{\def\@tempa{#1}%
  \ifx\@tempa\@empty\itemOld\else\boldmath\itemOld[#1]\unboldmath\fi}
\newcommand\Section[2][]{\def\@tempa{#1}
  \protect\boldmath\ifx\@tempa\@empty\section[#2]{#2}\else
    \section[#1]{#2}\fi\protect\unboldmath} \makeatother
\Section{Function $f(x)=x^2$}
\begin{description}
  \item[This is $y=f(x)$] only a demonstration.
  \item[And $z=f(x,y)$] another demonstration.
\end{description}
```

## 1 Function $f(x) = x^2$

04-18-11

This is $y = f(x)$ only a demonstration.

And $z = f(x, y)$ another demonstration.

## 4.19 Multiplication symbols

National differences do not only affect special characters (such as the decimal separator), but also some math symbols like the multiplication symbol. In European countries the dot (\cdot) is common, while Anglo-American countries tend to use the cross (\times). This is what B. N. Taylor has to say on the subject in his *Guide for the Use of the International System of Units (SI)*:

> When the dot is used as the decimal marker as in the United States, the preferred sign for the multiplication of numbers or values of quantities is a cross (that is, multiplication sign) ($\times$), not a half-high (that is, centred) dot ($\cdot$).
>
> [...]
>
> When the comma is used as the decimal marker, the preferred sign for the multiplication of numbers is the half-high dot. However, even when the comma is so used, this Guide prefers the cross for the multiplication of values of quantities.
>
> The multiplication of quantity symbols (or numbers in parentheses or values of quantities in parentheses) may be indicated in one of the following ways: $ab$, $a \cdot b$, $a \times b$. [25]

## 4.20 Further commands

### 4.20.1 \everymath and \everydisplay

\everymath lets you define an arbitrary expression that will be executed before **every** math equation in inline mode.

> \everymath{*argument*}

A fraction $\frac{3}{4}$ within a sentence.

A fraction $\frac{3}{4}$ within a sentence.

```
A fraction $\frac{3}{4}$ within a sentence.\\[5pt]
\everymath{\displaystyle}
A fraction $\frac{3}{4}$ within a sentence.
```

04-20-1

\everymath should be used with great care; remember that footnote numbers in the text *footnotes* are typeset by definition in math mode and will therefore also be affected by an \everymath command.

\everydisplay has the same purpose for displayed formulae. It does not affect footnote numbers, however.

$\boxed{\text{\everydisplay\{\textit{argument}\}}}$

04-20-2

Black text

$$f(x) = \oint_U (r) H \,\mathrm{d}r$$

Black text

```
\usepackage{color}

\everydisplay{\color{red}} Black text
\[ f(x)=\oint_U(r) H\,\mathrm{d}r \]
Black text
```

## 4.20.2 \underline

When underlining with the \underline command, the position of the line is determined solely by the width and depth of the box of space containing the elements that are to be underlined; it does not otherwise depend on the elements themselves. With display environments, the command must be used inside the environment.

```
$f(x)=\underline{x^2-1}$
\[ F(x)=\int \underline{f(x)\,\mathrm{d}x} \]
\[ \underline{F(x)=\int f(x)\,\mathrm{d}x} \]
```

04-20-3

$$f(x) = \underline{x^2 - 1}$$

$$F(x) = \int \underline{f(x)\,\mathrm{d}x}$$

$$\underline{F(x) = \int f(x)\,\mathrm{d}x}$$

## 4.20.3 \raisebox

The \raisebox command is not primarily associated with math, but can be very useful for translating certain parts of an equation vertically. \raisebox has several optional parameters, allowing the object to be translated or the surrounding box translated and scaled. This example summarizes the possibilities:

```
\newcommand*\Frac[2]{\displaystyle\frac{#1}{#2}}
left\rule{1cm}{.5pt}\fbox{$\Frac{A}{B}$}
\fbox{\raisebox{0pt}[0pt][0pt]{$\Frac{C}{D}$}}
\fbox{\raisebox{0pt}[0pt][1cm]{$\Frac{E}{F}$}}
\fbox{\raisebox{0pt}[1cm][0pt]{$\Frac{G}{H}$}}
\fbox{\raisebox{0pt}[1cm][1cm]{$\Frac{I}{J}$}}
\fbox{\raisebox{1cm}[0pt][0pt]{$\Frac{K}{L}$}}
\fbox{\raisebox{-1cm}{$\Frac{M}{N}$}}\rule{1cm}{.5pt}right
```

04-20-4

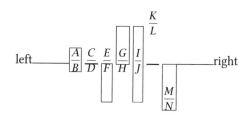

### 4.20.4  Differential coefficient

The differential coefficient is neither an operator nor a variable. Usually the *d* symbol is typeset upright and the following variable normally. It is best to define a new command so that you don't need to remember each time how to write it. This can then be used to display both the differential coefficient and the integration variable:

```
\newcommand*\dy{\,\mathrm{d}y}
\newcommand*\dx{\,\mathrm{d}x}\newcommand*\dyx{\,\frac{\mathrm{d}y}{\mathrm{d}x}}
\newcommand*\ds{\,\mathrm{d}s}\newcommand*\dt{\,\mathrm{d}t}
\newcommand*\dst{\,\frac{\mathrm{d}s}{\mathrm{d}t}}
\[ ay^\prime=a\dyx \quad c\times v = c\,\dst \quad
   a = \frac{\mathrm{d}^2t}{\dt^2} \quad \int\!\!\!\dy=\int\!\sin(x)\dx \]
```

$$ay' = a\,\frac{\mathrm{d}y}{\mathrm{d}x} \quad c \times v = c\,\frac{\mathrm{d}s}{\mathrm{d}t} \quad a = \frac{\mathrm{d}^2t}{\mathrm{d}t^2} \quad \int \mathrm{d}y = \int \sin(x)\,\mathrm{d}x$$

04-20-5

The space `\,` inserted before the differential coefficient in the definitions of the commands helps the layout of derivatives to look good, but it is not so useful when an integral is immediately preceding the expression. In these cases it looks better to have the integrand closer to the integral. So, as in the last example above, use `\int\!\!\!\dy` to manually cancel the additional space inserted by `\dy` and add an additional negative space.

Another option for typesetting derivatives by Luciano Battaia is to define a single command that then expects a parameter. This gives you even more flexibility to change the spacings. When only single characters are given as arguments, the argument's parentheses can be omitted as in `\diff x`.

```
\newcommand*\diff{\mathop{}\!\mathrm{d}}
\[ay^\prime=a\frac{\diff y}{\diff x}\quad c\times v = c\,\frac{\diff s}{\diff t}\quad
   a = \frac{\diff{^2t}}{\diff{t^2}} \quad \int\!\!\!\diff y=\int\!\sin(x)\diff x \]
```

$$ay' = a\,\frac{\mathrm{d}y}{\mathrm{d}x} \quad c \times v = c\,\frac{\mathrm{d}s}{\mathrm{d}t} \quad a = \frac{\mathrm{d}^2t}{\mathrm{d}t^2} \quad \int \mathrm{d}y = \int \sin(x)\,\mathrm{d}x$$

04-20-6

Short forms for commands may be defined at any time; avoid single-letter command names though as LaTeX and TeX frequently use these for commands themselves.

In all of the following examples short form commands have not been used; instead the long form is employed. This is just to make it easier for you to understand the coding without having to look up the definitions of the short forms.

Chapter **5**

# Colour in math expressions

To use colour in math expressions, you should need only the common `color` package or the extended `xcolor` package. They provide the predefined colours `black`, `red`, `green`, `blue`, `cyan`, `magenta`, `yellow` and, of course, `white`. You can also define additional colours with a `\definecolor` command. The `xcolor` package offers some significant advantages — it supports much more complex colour expressions, for example:

05-00-1

red
red!75
red!75!blue!100
red!75!blue!40
red!75!blue!40!yellow!50

-red
-red!75
-red!75!blue!100
-red!75!blue!40
-red!75!blue!40!yellow!50

```
\usepackage{xcolor}

\colorbox{red}{red}          \\
\colorbox{red!75}{red!75}\\
\colorbox{red!75!blue!100}{\color{white}red!75!blue!100}\\
\colorbox{red!75!blue!40}{red!75!blue!40} \\
\colorbox{red!75!blue!40!yellow!50}%
   {red!75!blue!40!yellow!50} \\[6pt]
%
\colorbox{-red}{-red} \\
\colorbox{-red!75}{-red!75} \\
\colorbox{-red!75!blue!100} {-red!75!blue!100} \\
\colorbox{-red!75!blue!40}{\color{white}-red!75!blue!40}\\
\colorbox{-red!75!blue!40!yellow!50}%
   {\color{white}-red!75!blue!40!yellow!50}
```

## 5.1 Partial colouring

You can change the colour of individual words with the \textcolor command; it can also be used to colour individual parts of a math formula.

> \textcolor{*colour*}{*text*}

Despite the confusing name of the command, \textcolor can be used in math mode without having to leave it. In Example 05-01-1 the variable $x$ is in italics so the math mode font style is still active:

$$\int_1^2 \frac{1}{x^2}\, dx = 0.5$$

```
\usepackage{xcolor}
\newcommand*\diff{\mathop{}\!\mathrm{d}}
```

05-01-1

```
\[
    \int_1^2\textcolor{red}{\frac{1}{x^2}}\diff x=0.5
\]
```

Even trivial math expressions can be made clearer by the use of colour, as in Example 05-01-2:

```
\usepackage{amsmath,xcolor}
\begin{align*}
y &= 2x^2 -3x +5\\\nonumber
  & \hphantom{= \ 2\left(x^2-\frac{3}{2}\,x\right. }%
       \textcolor{blue}{\overbrace{\hphantom{+\left(\frac{3}{4}\right)^2- %
       \left(\frac{3}{4}\right)^2}}^{=0}}\\[-11pt]
  &= 2\left(\textcolor{red}{%
     \underbrace{x^2-\frac{3}{2}\,x + \left(\frac{3}{4}\right)^2}}%
     \underbrace{-\left(\frac{3}{4}\right)^2 + \frac{5}{2}}\right)\\
  &= 2\left(\qquad\textcolor{red}{\left(x-\frac{3}{4}\right)^2}
     \qquad + \ \frac{31}{16}\qquad\right)\\
y\,\textcolor{blue}{\mathrel{-}\frac{31}{8}}
  &= 2\left(x\,\textcolor{cyan}{\mathrel{-}\frac{3}{4}}\right)^2%
\end{align*}
```

$$y = 2x^2 - 3x + 5$$

05-01-2

$$= 2\left(x^2 - \frac{3}{2}x + \overbrace{\left(\frac{3}{4}\right)^2 - \left(\frac{3}{4}\right)^2}^{=0} + \frac{5}{2}\right)$$

$$= 2\left(\left(x - \frac{3}{4}\right)^2 + \frac{31}{16}\right)$$

$$y - \frac{31}{8} = 2\left(x - \frac{3}{4}\right)^2$$

## 5.2 Complete colouring

In inline mode, entire math expressions can be typeset in colour by putting the whole expression as the argument in a \textcolor command. It does not matter whether the \textcolor command occurs inside math mode $\texttt{\textcolor\{}\textit{blue}\texttt{\}\{}\textit{y=x\textasciicircum2+px+q}\texttt{\}}$ or outside math mode $\texttt{\textcolor\{}\textit{blue}\texttt{\}\{}\textit{\$y=x\textasciicircum2+px+q\$}\texttt{\}}$.

Displayed formulae can be coloured completely as well. You can make a global switch to another colour with a $\texttt{\color\{}\textit{colour}\texttt{\}}$ declaration; simply put it at the beginning of the displayed environment. Example 05-02-1 illustrates the use of $\texttt{\color}$. The second expression is also partially coloured red as well.

```
\usepackage{xcolor,amsmath,amscd}
\[ \color{magenta}\left(\prod^n_{\, j=1}\hat{x}_j\right)H_c=%
    \frac{1}{2}\hat{k}_{ij}\det\widehat{\mathbf{K}}(i|i) \]

\begin{align}\color{blue}
\begin{CD}
  R\times S\times T @>\text{restriction}>> S\times T \\
        @VprojVV                          @VVprojV \\
  R\times S             @<<\text{\textcolor{red}{inclusion}}< S
\end{CD}
\end{align}
```

05-02-1

$$\left(\prod_{j=1}^{n} \hat{x}_j\right) H_c = \frac{1}{2}\hat{k}_{ij}\det\widehat{\mathbf{K}}(i|i)$$

$$R \times S \times T \xrightarrow{\text{restriction}} S \times T$$
$$proj\Big\downarrow \qquad\qquad \Big\downarrow proj \qquad\qquad (5.1)$$
$$R \times S \xleftarrow[\text{inclusion}]{} S$$

## 5.3 Coloured boxes

The $\texttt{\colorbox}$ and $\texttt{\fcolorbox}$ commands were mentioned in Section 2.2.4 on page 9. The $\texttt{\colorbox}$ command requires one colour to be specified, whereas the $\texttt{\fcolorbox}$ command needs an additional colour to be specified for the frame.

> $\texttt{\colorbox\{}\textit{colour}\texttt{\}\{}\textit{argument}\texttt{\}}$
> $\texttt{\fcolorbox\{}\textit{frame colour}\texttt{\}\{}\textit{box colour}\texttt{\}\{}\textit{argument}\texttt{\}}$

Both commands can be used to colour boxes in math mode as well as text mode However, unlike $\texttt{\textcolor}$, $\texttt{\colorbox}$ and $\texttt{\fcolorbox}$ expect the argument to be in text mode. If you want to use them after math mode has started, you have to deactivate it and activate it again afterwards (Example 05-03-1).

```
\usepackage{xcolor,amsmath}
\def\xstrut{\vphantom{\dfrac{(A)^1}{(B)^1}}}
\[ E = \colorbox{lightgray}{$\xstrut a_vA$} -
    \colorbox{magenta!30}{$\xstrut a_fA^{2/3}$} -
```

```
\colorbox{green!60}{$\xstrut a_c\dfrac{Z(Z-1)}{A^{1/3}}$} -
\colorbox{cyan}{$\xstrut a_s\dfrac{(A-2Z)^2}{A}$} +
\colorbox{yellow}{$\xstrut E_p$} \]
```

$$E = \boxed{a_v A} - \boxed{a_f A^{2/3}} - \boxed{a_c \frac{Z(Z-1)}{A^{1/3}}} - \boxed{a_s \frac{(A-2Z)^2}{A}} + E_p$$

05-03-1

This example uses a `\colorbox` for a formula in inline mode:

$\boxed{x_{n+1} \leftarrow x_n^2 + c}$ seen in the set of complex numbers, related to Benoît Mandelbrot.

```
\usepackage{xcolor}

\colorbox{yellow}{$x_{n+1}\leftarrow x_n^2+c$}
seen in the set of complex numbers,
related to Beno\^it Mandelbrot.
```

05-03-2

The `\fcolorbox` command can also be used for a formula in inline mode. It is equivalent to using `\fbox`: $\boxed{x_{n+1} \leftarrow x_n^2 + c}$.

Colouring displayed formulae that have an equation number is a bit more tricky as the number could appear inside (see Section 3.6 on page 26) or outside the coloured box. The latter is only possible if the displayed formula is inside a `\parbox`:

```
\usepackage{xcolor,amsmath,calc}
\colorbox{black!20}{%
\parbox{\linewidth-2\fboxsep}{%
  \begin{align}
    \frac{I(\alpha )}{I_{0}}=
    \begin{cases}
      \sqrt{1-\frac{\alpha}{\pi}+\frac{1}{2\pi}\sin 2\alpha} & \varphi =0\\[0.4cm]
      \sqrt{2\left(1-\frac{\alpha}{\pi}\right)\left(2+\cos 2\alpha \right)+
        \frac{3}{\pi}\sin 2\alpha} & \varphi =\frac{\pi}{2}
    \end{cases}
  \end{align}}}
```

05-03-3

$$\frac{I(\alpha)}{I_0} = \begin{cases} \sqrt{1 - \frac{\alpha}{\pi} + \frac{1}{2\pi}\sin 2\alpha} & \varphi = 0 \\ \sqrt{2\left(1 - \frac{\alpha}{\pi}\right)(2 + \cos 2\alpha) + \frac{3}{\pi}\sin 2\alpha} & \varphi = \frac{\pi}{2} \end{cases} \qquad (5.2)$$

If on the other hand there is no equation number, and you only want to colour the area that the formula really occupies, you will need to know the width of the `\parbox` when defining it. The empheq package by Morten Høgholm will help you determine the width of the box. It does this independent of any equation numbers. It is described in detail in Section 9.12 on page 176.

$$\boxed{\begin{aligned} x(t) &= \frac{a\left(t^2 - 1\right)}{t^2 + 1} \\ y(t) &= \frac{at\left(t^2 - 1\right)}{t^2 + 1} \end{aligned}}$$

```
\usepackage{xcolor,empheq}

\begin{empheq}[box=\fcolorbox{blue}{cyan!40}]{align*}
        x(t) & =\frac{a\left(t^{2}-1\right)}{t^2+1}\\
        y(t) & =\frac{at\left(t^{2}-1\right)}{t^2+1}
\end{empheq}
```

05-03-4

For equations with a coloured background, one can use also the `framed` package by Donald Arseneau. It defines the `shaded` environment, which can then be used to encompass several different math environments and which works across page breaks.

```
\usepackage{xcolor,framed,amsmath,esint}
\newcommand*\Q[2]{\frac{\partial #1}{\partial #2}}
\definecolor{shadecolor}{cmyk}{0.2,0.1,0.9,0.1}
\begin{shaded}
\begin{align}
\underset{\mathcal{G}\quad}\iiint\!%
        \left[u\nabla^{2}v+\left(\nabla u,\nabla v\right)\right]\mathrm{d}^{3}V%
        =\underset{\mathcal{S}\quad}\oiint u\Q{v}{n}\mathrm{d}^{2}A
\end{align}
\begin{align}
\underset{{\mathcal{G}\quad}}\iiint\!%
        \left[u\nabla^{2}v-v\nabla^{2}u\right]\mathrm{d}^{3}V%
        =\underset{\mathcal{S}\quad}\oiint%
        \left(u\Q{v}{n}-v\Q{u}{n}\right)\mathrm{d}^{2}A
\end{align}
\end{shaded}
```

05-03-5

$$\iiint_{\mathcal{G}} \left[ u\nabla^2 v + (\nabla u, \nabla v) \right] \mathrm{d}^3 V = \oiint_{S} u \frac{\partial v}{\partial n} \mathrm{d}^2 A \qquad (5.3)$$

$$\iiint_{\mathcal{G}} \left[ u\nabla^2 v - v\nabla^2 u \right] \mathrm{d}^3 V = \oiint_{S} \left( u \frac{\partial v}{\partial n} - v \frac{\partial u}{\partial n} \right) \mathrm{d}^2 A \qquad (5.4)$$

## 5.4 Coloured tables or arrays

In principle there is no difference between colouring the individual cells of a table or a matrix.

05-04-1

$$\underline{A} = \begin{pmatrix} A & B & C \\ A & BBB & C \\ A & B & C \end{pmatrix}$$

```
\usepackage[table]{xcolor}

\[ \underline{A} =
    \left(\begin{array}{c>{\columncolor{magenta}}cc}
        A & B & C\\
        \rowcolor{cyan} A & \cellcolor{white} BBB & C\\
        A & B & C
    \end{array}\right) \]
```

The `\columncolor`, `\rowcolor`, and `\cellcolor` commands can be used to create any design. These require the `colortbl` package by David Carlisle or the `xcolor` package by Uwe Kern, which has to be loaded with the `table` option.

An alternative option to using \cellcolor is to define a single cell with \multicolumn in order to be able to use \columncolor for it.

```
\usepackage{array}
\usepackage[table]{xcolor}
\definecolor{umbra}{rgb}{0.8,0.8,0.5}
\newcommand*\zero{\multicolumn{1}{>{\columncolor{white}}c}{0}}
\newcommand\colCell[2]{\multicolumn{1}{>{\columncolor{#1}}c}{#2}}
\delimitershortfall=-3pt
\[
\left[\,
\begin{array}{*{5}{>{\columncolor[gray]{0.95}}c}}
  h_{k,1,0}(n) & h_{k,1,1}(n) & h_{k,1,2}(n) & \zero & \zero \\
  h_{k,2,0}(n) & h_{k,2,1}(n) & h_{k,2,2}(n) & \zero & \zero \\
  h_{k,3,0}(n) & h_{k,3,1}(n) & h_{k,3,2}(n) & \zero & \zero \\
  h_{k,4,0}(n) & \colCell{umbra}{h_{k,4,1}(n)} & h_{k,4,2}(n) & \zero & \zero \\
  \zero & h_{k,1,0}(n-1) & h_{k,1,1}(n-1) & h_{k,1,2}(n-1) & \zero \\
  \zero & h_{k,2,0}(n-1) & h_{k,2,1}(n-1) & h_{k,2,2}(n-1) & \zero \\
  \zero & h_{k,3,0}(n-1) & h_{k,3,1}(n-1) & h_{k,3,2}(n-1) & \zero \\
  \zero & \colCell{umbra}{h_{k,4,0}(n-1)} & h_{k,4,1}(n-1) & h_{k,4,2}(n-1) & \zero \\
  \zero & \zero & h_{k,1,0}(n-2) & h_{k,1,1}(n-2) & h_{k,1,2}(n-2)\\
\rowcolor[gray]{0.75}%
  \zero & \zero & h_{k,2,0}(n-2) & h_{k,2,1}(n-2) & h_{k,2,2}(n-2)\\
  \zero & \zero & h_{k,3,0}(n-2) & h_{k,3,1}(n-2) & h_{k,3,2}(n-2)\\
  \zero & \zero & h_{k,4,0}(n-2) & h_{k,4,1}(n-2) & h_{k,4,2}(n-2)
\end{array} \,\right]_{12\times 5}\]
```

$$
\left[
\begin{array}{ccccc}
h_{k,1,0}(n) & h_{k,1,1}(n) & h_{k,1,2}(n) & 0 & 0 \\
h_{k,2,0}(n) & h_{k,2,1}(n) & h_{k,2,2}(n) & 0 & 0 \\
h_{k,3,0}(n) & h_{k,3,1}(n) & h_{k,3,2}(n) & 0 & 0 \\
h_{k,4,0}(n) & h_{k,4,1}(n) & h_{k,4,2}(n) & 0 & 0 \\
0 & h_{k,1,0}(n-1) & h_{k,1,1}(n-1) & h_{k,1,2}(n-1) & 0 \\
0 & h_{k,2,0}(n-1) & h_{k,2,1}(n-1) & h_{k,2,2}(n-1) & 0 \\
0 & h_{k,3,0}(n-1) & h_{k,3,1}(n-1) & h_{k,3,2}(n-1) & 0 \\
0 & h_{k,4,0}(n-1) & h_{k,4,1}(n-1) & h_{k,4,2}(n-1) & 0 \\
0 & 0 & h_{k,1,0}(n-2) & h_{k,1,1}(n-2) & h_{k,1,2}(n-2) \\
0 & 0 & h_{k,2,0}(n-2) & h_{k,2,1}(n-2) & h_{k,2,2}(n-2) \\
0 & 0 & h_{k,3,0}(n-2) & h_{k,3,1}(n-2) & h_{k,3,2}(n-2) \\
0 & 0 & h_{k,4,0}(n-2) & h_{k,4,1}(n-2) & h_{k,4,2}(n-2)
\end{array}
\right]_{12\times5}
$$

05-04-2

Chapter 6

# $\mathcal{AMS}$ packages

## 6.1 Introduction

The American Mathematical Society (AMS) recognised early on the significant benefits of TeX and later LaTeX and quickly started to publish extensions.[1] This productive phase resulted in many packages and classes; this can cause confusion now, with users unsure which class or which package to use. We only describe packages here; the document classes refer to them and they are not the primary topic of this book. This is no disadvantage however; the document classes amsbook and amsart use the amsmath packages by definition and differ from the standard LaTeX classes only in small details. $\mathcal{A}_{\mathcal{M}}\mathcal{S}$ here means using the packages, not the classes.

This chapter primarily deals with the amsmath package. Loading this package also loads the amstext, amsbsy, and amsopn packages, plus the amsgen package that is required by amstext.

```
\RequirePackage{amstext}[1995/01/25] \RequirePackage{amsbsy}[1995/01/20]
\RequirePackage{amsopn}[1995/01/20]  \RequirePackage{amsgen}
```

Table 6.1 lists a selection of the packages and describes their significance:

Table 6.1: Summary of the most important packages from the project of the $\mathcal{A}_{\mathcal{M}}\mathcal{S}$.

| name | description |
| --- | --- |
| amsmath | This is a LaTeX package that provides a variety of extra math features, largely derived from $\mathcal{A}_{\mathcal{M}}\mathcal{S}$TeX. |
| amsbsy | This is a LaTeX package that provides a command for producing bold math symbols when appropriate fonts exist, and a "poor man's bold" command that can be applied when no appropriate bold font is available. |
| amsopn | This is a LaTeX package that provides a DeclareMathOperator command for defining named operators like sin and lim. |
| amstext | This is a LaTeX package that defines a \text command, which makes it easy to incorporate fragments of text inside a displayed equation or a sub- or superscript. Font sizes are automatically scaled in sub-/superscripts. |
| amsgen | It contains some general internal commands shared by several different files in $\mathcal{A}_{\mathcal{M}}\mathcal{S}$LaTeX. |
| amscd | This is a LaTeX package that adapts the commutative diagram commands of $\mathcal{A}_{\mathcal{M}}\mathcal{S}$TeX for use in LaTeX. |
| amsrefs | This is a LaTeX package that permits bibliography style to be controlled completely from the LaTeX side instead of being determined chiefly by the BibTeX style file. |
| amsthm | This is the source for the amsthm package and three $\mathcal{A}_{\mathcal{M}}\mathcal{S}$document classes: amsart, amsproc and amsbook. They are designed for use with LaTeX. |
| upref | This is a LaTeX package that provides printing of \refs in an upright font even if the current context is italic or slanted. |

The most important package, amsmath, can be loaded with a choice of options listed in Table 6.2.

---

[1]Also see http://www.ams.org/tex/author-info.html

Table 6.2: Summary of the package options for the amsmath package

| name | description |
| --- | --- |
| centertags | (default) For a split equation, centre equation numbers vertically on the total height of the equation. |
| tbtags | "Top-or-bottom tags" For a split equation, place equation numbers level with the last line if numbers are on the right, or level with the first line if numbers are on the left. |
| sumlimits | (default) In displayed equations, place the subscripts and superscripts of summation symbols above and below. This option also applies to other symbols of the same type – $\prod$, $\bigsqcup$, $\otimes$, $\oplus$, and so forth – but does not apply to integrals (see below). |
| nosumlimits | Always place the subscripts and superscripts of summation-type symbols to the side, even in displayed equations. |
| intlimits | Like sumlimits, but for integral symbols. |
| nointlimits | (default) Opposite of intlimits. |
| namelimits | (default) Like sumlimits, but for certain 'operator names' such as \det, \inf, \lim, \max, \min, that traditionally have subscripts placed underneath when they occur in a displayed equation. |
| nonamelimits | Opposite of namelimits. |

The activation or the loading of the options is done in the usual LaTeX way:

\usepackage [settings] {amsmath}

Alternatively, the options can be given through the optional field of the document class and then passed on to the amsmath package. In principle there is no difference between the two methods; however, for options that may be evaluated by other packages as well it is advisable to pass them through the document class. General options that are also evaluated by LaTeX are summarised in Table 6.3.

Table 6.3: Summary of the important document class options

| name | description |
| --- | --- |
| leqno | Left-align equation numbers. |
| reqno | (default) Right-align equation numbers. |
| fleqn | Place equations with a fixed space to the left margin. amsmath defines the rubber length \mathindent and takes this length into account only on the left side of the equation. |

Remember that the amsmath package offers extensions only for typesetting displayed formulae. It does not extend functionality in inline mode as shown in Chapter 2 on page 5.

*only displayed formulae*

## 6.2 `align` environments

In the `amsmath` package there are in principle three different alignment environments, which are summarised in examples 06-02-1 to 06-02-3 and explained in detail below. The code for the graphical display follows the following principle:

```
\begin{<name>}
   <name> & = x & x & = x\\
   <name> & = x & x & = x
\end{<name>}
```

06-02-1

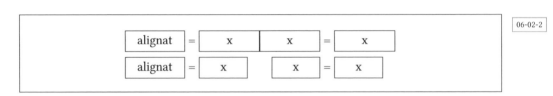

06-02-2

06-02-3

These align environments are different to the `eqnarray` environment from standard LaTeX (cf. page 3.2.3 on page 19) in that the "three" parts of an equation (expression — symbol — expression) require just a *single* ampersand (tabbing character).

    &    Always put the column separator & *in front of* the math symbol (y &= x). Otherwise the horizontal spacing may be wrong.
           There is no limit to the number of & signs.
           The & separator does not have to be followed by a math symbol. It can also be used as a normal column separator like in a table.

    The importance of placing the column separator correctly is shown in the following three equations. The & is placed correctly in the first equation. However, in the second equation y =& x the & is in the wrong place, after the equals sign. This prevents the correct horizontal spacing; LaTeX looks at the type of the following character, the $x$, which is not a math symbol, so it inserts normal character spacing. The result is that the spacing between the $y$ and the equals sign is correct, but the space before the $x$ is too small. In the last equation, the curly braces prevent the spacing allocated after the equals sign from being changed.

06-02-4

$$y = x$$
$$y = x$$
$$y = x$$

```
\usepackage{amsmath}

\begin{align*}  y &= x    \end{align*}
\begin{align*}  y =& x    \end{align*}
\begin{align*}  y ={}& x \end{align*}
```

The equations in Example 03-05-1 already demonstrated that using an eqnarray environment results in too much horizontal spacing. Using a standard align environment for the same equations shows instantly that the amsmath package has significant advantages over standard LaTeX:

06-02-5

$$y = d \tag{6.1}$$
$$y = cx + d \tag{6.2}$$
$$y_{12} = bx^2 + cx + d \tag{6.3}$$
$$y(x) = ax^3 + bx^2 + cx + d \tag{6.4}$$

```
\usepackage{amsmath}

\begin{align}
y       &= d\\
y       &= cx+d\\
y_{12}  &= bx^2+cx+d\\
y(x)    &= ax^3+bx^2+cx+d
\end{align}
```

Unlike the eqnarray environment, the align environment consists of right-left sequences ($\lfloor rlrl\ldots\rfloor$) with corresponding vertical alignment for all lines; this can be seen in Example 06-02-1 on the preceding page, where the first element is right-aligned, the second is left-aligned, the third is right-aligned and the fourth is left-aligned. This means that if there are only three parts to the equation the result is strange, with the third element spaced along way apart from the first two:

06-02-6

$$12 \qquad\qquad 3$$
$$AB \qquad\qquad C$$

```
\usepackage{amsmath}

\begin{align*}
  1 & 2 & 3\\  A & B & C
\end{align*}
```

```
\begin{align * }
    left & right & left & right...\\
    left & right & left & right...\\
    ...
    left & right & left & right...
\end{align * }
```

The individual parts of an equation form *left–right* blocks, which are distributed symmetrically and evenly across the line such that they are usually horizontally centred. The individual parts are always aligned symmetrically to the separator on a line or, to put it another way, only the "invisible" separator characters are spaced evenly and symetrically across the line and the visible parts of the equation are placed after that.

We can improve the layout of the example above by adding an even number of extra separation characters at both the start and the end of each line. This increases the number of *left–right* blocks and therefore reduces the distance between them, pushing the third part "closer" to the previous parts of the equation (number of separator characters) while keeping the equation symmetric to the centre of the line (symmetric extension):

$$12 \qquad 3$$

$$AB \qquad C$$

```
\usepackage{amsmath}

\begin{align*}
    &&&&  1 & 2 & 3 &&&&\\
    &&&&  A & B & C &&&&
\end{align*}
```

06-02-7

The sequence above then corresponds to

*L & R & L & R & 1 & 2 & 3 & R & L & R & L*

The "1" is therefore a left-hand side expression and the "2" a right-hand side expression which are arranged together. The following "3" belongs to the next block and is therefore translated horizontally until the whole line is symmetric. This example is a bit contrived, but it does illustrate the main principle of column separators and their behaviour in a `align` environment. Both can be transferred to the environments described below.

Equation numbers are put on the right-hand side by default unless the user specified otherwise when loading the `amsmath` package or in the document class options. This default placing is shown below for the `flalign` environment, which only differs from `align` in that it is left- and right-aligned on the line if at least two *left–right* sequences exist:

| flalign | = | x | | x | = | x | (6.5) |

06-02-8

The same equations created with the document class option `leqno` (i.e./ left-aligned equation numbers) result in a comparable arrangement. If there is not enough space on the line for the equation number and the individual *left–right* blocks, the equation is placed underneath the equation number.

The fourth line in Example 06-02-9 has pushed the equation number onto the next line. This shows that the space between the blocks can't fall below a fixed amount, called the `\minalignsep`. The default for this space is 10pt. This length is defined as a command, so to change it you must use a `\renewcommand`. In the example, `\minalignsep` was reduced just before the last equation line, which then allowed the equation number to fit on the same line:

\minalignsep

*command*

```
\documentclass[leqno]{article}
\usepackage{amsmath,calc}
\fbox{\parbox{\linewidth-2\fboxsep-2\fboxrule}{%
\begin{align} y &= f(x) \end{align}
\begin{align} y &= f(x) & y &= f(x) & y &= f(x) \end{align}
\begin{align} y &= f(x) & y &= f(x) & y &= f(x) & y &= f(x) & y &= f(x) \end{align}
\begin{align}
y &= f(x) & y &= f(x) & y &= f(x) & y &= f(x) & y &= f(x) & y &= f(x)  & y &= f(x)
\end{align}
\renewcommand\minalignsep{2pt}% reduction of the minimum space
\begin{align}
y &= f(x) & y &= f(x) & y &= f(x) & y &= f(x) & y &= f(x) & y &= f(x)  & y &= f(x)
\end{align}}}
```

$$(6.6) \qquad\qquad\qquad y = f(x)$$

$$(6.7) \qquad y = f(x) \qquad\qquad y = f(x) \qquad\qquad y = f(x)$$

$$(6.8) \qquad y = f(x) \qquad y = f(x) \qquad y = f(x) \qquad y = f(x) \qquad y = f(x)$$

$$(6.9) \quad y = f(x) \quad y = f(x) \quad y = f(x) \quad y = f(x) \quad y = f(x) \quad y = f(x) \quad y = f(x)$$

$$(6.10) \quad y = f(x) \quad y = f(x) \quad y = f(x) \quad y = f(x) \quad y = f(x) \quad y = f(x) \quad y = f(x)$$

The obsolete, but supported xxalignat environment does not have an equation number by default (and therefore has no starred version to suppress its output). For more information on this environment see Section 6.20 on page 122.

In the in examples 06-02-1 to 06-02-3 on page 80, all environments were used as their starred version (i.e. with equation numbers suppressed). Alternatively, if you just want to exclude certain lines from the numbering you can use the \nonumber command as usual (cf. Section 3.4 on page 21).

```
\usepackage{amsmath}
\begin{align}
y       &= d              & z &= 1                            \\
y       &= cx+d           & z &= x+1             \nonumber\\
y_{12}  &= bx^2+cx+d      & z &= x^2+x+1         \nonumber\\
y(x)    &= ax^3+bx^2+cx+d & z &= x^3+x^2+x+1
\end{align}
```

$$y = d \qquad\qquad\qquad z = 1 \qquad\qquad\qquad (6.11)$$
$$y = cx + d \qquad\qquad\qquad z = x + 1$$
$$y_{12} = bx^2 + cx + d \qquad\qquad z = x^2 + x + 1$$
$$y(x) = ax^3 + bx^2 + cx + d \qquad z = x^3 + x^2 + x + 1 \qquad (6.12)$$

Whichever environment you are using, don't put an end-of-line character \\ on the last line because otherwise the following (empty) line will also get an equation number (6.14 in the following example): *Don't end last line with \\!*

$$2x + 3 = 7 \qquad\qquad (6.13)$$
$$(6.14)$$
$$2x + 3 = 7 \qquad\qquad (6.15)$$

```
\usepackage{amsmath}

\begin{align}
    2x+3 &= 7\\% <=== wrong
\end{align}
\begin{align}
    2x+3 &= 7   % <=== right
\end{align}
```

You can use the align environment within a custom theorem environment:

```
\usepackage{amsmath} \usepackage[amsmath,thmmarks]{ntheorem}
\theoremsymbol{\ensuremath{\triangle}}
\theoremprework{\bigskip\hrule}  \theorempostwork{\hrule\bigskip}
\newtheorem{example}{example}[section]
Normal text before the new environment.
\begin{example} Use of the transformation theorem:
\begin{align*}
  [z^n]C(z) &= [z^n]\biggl[\frac{e^{3/4}}{\sqrt{1-z}}+e^{-3/4}(1-z)^{1/2}+
             \frac{e^{-3/4}}{4}(1-z)^{3/2}+O\Bigl((1-z)^{5/2}\Bigr)\biggr]\\[5pt]
          &= \frac{e^{-3/4}}{\sqrt{\pi n}}-\frac{5e^{-3/4}}{8\sqrt{\pi n^3}}+
             \frac{e^{-3/4}}{128\sqrt{\pi n^5}}+O\biggl(\frac{1}{\sqrt{\pi n^7}}\biggr)
\end{align*}
\end{example}
```

Normal text before the new environment.

06-02-12

---

**example 6.1** *Use of the transformation theorem:*

$$[z^n]C(z) = [z^n]\left[ \frac{e^{3/4}}{\sqrt{1-z}} + e^{-3/4}(1-z)^{1/2} + \frac{e^{-3/4}}{4}(1-z)^{3/2} + O\left((1-z)^{5/2}\right)\right]$$

$$= \frac{e^{-3/4}}{\sqrt{\pi n}} - \frac{5e^{-3/4}}{8\sqrt{\pi n^3}} + \frac{e^{-3/4}}{128\sqrt{\pi n^5}} + O\left(\frac{1}{\sqrt{\pi n^7}}\right) \qquad \triangle$$

---

## 6.2.1 `alignat` environment

The name of the `alignat` environment is derived from "align at several places". It corresponds to two separate `align` environments side by side. The main difference is the parameter, which specifies the number of these side by side objects (blocks) which are separated by another & column separator. A parameter value of *2* implies that 3 column separators have to occur on a line. In general $2n - 1$ column separators must occur for a parameter value of $n$. If the specification of the number of blocks is missing, LaTeX does not output an error message.

The syntax is:

```
\begin{alignat * }{number of blocks}
      left & right & left & right...\\
   ...
      left & right & left & right...
\end{alignat * }
```

If we use the `alignat` environment to typeset the equations in Example 06-02-10 on the preceding page, the result is only slightly different but the blocks collide in the middle.

```
\usepackage{amsmath}
\begin{alignat}{2}
y        &= d              & z &= 1                        \\ % <=== 3 times &
y        &= cx+d           & z &= x+1           \nonumber\\
y_{12} &= bx^2+cx+d      & z &= x^2+x+1       \nonumber\\
y(x)   &= ax^3+bx^2+cx+d & z &= x^3+x^2+x+1
\end{alignat}
```

06-02-13

$$y = d \qquad\qquad z = 1 \qquad\qquad\qquad (6.16)$$

$$y = cx + d \qquad\qquad z = x + 1$$

$$y_{12} = bx^2 + cx + d \qquad z = x^2 + x + 1$$

$$y(x) = ax^3 + bx^2 + cx + dz = x^3 + x^2 + x + 1 \qquad (6.17)$$

On each line of the next example there are three different parts of equations, separated by five & symbols ($2 \times 3 - 1 = 5$):

```
\usepackage{amsmath}
\begin{alignat}{3}
  i_{11} &= 0.25 & i_{12} & =i_{21} & i_{13} & =i_{23}\nonumber \\
  i_{21} &= \frac{1}{3}i_{11} & i_{22} & =0.5i_{12} & i_{23} & =i_{31}\\
  i_{31} &= 0.33i_{22}\quad & i_{32} & =0.15i_{32}\quad & i_{33} & =i_{11}
\end{alignat}
```

06-02-14

$$i_{11} = 0.25 \qquad i_{12} = i_{21} \qquad i_{13} = i_{23}$$

$$i_{21} = \frac{1}{3}i_{11} \qquad i_{22} = 0.5i_{12} \qquad i_{23} = i_{31} \qquad (6.18)$$

$$i_{31} = 0.33i_{22} \quad i_{32} = 0.15i_{32} \quad i_{33} = i_{11} \qquad (6.19)$$

This shows the usefulness of the `alignat` environment for aligning several equations vertically at different separators. This still works even if the expressions are of quite different lengths:

```
\usepackage{amsmath}
\begin{alignat}{3}
  abc &= xxx                 &&= xxxxxxxxxxxx &&= aaaaaaaaa \\
  ab  &= yyyyyyyyyyyyyyyy &&= yyyy             &&= ab
\end{alignat}
```

06-02-15

$$abc = xxx \qquad\qquad\qquad = xxxxxxxxxxxx = aaaaaaaaa \qquad (6.20)$$

$$ab = yyyyyyyyyyyyyyyy = yyyy \qquad\qquad = ab \qquad (6.21)$$

### 6.2.2 **flalign** environment

The `flalign` environment replaces the obsolete `xalignat` and `xxalignat` environments (cf. Section 6.20 on page 122). `flalign` is not much different except that the first block on each line is left-aligned and the blocks are spaced further apart.

Completely normal and pointless text before the flalign environment.

$$i_{11} = 0.25$$

$$i_{21} = \frac{1}{3}i_{11} \tag{6.22}$$

$$i_{31} = 0.33i_{22} \tag{6.23}$$

```
\usepackage{amsmath}

Completely normal and pointless text before
the \texttt{flalign} environment.
\begin{flalign}
    i_{11} &= 0.25      \nonumber \\
    i_{21} &= \frac{1}{3}i_{11}\\
    i_{31} &= 0.33i_{22}
\end{flalign}
```

06-02-16

You can see in this example that the equations are not left-aligned (although they were in the previous example of the flalign environment in Equation 06-02-8 on page 82). The reason is that there was only one column separator in this example; when this is the case, the equations are centred and the layout is identical to the standard align environment.

However, as soon as more than one column separator (&) is used, the first block becomes left-aligned. Therefore inserting additional column separators at the end of the lines in a flalign environment is a simple way of left-aligning equations. This is done below to left-align the previous example, and can also be seen in Example 06-02-19.

Completely normal and pointless text before the flalign environment.

$$i_{11} = 0.25$$

$$i_{21} = \frac{1}{3}i_{11} \tag{6.24}$$

$$i_{31} = 0.33i_{22} \tag{6.25}$$

```
\usepackage{amsmath}

Completely normal and pointless text before
the \texttt{flalign} environment.
\begin{flalign}
    i_{11} &= 0.25                  & \nonumber \\
    i_{21} &= \frac{1}{3}i_{11} & \\
    i_{31} &= 0.33i_{22}            &
\end{flalign}
```

06-02-17

In contrast to the alignat environment, multiple groups can be placed side by side without specifying the number of blocks. Take care though to place the column separators correctly:

```
\usepackage{amsmath}
\begin{flalign}
i_{11} & =0.25 & i_{12} & =i_{21} & i_{13} & =i_{23}\nonumber \\
i_{21} & =\frac{1}{3}i_{11} & i_{22} & =0.5i_{12} & i_{23} & =i_{31}\\
i_{31} & =0.33i_{22} & i_{32} & =0.15i_{32} & i_{33} & =i_{11}
\end{flalign}
```

$$i_{11} = 0.25 \qquad\qquad i_{12} = i_{21} \qquad\qquad i_{13} = i_{23}$$

$$i_{21} = \frac{1}{3}i_{11} \qquad\qquad i_{22} = 0.5i_{12} \qquad\qquad i_{23} = i_{31} \tag{6.26}$$

$$i_{31} = 0.33i_{22} \qquad\qquad i_{32} = 0.15i_{32} \qquad\qquad i_{33} = i_{11} \tag{6.27}$$

06-02-18

On occasion you might want to mix left-aligned and centred equations. One option, of course, is to use the global document option fleqn, but if you can't use this for some reason, the desired effect can be achieved with a flalign environment:

06-02-19

$$f(x) = \int \frac{1}{x^2}\, \mathrm{d}x \qquad (6.28)$$

```
\usepackage{amsmath}
\newcommand*\diff{\mathop{}\!\mathrm{d}}
```

$$f(x) = \int \frac{1}{x^2}\, \mathrm{d}x \qquad (6.29)$$

```
\begin{flalign}
    f(x) & = \int\frac{1}{x^2}\diff x
\end{flalign}
\begin{flalign}
    f(x) & = \int\frac{1}{x^2}\diff x & %<== dummy
\end{flalign}
```

Another occasion that a `flalign` environment comes in useful is to left-align some additional text while right-aligning the equations. This is also achieved by inserting dummy column separators. Adding an equation number at the same time does not make sense in this case.

```
\usepackage{amsmath}
\begin{flalign*}
                        && 12(x-1)+20(y-3)+14(z-2) &= 0\\
\text{expand and transpose} &&  6x+10y+7z-50 &= 0
\end{flalign*}
```

06-02-20

$$12(x-1) + 20(y-3) + 14(z-2) = 0$$
$$\text{expand and transpose} \qquad 6x + 10y + 7z - 50 = 0$$

### 6.2.3 **aligned** environment

The `aligned` environment is the better alternative to the `array` environment. It caters for multiline formulae, giving one equation number to the whole set, but has much better horizontal alignment of the individual parts than an `array`. It, too, does not start math mode itself, so it must be part of another math environment. LaTeX outputs the following error message if the `aligned` environment is not used inside another math environment:

```
! Package amsmath Error: \begin{aligned} allowed only in math mode.

See the amsmath package documentation for explanation.
```

The requirement to be part of another math environment is also the reason why there is no starred version of the `aligned` environment; it is the outer environment that determines whether there is an equation number or not.

```
\begin{aligned}
    left & right & left & right...\\
    ...
    left & right & left & right...
\end{aligned}
```

Unlike the otherwise similar `split` environment (cf. Section 6.3.3 on page 92), `aligned` can have multiple column separators.

A normal line of text without any point before the formula.

$$2x + 3 = 7 \quad 2x + 3 - 3 = 7 - 3$$
$$2x = 4 \qquad \frac{2x}{2} = \frac{4}{2} \qquad (6.30)$$
$$x = 2$$

```
\usepackage{amsmath}

A normal line of text without any point
before the formula.
\begin{align}
\begin{aligned}
  2x+3 &= 7 &      2x+3-3   &= 7-3    \\
  2x   &= 4 & \frac{2x}{2} &= \frac42\\
  x    &= 2
\end{aligned}
\end{align}
```

06-02-21

## 6.3 Further environments for displayed formulae

The environments described so far cover a large part of what is needed for math typesetting in LaTeX and extend the functionality of standard LaTeX considerably. Nevertheless there are some interesting special cases where specific things can be done better.

### 6.3.1 gather and gathered environments

Use the gather environment if equations need to simply be centred horizontally. There are no column separators for this environment; you just need to specify the sequence of equations. The gathered environment has identical syntax, but like the aligned environment, it is used within another math environment (so also has no starred version) and results in a single equation number.

```
\begin{gather * }              \begin{gathered}
 expression\\                    expression\\
 expression\\                    expression\\
...                            ...
 expression                     expression
\end{gather * }                \end{gathered}
```

A normal pointless line of text before the formula.

$$i_{11} = 0.25 \qquad (6.31)$$
$$i_{21} = \frac{1}{3}i_{11}$$
$$i_{31} = 0.33i_{22}$$

```
\usepackage{amsmath}

A normal pointless line of text before
the formula.
\begin{gather}
 i_{11}=0.25\\
 i_{21}=\frac{1}{3}i_{11}\nonumber \\
 i_{31}=0.33i_{22}          \nonumber
\end{gather}
```

06-03-1

The following example shows a situation where you would want to use a gather environment (here used in its starred version). However, the equals sign in the first line is still not at the outer border of all equations. A better environment is described in the following section.

```
\usepackage{amsmath}
\begin{gather*}
  A = \lim _{n\rightarrow \infty }\Delta x\left( a^{2}+\left( a^{2}+2a\Delta x+\left(
      \Delta x\right) ^{2}\right)\right.\\
  +\left( a^{2}+2\times 2a\Delta x+2^{2}\left( \Delta x\right) ^{2}\right)
      +\left( a^{2}+2\times 3a\Delta x+3^{2}\left( \Delta x\right) ^{2}\right)\\
  + \ldots\\
  \left.+\left( a^{2}+2\times (n-1)a\Delta x +(n-1)^{2}
      \left( \Delta x\right) ^{2}\right) \right)\\
  = \frac{1}{3}\left( b^{3}-a^{3}\right)
\end{gather*}
```

06-03-2

$$A = \lim_{n \to \infty} \Delta x \left( a^2 + \left( a^2 + 2a\Delta x + (\Delta x)^2 \right) \right.$$
$$+ \left( a^2 + 2 \times 2a\Delta x + 2^2 (\Delta x)^2 \right) + \left( a^2 + 2 \times 3a\Delta x + 3^2 (\Delta x)^2 \right)$$
$$+ \ldots$$
$$\left. + \left( a^2 + 2 \times (n-1)a\Delta x + (n-1)^2 (\Delta x)^2 \right) \right)$$
$$= \frac{1}{3} \left( b^3 - a^3 \right)$$

Here is a gathered environment set within a gather environment. Note that there is only one equation number, vertically centred.

06-03-3

$$i_{11} = 0.25$$
$$i_{21} = \frac{1}{3} i_{11}$$
$$i_{31} = 0.33 i_{22} \tag{6.32}$$
$$i_{41} = i_{11} + i_{22}$$

A reference to Equation (6.32).

```
\usepackage{amsmath}

\begin{gather}\label{eq:gather}
\begin{gathered}
  i_{11}=0.25\\
  i_{21}=\frac{1}{3}i_{11}\\
  i_{31}=0.33i_{22}\\
  i_{41}=i_{11}+i_{22}
\end{gathered}
\end{gather}
%
A reference to Equation~\eqref{eq:gather}.
```

### 6.3.2 `multline` environment

The multline environment[2] is another multiline environments; it has advantages especially for very long equations because its internal composition follows the sequence left — centred — ... — centred — right, as shown in Example 06-03-4. Only the last line gets an equation number, which can be suppressed with the starred version.

---

[2]The first "missing" i is not a typo.

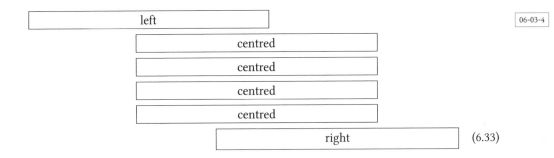

06-03-4

(6.33)

The example from the previous section can now be typeset more attractively with the first equation almost aligned with the left margin and the last one almost right-aligned.

```
\usepackage{amsmath}
\begin{multline}
  A = \lim_{n\rightarrow\infty}\Delta x\left(a^{2}+\left(a^{2}+2a \Delta x
      +\left(\Delta x\right)^{2}\right)\right.\\
  +\left(a^{2}+2\times2a\Delta x+2^{2}\left(\Delta x\right)^{2}\right)\\
  +\left(a^{2}+2\times3a\Delta x+3^{2}\left(\Delta x\right)^{2}\right)\\
  + \ldots\\
  \left.+\left(a^{2}+2\times(n-2)a\Delta x+(n-2)^{2}
      \left(\Delta x\right)^{2}\right)\right)\\
  \left.+\left(a^{2}+2\times(n-1)a\Delta x+(n-1)^{2}
      \left(\Delta x\right)^{2}\right)\right)\\
  = \frac{1}{3}\left( b^{3}-a^{3}\right)
\end{multline}
```

$$
\begin{aligned}
A = \lim_{n\to\infty} \Delta x \left(a^2 + \left(a^2 + 2a\Delta x + (\Delta x)^2\right)\right. \\
+ \left(a^2 + 2\times 2a\Delta x + 2^2 (\Delta x)^2\right) \\
+ \left(a^2 + 2\times 3a\Delta x + 3^2 (\Delta x)^2\right) \\
+ \ldots \\
+ \left(a^2 + 2\times (n-2)a\Delta x + (n-2)^2 (\Delta x)^2\right)\Big) \\
+ \left(a^2 + 2\times (n-1)a\Delta x + (n-1)^2 (\Delta x)^2\right)\Big) \\
\left. = \frac{1}{3}\left(b^3 - a^3\right)\right. \quad (6.34)
\end{aligned}
$$

06-03-5

The first equation is inset from the left margin by a gap of length \multlinegap, while the last equation is inset from the equation number (or from the right margin if using the starred version) by a gap of length \multlinetaggap. The default values are \multlinegap=10.0pt and \multlinetaggap=10.0pt.

The following Example 06-03-6 shows how different values for these gaps affect the layout of multline and the starred version multline*.

06-03-6

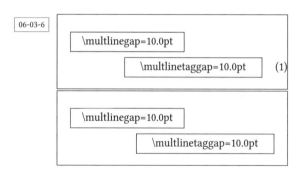
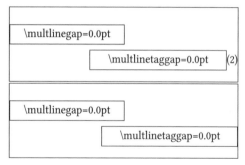

You can change the default arrangement of the individual lines ("lc...cr")[3] by using the \shoveleft or \shoveright commands.

\shoveleft{*content*} \shoveright{*content*}

These commands let you shift individual lines to the left or the right, as in this example:

06-03-7

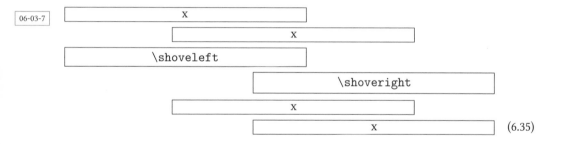

Using \shoveright on the first line, or \shoveleft on the last one, would be contrary to the concept of what the multline environment is designed to do. Therefore the fact that doing this has a strange effect on the layout, including relocating the equation number (as shown in the next example), should never actually cause the user a problem.

```
\usepackage{amsmath}
\newcommand*\CMD[1]{\texttt{\textbackslash#1}}
\begin{multline}
  \shoveright{\framebox[0.5\columnwidth]{\CMD{shoveright}}}\\
  \framebox[0.5\columnwidth]{x}\\
  \framebox[0.5\columnwidth]{x}\\
  \framebox[0.5\columnwidth]{x}\\
  \shoveleft{\framebox[0.5\columnwidth]{\CMD{shoveleft}}}
\end{multline}
```

---

[3]left — centre —... — centre — right

| | |
|---|---|
| \shoveright | |

| |
|---|
| X |

| |
|---|
| X |

| |
|---|
| X |

| | |
|---|---|
| \shoveleft | (6.36) |

06-03-8

In Section 4.5.1 on page 41 we discussed how to layout equations that span several lines. Let's rewrite the equation given there using the `multline` environment:

```
\usepackage{amsmath}
\begin{multline}
  \frac{1}{2}\Delta(f_{ij}f^{ij})=2\left(\sum_{i<j}\chi_{ij}(\sigma_{i}
      -\sigma_{j})^{2}+f^{ij}\nabla_{j}\nabla_{i}(\Delta f)+\right.\\
  +\left.\nabla_{k}f_{ij}\nabla^{k}f^{ij}+f^{ij}f^{k}
      \left[2\nabla_{i}R_{jk}-\nabla_{k}R_{ij}\right]\right)
\end{multline}
```

$$\frac{1}{2}\Delta(f_{ij}f^{ij}) = 2\left(\sum_{i<j}\chi_{ij}(\sigma_i - \sigma_j)^2 + f^{ij}\nabla_j\nabla_i(\Delta f)+ \right.$$
$$\left. + \nabla_k f_{ij}\nabla^k f^{ij} + f^{ij}f^k\left[2\nabla_i R_{jk} - \nabla_k R_{ij}\right]\right) \quad (6.37)$$

06-03-9

Again we have the problem of a pair of opening and closing parentheses on separate lines that do not match, but the solutions given earlier in Section 4.5.1 still work. One of the possibilities is to insert a \vphantom command containing the summation element on the second line:

```
\usepackage{amsmath}
\begin{multline}
  \frac{1}{2}\Delta(f_{ij}f^{ij})=2\left(\sum_{i<j}\chi_{ij}(\sigma_{i}
      -\sigma_{j})^{2}+f^{ij}\nabla_{j}\nabla_{i}(\Delta f)+\right.\\
  \left.+\nabla_{k}f_{ij}\nabla^{k}f^{ij}+f^{ij}f^{k}
      \left[2\nabla_{i}R_{jk}-\nabla_{k}R_{ij}\right]\vphantom{\sum_{i<j}}\right)
\end{multline}
```

$$\frac{1}{2}\Delta(f_{ij}f^{ij}) = 2\left(\sum_{i<j}\chi_{ij}(\sigma_i - \sigma_j)^2 + f^{ij}\nabla_j\nabla_i(\Delta f)+ \right.$$
$$\left. +\nabla_k f_{ij}\nabla^k f^{ij} + f^{ij}f^k\left[2\nabla_i R_{jk} - \nabla_k R_{ij}\right]\right) \quad (6.38)$$

06-03-10

### 6.3.3 `split` environment

The `split` environment is similar to the `multline` environment and is used when the actual equation is too long for a single line. Like the `aligned` environment, `split` can only be used within other math environments (so does not number equations itself). If the `split` environment is not used inside another one, an error message is given:

```
! Package amsmath Error: \begin{split} won't work here.
See the amsmath package documentation for explanation.
```

If you use `split` without any column separator, all lines are right-aligned:

06-03-11

With a column separator they are centred, but without any spacing in the centre unless there is a math comparison operator after the column separator.

06-03-12

06-03-13

$$f(x) = x$$
$$g(x) = \sinh\left(x^2 - 1\right)$$
$$h(x) = \mathrm{atan}\left(\frac{x}{\sqrt{1+x^2}}\right)$$

(6.39)

```
\usepackage{amsmath}

\begin{align}
\begin{split}
  f(x) &= x                         \\
  g(x) &= \sinh\left(x^2-1\right)\\
  h(x) &= \mathrm{atan}\left(\frac{x}
            {\sqrt{1+x^2}}\right)
\end{split}
\end{align}
```

> split   In contrast to `aligned` (cf. Section 6.2.3 on page 87), `split` only allows a single
> column separator. Only use it as part of one of the other `amsmath` environments
> described here, and not as part of an `eqnarray` environment from standard LaTeX
> because that allows several separators for `split`, which can lead to confusion.

The `split` environment is effective for long chains of equations with the same alignment, but its use is not necessary as the same can be achieved with the `align` environment.

```
\usepackage{amsmath} \newcommand*\diff{\mathop{}\!\mathrm{d}}
\begin{align}
\begin{split}
A_1 &= \left|\int_0^1 (f(x)-g(x))\diff x\right|+\left|\int_1^2
       (g(x)-h(x))\diff x\right|\\
    &= \left|\int_0^1 (x^2-3x)\diff x\right|+\left|\int_1^2(x^2-5x+6)
       \diff x\right|\\
    &= \left|\frac{x^3}{3}-\frac{3}{2}x^2\right|_0^1+\left|\frac{x^3}{3}
       -\frac{5}{2}x^2+6x\right|_1^2\\
```

```
&= \left|\frac{1}{3}-\frac{3}{2}\right|+\left|\frac{8}{3}-\frac{20}{2}+12
    -\left(\frac{1}{3}-\frac{5}{2}+6\right)\right|\\
&= \left|-\frac{7}{6}\right|+\left|\frac{28}{6}-\frac{23}{6}\right|
    =\frac{7}{6}+\frac{5}{6}=2\,\text{AU}
\end{split}
\end{align}
```

$$A_1 = \left| \int_0^1 (f(x) - g(x))\,\mathrm{d}x \right| + \left| \int_1^2 (g(x) - h(x))\,\mathrm{d}x \right|$$

$$= \left| \int_0^1 (x^2 - 3x)\,\mathrm{d}x \right| + \left| \int_1^2 (x^2 - 5x + 6)\,\mathrm{d}x \right|$$

$$= \left| \frac{x^3}{3} - \frac{3}{2}x^2 \right|_0^1 + \left| \frac{x^3}{3} - \frac{5}{2}x^2 + 6x \right|_1^2 \tag{6.40}$$

$$= \left| \frac{1}{3} - \frac{3}{2} \right| + \left| \frac{8}{3} - \frac{20}{2} + 12 - \left( \frac{1}{3} - \frac{5}{2} + 6 \right) \right|$$

$$= \left| -\frac{7}{6} \right| + \left| \frac{28}{6} - \frac{23}{6} \right| = \frac{7}{6} + \frac{5}{6} = 2\,\text{AU}$$

06-03-14

Comparing the example above to the following where the array environment is used, the superiority of the amsmath environments can be clearly seen again. The appearance of the array environment could be improved through application of \displaystyle (cf. Section 4.9 on page 55), \arraystretch, etc. However, it is easier to use the more sophisticated environment.

```
\newcommand*\diff{\mathop{}\!\mathrm{d}}
\begin{equation}\begin{array}{rl}
A_1 &= \left|\int_0^1(f(x)-g(x))\diff x\right|+\left|\int_1^2(g(x)-h(x))
    \diff x\right|\\
&= \left|\int_0^1 (x^2 -3x)\diff x\right|+\left|\int_1^2(x^2-5x+6)
    \diff x\right|\\
&= \left|\frac{x^3}{3}-\frac{3}{2}x^2\right|_0^1+\left|\frac{x^3}{3}
    -\frac{5}{2}x^2+6x\right|_1^2\\
&= \left|\frac{1}{3}-\frac{3}{2}\right|+\left|\frac{8}{3}-\frac{20}{2}+12
    -\left(\frac{1}{3}-\frac{5}{2}+6\right)\right|\\
&= \left|-\frac{7}{6}\right|+\left|\frac{28}{6}-\frac{23}{6}\right|
    =\frac{7}{6}+\frac{5}{6}=2\,\textrm{AU}
\end{array}\end{equation}
```

$$A_1 \quad = \left| \int_0^1 (f(x) - g(x))\,\mathrm{d}x \right| + \left| \int_1^2 (g(x) - h(x))\,\mathrm{d}x \right|$$
$$= \left| \int_0^1 (x^2 - 3x)\,\mathrm{d}x \right| + \left| \int_1^2 (x^2 - 5x + 6)\,\mathrm{d}x \right|$$
$$= \left| \frac{x^3}{3} - \frac{3}{2}x^2 \right|_0^1 + \left| \frac{x^3}{3} - \frac{5}{2}x^2 + 6x \right|_1^2 \tag{6.41}$$
$$= \left| \frac{1}{3} - \frac{3}{2} \right| + \left| \frac{8}{3} - \frac{20}{2} + 12 - \left( \frac{1}{3} - \frac{5}{2} + 6 \right) \right|$$
$$= \left| -\frac{7}{6} \right| + \left| \frac{28}{6} - \frac{23}{6} \right| = \frac{7}{6} + \frac{5}{6} = 2\,\text{AU}$$

06-03-15

### 6.3.4 cases environment

Example 04-01-5 on page 33 offered a way of typesetting cases with the standard array environment, but the cases environment offers much better support for this kind of construction. The syntax is similar to a $n \times 2$ matrix:

```
\begin{cases}
    ...&...\\
    ...&...\\
    ...
    ...&...
\end{cases}
```

The cases environment can only be used in math mode as part of another environment or a command. In the following example, a cases environment is set within an align environment; we've also used the \text command, which will be described later (Section 6.14 on page 114).

```
\usepackage{amsmath}
\begin{align}
x &= \begin{cases}
        0 & \text{if } a=\ldots\\
        1 & \text{if } b=\ldots\\
        x & \text{This text may be as long as it wants to be, but sometimes there
                    are problems\ldots}
    \end{cases}
\end{align}
```

06-03-16

$$x = \begin{cases} 0 & \text{if } a = \ldots \\ 1 & \text{if } b = \ldots \\ x & \text{This text may be as long as it wants to be, but sometimes there are problems} \ldots \end{cases}$$

$$(6.42)$$

Use a \parbox to wrap text that is too long to fit on one line. However, it requires a fixed size to be specified, and it should be typeset with \raggedright for better inter-word spacing.

```
\usepackage{amsmath}
\begin{align}
x &= \begin{cases}
        0 & \text{if } a=\ldots\\
        1 & \text{if } b=\ldots\\
        x & \parbox{5cm}{\raggedright This text may be as long as it wants to be,
            but sometimes there are problems\ldots\ not anymore :)}
    \end{cases}
\end{align}
```

06-03-17

$$x = \begin{cases} 0 & \text{if } a = \ldots \\ 1 & \text{if } b = \ldots \\ x & \text{This text may be as long as it wants to be, but sometimes there are problems}\ldots \text{ not anymore :)} \end{cases}$$

$$(6.43)$$

### 6.3.5 Matrix environments

In standard LATEX the separators of a matrix have to be specified, whereas amsmath defines differently named environments for each type of separator. They are all summarised in Example 06-03-18.

| environment | example | environment | example | environment | example | |
|---|---|---|---|---|---|---|
| Vmatrix | $\begin{Vmatrix} a & b \\ c & d \end{Vmatrix}$ | Bmatrix | $\begin{Bmatrix} a & b \\ c & d \end{Bmatrix}$ | matrix | $\begin{matrix} a & b \\ c & d \end{matrix}$ | 06-03-18 |
| vmatrix | $\begin{vmatrix} a & b \\ c & d \end{vmatrix}$ | bmatrix | $\begin{bmatrix} a & b \\ c & d \end{bmatrix}$ | pmatrix | $\begin{pmatrix} a & b \\ c & d \end{pmatrix}$ | |
| smallmatrix | $\begin{smallmatrix} a & b \\ c & d \end{smallmatrix}$ | | | | | |

Matrix environments can be nested arbitrarily; any arrangement is theoretically possible, however complex it may be. They don't activate math mode themselves so have to be part of another math environment or a command.

$$\begin{bmatrix} & a & & b \\ \begin{Vmatrix} A & B \\ C & D \end{Vmatrix} & c & & d \end{bmatrix} \qquad (6.44)$$

```
\usepackage{amsmath}

\begin{align}
\begin{bmatrix}
 a & b\\
 \begin{Vmatrix} A & B\\ C & D \end{Vmatrix}
 c & d
\end{bmatrix}
\end{align}
```
06-03-19

This example again shows that LATEX does not automatically size nested pairs of parentheses such that the outer ones encompass the inner ones. The solution is the same as before; modify the length \delimitershortfall. By setting it to a negative length of −2pt, the enclosing brace will be able to stretch by up to 2pt. The resulting matrix looks much better:

$$\begin{bmatrix} & a & & b \\ \begin{Vmatrix} A & B \\ C & D \end{Vmatrix} & c & & d \end{bmatrix} \qquad (6.45)$$

```
\usepackage{amsmath}

\setlength\delimitershortfall{-2pt}
\begin{align}
\begin{bmatrix}
 a & b\\
 \begin{Vmatrix} A & B\\ C & D\end{Vmatrix}
 c & d
\end{bmatrix}
\end{align}
```
06-03-20

For all matrix environments the arrangement of the individual elements is centred by default. However, this is not always the best layout, for example for columns of negative numbers:

06-03-21

$$\begin{pmatrix} 1 & -2 & 1 & 2 \\ 2 & 3 & -2 & 3 \\ 4 & -1 & 3 & -1 \\ 3 & 2 & -4 & 5 \end{pmatrix} \qquad (6.46)$$

```
\usepackage{amsmath}

\begin{align}
\begin{pmatrix}
 1 &-2 & 1 & 2\\ 2 & 3 &-2 & 3\\
 4 &-1 & 3 &-1\\ 3 & 2 &-4 & 5
\end{pmatrix}
\end{align}
```

We need to replace the centred arrangement with right-alignment. This is done by making a small modification to the matrix environment or by redefining the `smallmatrix` environment. The same matrix now looks much better:

(The corresponding code was put into the preamble, which is not visible here for space reasons.)

06-03-22

$$\begin{pmatrix} 1 & -2 & 1 & 2 \\ 2 & 3 & -2 & 3 \\ 4 & -1 & 3 & -1 \\ 3 & 2 & -4 & 5 \end{pmatrix} \qquad (6.47)$$

$$\begin{pmatrix} 1 & -2 & 1 & 2 \\ +2 & 3 & -2 & 3 \\ 4 & -1 & 3 & -1 \\ +3 & 2 & -4 & 5 \end{pmatrix} \qquad (6.48)$$

```
\usepackage{amsmath}

\begin{align}
\begin{pmatrix}
 1 &-2 & 1 & 2\\ 2 & 3 &-2 & 3\\
 4 &-1 & 3 &-1\\ 3 & 2 &-4 & 5
\end{pmatrix}
\end{align}
\begin{align}
\left(\begin{smallmatrix}
 1 &-2 & 1 & 2\\ +2 & 3 &-2 & 3\\
 4 &-1 & 3 &-1\\ +3 & 2 &-4 & 5
\end{smallmatrix}\right)
\end{align}
```

The `smallmatrix` environment is primarily intended for inline mode; it is typeset in `\scriptstyle` by default, which reduces its impact on the line spacing of the text (cf. Section 2.2 on page 5).

Dotted lines can be inserted across several columns with the `\hdotsfor` command; this is discussed in Section 6.6 on the following page. The maximum number of columns is limited to 10 for matrices that were created without explicit column specification. You can *MaxMatrixCols* change this limit at any time by modifying the internal counter `MaxMatrixCols`; for example, `\setcounter{MaxMatrixCols}{20}`.

## 6.4  Vertical space

amsmath does not define commands to deal with vertical whitespace; more information can be found in Section 4.8.4 on page 52, where this topic was covered for standard LATEX.

## 6.5  Page breaks

The `\allowdisplaybreaks` and `\displaybreak` commands let you control page breaks between the lines of displayed equations. The first command affects all equations within the current or nested groups, the second one only the line break (\\) following the command.

```
\allowdisplaybreaks [value]
\displaybreak [value]
```

The syntax of \displaybreak follows that of \pagebreak; a parameter value of 0 makes a page break possible without suggesting it, a parameter value of 4 forces a page break. The default value is 4 is the parameter is not specified.

The opposite applies for \allowdisplaybreaks; a parameter value of 0 prevents any page breaks, which corresponds to the default behaviour after loading the package. A parameter value of 4 on the other hand leaves it up to LaTeX or the amsmath commands to determine the optimal location for a page break.

## 6.6 Dots

In addition to the dot commands given in Section 4.10 on page 58, amsmath defines two more commands; \dddot[4] and \ddddot. Both expect a single-character argument and only work in math mode.

$$\dddot{y} \qquad \ddddot{y}$$

```
\usepackage{amsmath}

\Large $\dddot{y}$ \qquad $\ddddot{y}$
```

06-06-1

An interesting and useful command is \hdotsfor, which lets you put a dotted line within a matrix:

```
\hdotsfor [step size] {number of columns}
```

The optional step size, which should be thought of as a factor, can be used to stretch or compress the line of dots. The following example shows how to typeset a tri-diagonal matrix:

```
\usepackage{amsmath,array}
\setlength\delimitershortfall{-1pt}
\begin{align}
\underline{A} &= \left[\begin{array}{*7c}
    a_{11} & a_{12} & 0 & \ldots & \ldots & \ldots & 0\\
    a_{21} & a_{22} & a_{23} & 0 & \ldots & \ldots & 0\\
    0 & a_{32} & a_{33} & a_{34} & 0 & \ldots & 0\\
    \vdots & \vdots & \vdots & \vdots & \vdots & \vdots & \vdots\\
    \hdotsfor{7}\\
    \vdots & \vdots & \vdots & \vdots & \vdots & \vdots & \vdots\\
    0 & \ldots & 0 & a_{n-2,n-3} & a_{n-2,n-2} & a_{n-2,n-1} & 0\\
    0 & \ldots & \ldots & 0 & q_{n-1,n-2} & a_{n-1,n-1} & a_{n-1,n}\\
    0 & \ldots & \ldots & \ldots & 0 & a_{n,n-1} & a_{nn}
  \end{array}\right]
\end{align}
```

---

[4] Already mentioned in Section 4.11 on page 59.

06-06-2

$$
\underline{A} = \begin{bmatrix}
a_{11} & a_{12} & 0 & \dots & \dots & \dots & 0 \\
a_{21} & a_{22} & a_{23} & 0 & \dots & \dots & 0 \\
0 & a_{32} & a_{33} & a_{34} & 0 & \dots & 0 \\
\vdots & \vdots & \vdots & \vdots & \vdots & \vdots & \vdots \\
\dots\dots\dots\dots\dots\dots\dots\dots\dots\dots\dots\dots\dots\dots\dots\dots\dots \\
\vdots & \vdots & \vdots & \vdots & \vdots & \vdots & \vdots \\
0 & \dots & 0 & a_{n-2,n-3} & a_{n-2,n-2} & a_{n-2,n-1} & 0 \\
0 & \dots & \dots & 0 & q_{n-1,n-2} & a_{n-1,n-1} & a_{n-1,n} \\
0 & \dots & \dots & \dots & 0 & a_{n,n-1} & a_{nn}
\end{bmatrix}
\tag{6.49}
$$

## 6.7  Fractions and binomials

In addition to the problem with the font size already mentioned in Section 2.2.1 on page 6 for fractions, amsmath defines an own global command which intended to be used to define the different fraction commands. The \frac command as described in [7] does not exist in the amsmath package. Instead, all fractions, binomials and similar constructs are defined through the \genfrac command.[5] It expects six parameters; some can be empty, but none may be omitted.

$$\boxed{\texttt{\textbackslash genfrac}\{\textit{left}\}\{\textit{right}\}\{\textit{line width}\}\{\textit{math style}\}\{\textit{numerator}\}\{\textit{denominator}\}}$$

| | |
|---|---|
| *left* | The left delimiter symbol; nothing for a fraction, an opening parenthesis for a binomial. |
| *right* | The right delimiter symbol. |
| *line width* | The thickness of the fraction bar for fractions in any valid TeX unit. For a binomial this value *must* be set to 0pt. An empty value does not lead to no line, but to a line with the standard thickness. |

*math style*   One of the math styles described in Section 4.9 on page 55, encoded as a digit:

**0** \displaystyle
**1** \textstyle
**2** \scriptstyle
**3** \scriptscriptstyle

An empty value continues the current math style. If this is \scriptstyle, it is also selected as style for \genfrac.

| | |
|---|---|
| *numerator* | The numerator of a fraction or the upper part of a binomial. |
| *denominator* | The denominator or the lower part. |

If the parameters of the delimiters are only single characters, you don't need to enclose them in curly braces — TeX will assume that every character is a separate parameter.

---

[5]generalised fraction

$$\frac{x^2+x+1}{3x-2} \qquad \tfrac{x^2+x+1}{3x-2}$$

```
\usepackage{amsmath}
```
06-07-1

```
\[ \genfrac{}{}{0.1pt}{}{x^2+x+1}{3x-2}\quad
   \genfrac{}{}{0.1pt}{3}{x^2+x+1}{3x-2} \]
```

$$\frac{x^2+x+1}{3x-2} \qquad \left[\frac{x^2+x+1}{3x-2}\right]$$

```
\[ \genfrac{}{}{1pt}{}{x^2+x+1}{3x-2} \quad
   \genfrac[]{1pt}{1}{x^2+x+1}{3x-2}        \]
```

$$\left\{\frac{x^2+x+1}{3x-2}\right\} \qquad \binom{n+1}{n}$$

```
\[ \genfrac\{\}{0.1pt}{1}{x^2+x+1}{3x-2}\quad
   \genfrac(){0pt}{}{n+1}{n}                \]
```

$$\left\{{n+1 \atop n}\right\}$$

```
\[ \genfrac{\{}{\}}{0pt}{2}{n+1}{n}        \]
```

amsmath itself uses \genfrac to define several further fraction and binomial commands. Each of these commands has four predetermined parameters to pass to \genfrac for delimiters, bar width and math style. Then you only need to give the commands a further two parameters, which are passed straight on to \genfrac for the dividend and divisor.

```
\newcommand\dfrac{\genfrac{}{}{}{}0}
\newcommand\tfrac{\genfrac{}{}{}{}1}
\DeclareRobustCommand\binom{\genfrac()\z@{}}
\newcommand\dbinom{\genfrac(){0pt}0}
\newcommand\tbinom{\genfrac(){0pt}1}
```

| | |
|---|---|
| \dfrac | Outputs large fractions, useful in inline mode (no delimiters, a standard fraction bar and activates \displaystyle). |
| \tfrac | Outputs smaller fractions (activates \textstyle). |
| \binom | Takes ( ) as delimiters, sets the fraction bar width to the length \z@, which corresponds to 0pt[6] and continues the current math style. |
| \dbinom | Outputs large binomials (activates \displaystyle). |
| \tbinom | Outputs smaller binomials (activates \textstyle). |

Another command, \cfrac, is useful for displaying continued fractions. The fraction is always typeset in \displaystyle and not shrunk to \scriptscriptstyle as with the normal \frac command. The following examples show the difference:

$$\cfrac{1}{\sqrt{2}+\cfrac{1}{\sqrt{3}+\cfrac{1}{\sqrt{4}+\cfrac{1}{\ldots}}}}$$

```
\usepackage{amsmath}
```
06-07-2

```
\begin{align*}
\cfrac{1}{\sqrt{2}+\cfrac{1}{\sqrt{3}+
   \cfrac{1}{\sqrt{4}+ \cfrac{1}{\ldots}}}}
\end{align*}
```

$$\frac{1}{\sqrt{2}+\frac{1}{\sqrt{3}+\frac{1}{\sqrt{4}+\frac{1}{\ldots}}}}$$

```
\usepackage{amsmath}
```
06-07-3

```
\begin{align*}
\frac{1}{\sqrt{2}+\frac{1}{\sqrt{3}+
   \frac{1}{\sqrt{4}+ \frac{1}{\ldots}}}}
\end{align*}
```

---

[6] \z@ provides a minimal speedup because it is part of the precompiled LaTeX format.

\cfrac has an optional parameter that specifies the horizontal alignment of the divisor. The default is c (centre), as in Example 06-07-2. l (left) and r (right) are illustrated below:

```
\usepackage{amsmath}
\begin{gather}
\cfrac[l]{1}{\sqrt{2}+\cfrac[l]{1}{\sqrt{3}+\cfrac[l]{1}{\sqrt{4}+
  \cfrac[l]{1}{\ldots}}}} \qquad \cfrac[r]{1}{\sqrt{2}+\cfrac[r]{1}{\sqrt{3}+
  \cfrac[r]{1}{\sqrt{4}+\cfrac[r]{1}{\ldots}}}}
\end{gather}
```

06-07-4

$$\cfrac{1}{\sqrt{2}+\cfrac{1}{\sqrt{3}+\cfrac{1}{\sqrt{4}+\cfrac{1}{\dots}}}} \qquad \cfrac{1}{\sqrt{2}+\cfrac{1}{\sqrt{3}+\cfrac{1}{\sqrt{4}+\cfrac{1}{\dots}}}} \tag{6.50}$$

Binomials are in principle just fractions with delimiters and no fraction bar. The placing of the upper and the lower part follows the usual rules for fractions. The syntax of \binom is different to the one of \choose from standard LaTeX (cf. Section 2.2.1 on page 6). We have already mentioned the three different commands to create binomials provided by amsmath, but they are summarised again below:

06-07-5

| command | inline mode | displayed |
|---|---|---|
| \binom{m}{n} | $\binom{m}{n}$ | $\binom{m}{n}$ |
| \dbinom{m}{n} | $\binom{m}{n}$ | $\binom{m}{n}$ |
| \tbinom{m}{n} | $\binom{m}{n}$ | $\binom{m}{n}$ |

## 6.8  Roots

Getting different root expressions to look good, especially when positioned next to each other, isn't always easy. We described some possibilities for improving their display in Section 4.3 on page 35. amsmath defines two additional auxiliary commands to help with typesetting roots. They are used in conjunction with the actual \sqrt command.

\sqrt [\leftroot{*value*}  \uproot{*value*}*n*] {*radicand*}

Here *value* is the magnitude of the horizontal or vertical translation (respectively) of the root exponent (*n*) in mu.[7] *Value* must be an integer number without unit. The radicand is not affected by the translation. The following example has been enlarged with the \Large command to make the effects of \leftroot and \uproot clearer.

---

[7] "math unit" — $1\mathrm{mu} = \frac{1}{18}\mathrm{em}$.

$$\sqrt[k_n]{a} \qquad \sqrt[k_n]{a} \qquad \sqrt[k_n]{a}$$

```
\usepackage{amsmath}

\Large $\sqrt[k_n]{a}$ \quad
$\sqrt[\uproot{2}k_n]{a}$ \quad
$\sqrt[\leftroot{2}\uproot{2}k_n]{a}$
```

06-08-1

That's dealt with repositioning the exponent, but you may also need to reposition the radicand, especially if it contains an index or stacked indices. The \smash command lets you modify the base line to which the root refers for its position.

> \smash [position] {*radicand*}

The optional argument can take three values; the default is tb.

**t** Leaves the base and changes the height.
**b** Leaves the height and changes the base.
**tb** Changes the base as well as the height.

Only the radicand is affected, the root exponent remains untouched.

```
\usepackage{amsmath}\newcommand\rV{\rule{5mm}{0.4pt}}
\Large A\rV$\sqrt{\frac{x}{\lambda}}\rV\sqrt{\smash[b]{\frac{b}{\lambda}}}
 \rV\sqrt{\smash[t]{\frac{t}{\lambda}}}\rV\sqrt{\smash[tb]{\frac{tb}{\lambda}}}
 \rV\sqrt{\frac{b}{\lambda_{k_i}}}\rV\sqrt{\frac{b}{\smash[b]{\lambda_{k_i}}}}$\rV B
```

$$\text{A}\rule{4mm}{0.4pt}\sqrt{\frac{x}{\lambda}}\rule{4mm}{0.4pt}\sqrt{\frac{b}{\lambda}}\rule{4mm}{0.4pt}\sqrt{\frac{t}{\lambda}}\rule{4mm}{0.4pt}\sqrt{\frac{tb}{\lambda}}\rule{4mm}{0.4pt}\sqrt{\frac{b}{\lambda_{k_i}}}\rule{4mm}{0.4pt}\sqrt{\frac{b}{\lambda_{k_i}}}\rule{4mm}{0.4pt}\text{B}$$

06-08-2

> \smash   In standard LaTeX, the command is defined without an optional parameter, so is less flexible.
>
> The command can lead to unexpected results, especially when used in conjunction with fractions, stacked indices or double exponents.

## 6.9 Accents

The \mathaccent command lets you define new characters with accents, for example:

$$\dot\cup$$

```
\usepackage{amsmath}
\newcommand\dotcup{$\mathaccent\cdot\cup$}

\Huge $\mathaccent\cdot\cup$
```

06-09-1

The example demonstrates the syntax:

> \mathaccent{*accent character*}{*base character*}

The command may only be used in math mode, which can be activated when defining new characters with the \ensuremath command.

```
\usepackage{amsmath,txfonts}
\newcommand\CurveArrowLeftRight{%
  \ensuremath{\mathaccent\curvearrowright{\mkern-5mu\curvearrowleft}}}
{ \huge $\curvearrowleft\curvearrowright\CurveArrowLeftRight$ }
        $\curvearrowleft\curvearrowright\CurveArrowLeftRight$
\footnotesize $\curvearrowleft\curvearrowright\CurveArrowLeftRight$
```

06-09-2

This example does not work with the MnSymbol package; the individual arrows in that package have different heights so superimposing them does not have the desired effect. To superimpose the two characters, the second character in the example above was translated by 5 mu (math unit) to the left to place it on top of the first character. This value was found through trial and error. Math units are proportionate to the dynamic length 1em[8] so there is no problem using the new character if the font size is enlarged.[9] Other ways of creating characters with accents are shown in Section 9.1 on page 167.

## 6.10  Modulo command

In standard LaTeX, the commonly used \mod is not an operator (in contrast to \pmod and \bmod) and if it is used as a command, LaTeX outputs an error message when using (\$a\mod b\$). amsmath defines several types of the modulo operator; they are summarised in Table 6.4. The usual operator spacing applies before and after the commands, as \mod was defined as a binary operator by amsmath.

Table 6.4: Summary of the different modulo commands

06-10-1

| input → output | input → output |
|---|---|
| a\mod{n^2}=b → $a \mod n^2 = b$ | a\bmod{n^2}=b → $a \bmod n^2 = b$ |
| a\pmod{n^2}=b → $a \pmod{n^2} = b$ | a\pod{n^2}=b → $a \ (n^2) = b$ |

## 6.11  Equation numbering

Most of the ways to influence the form of equation numbers were covered in Section 3.4 on page 21. The only extra command defined by amsmath is \numberwithin. It is used to specify which levels should be part of the counter.

\numberwithin{*counter*}level

Its effect is shown in the second equation of Example 06-11-2 on the following page; for the last equation the chapter and section numbers are removed as an example (cf. Section 3.4 on page 21). The command \theequation always defines the appearance of the numbering.

---

[8]The width of an "M" in the current font, here only 8.0pt because the font of the footnote is smaller. For \normalsize 1em=10.0pt.

[9]Dynamic lengths refer to the current size of characters and not a fixed value.

> \theequation{*definition*}

The following examples show the setting of equation numbers and how the can be modified independent from the sense of the defined numbering.

```
\usepackage{amsmath,blkarray}
\renewcommand\theequation{eq. \roman{section}-\Roman{equation}}
\begin{align}\begin{blockarray}{*{4}{c}}
  \begin{block}{[ccc]c}
    \bigstrutht 1-\lambda x & 0 & 0 & \ell_1 \\  0 & 1-\lambda x & 0 & \ell_2 \\
    \bigstrutdp 0 & 0 & 1-\lambda x & \ell_3 \\
  \end{block}
  c_1 & c_2 & c_3
\end{blockarray}\end{align}
```

$$
\begin{bmatrix} 1-\lambda x & 0 & 0 \\ 0 & 1-\lambda x & 0 \\ 0 & 0 & 1-\lambda x \end{bmatrix} \begin{matrix} \ell_1 \\ \ell_2 \\ \ell_3 \end{matrix} \qquad\qquad \text{(eq. xi-LI)}
$$
$$
\begin{matrix} c_1 & c_2 & c_3 \end{matrix}
$$

06-11-1

The documentation class book numbers the equations in the form "chapter.number" by default. To change this to, for example, "chapter.section.number", we can use the \numberwithin command. (There is no corresponding command \numberwithout.) To reset it again, redefine \theequation.

```
\usepackage{amsmath}
\begin{flalign} z &= f(x,y) \end{flalign} \numberwithin{equation}{section}
\begin{align}\bordermatrix{%
   & c & o & l & u & m & n \cr
1 & 1 & 0 & 0 & 0 & \cos \phi & \sin\phi\cr
i & 0 & 1 & 0 & 0 & -\sin\phi & \cos\phi\cr
n & 0 & 0 & 1 & 0 &    0    &    0    \cr
e & 0 & 0 & 0 & 1 &    0    &    0    \cr} \end{align}
\renewcommand\theequation{\arabic{equation}} \begin{flalign} y &= f(x) \end{flalign}
```

$$
z = f(x,y) \tag{6.52}
$$

06-11-2

$$
\begin{matrix} & c & o & l & u & m & n \\ l & 1 & 0 & 0 & 0 & \cos\phi & \sin\phi \\ i & 0 & 1 & 0 & 0 & -\sin\phi & \cos\phi \\ n & 0 & 0 & 1 & 0 & 0 & 0 \\ e & 0 & 0 & 0 & 1 & 0 & 0 \end{matrix} \tag{6.11.53}
$$

$$
y = f(x) \tag{54}
$$

amsmath uses the term "tag" in connection with the equation number, and has useful commands with the same and similar names. Each equation is assigned a unique "number" (which is frequently the star). This is used to highlight individual unnumbered equations. You can put any arbitrary material as the parameter in the \tag command.

```
\usepackage{amsmath}
\begin{align}
  H_c &= \frac{1}{2n} \sum^n_{l=0} (-1)^l (n-l)^{p-2} \sum_{l_1 + \dots + l_p=l}
        \prod^p_{i=1} \binom{n_i}{l_i} \tag{*}\label{eq:tagDemo}\\
      & \quad\times [(n-l)-(n_i-l_i)]^{n_i-l_i}\tag{II-AZ}\label{eq:tagDemo2}
\end{align}
You can also reference tags, as the label~\eqref{eq:tagDemo}
or~\ref{eq:tagDemo2}.
```

06-11-3

$$H_c = \frac{1}{2n} \sum_{l=0}^{n} (-1)^l (n-l)^{p-2} \sum_{l_1+\cdots+l_p=l} \prod_{i=1}^{p} \binom{n_i}{l_i} \tag{*}$$

$$\times [(n-l)-(n_i-l_i)]^{n_i-l_i} \tag{II-AZ}$$

You can also reference tags, as the label (*) or II-AZ.

You can also use the \tag command for single equations; this is discussed in the next section.

Here the form of the equation numbering refers only to its enclosing in parentheses. This can be redefined in the usual LaTeX way. However, for amsmath, this is different to Section 3.4.1 on page 21 where it was shown for standard LaTeX. CJ Well is it the same or not? What do you mean - ...it can be redefined in the usual way... but... it is different for amsmath... ? In amsmath, you need to use the \tagform@ command. There are individual packages that assist in making these changes, by avoiding the need to use \makeatletter ... \makeatother.

```
\usepackage{amsmath}
\makeatletter
\let\oldTagForm\tagform@
\renewcommand\tagform@[1]{\maketag@@@{\ignorespaces*#1*\unskip\@@italiccorr}}
\makeatother
\begin{align}
  c^2 &\le a^2 + b^2\label{eq:pyth0}\\
  \left|\sum_{i=1}^n a_ib_i\right| &\le \left(\sum_{i=1}^n a_i^2\right)^{1/2}
    \left(\sum_{i=1}^n b_i^2\right)^{1/2}\label{eq:pyth1}
\end{align}
\makeatletter\let\tagform@\oldTagForm\makeatother
\begin{align}\label{eq:pyth2} c^2 &\le a^2 + b^2 \end{align}

A reference to eq.~\eqref{eq:pyth0}, eq.~\eqref{eq:pyth1} and
eq.~\eqref{eq:pyth2}.
```

$$c^2 \le a^2 + b^2 \qquad\qquad *6.55*$$

06-11-4

$$\left| \sum_{i=1}^{n} a_i b_i \right| \le \left( \sum_{i=1}^{n} a_i^2 \right)^{1/2} \left( \sum_{i=1}^{n} b_i^2 \right)^{1/2} \qquad *6.56*$$

$$c^2 \le a^2 + b^2 \qquad\qquad (6.57)$$

A reference to eq. (6.55), eq. (6.56) and eq. (6.57).

### 6.11.1 Labels and tags

For single equations, use the starred version of the \tag command. It ignores any parentheses and simply outputs its argument. This can also be "abused" for simple descriptions:

```
\usepackage{amsmath}
\begin{align}
  f(x) &= cx+b            \tag*{linear}\\
  g(x) &= dx^2+cx+b       \tag*{quadratic}\\
  h(x) &= ex^3+dx^2+cx+b \tag*{cubic}
\end{align}
```

06-11-5

$$
\begin{aligned}
f(x) &= cx + b & \text{linear}\\
g(x) &= dx^2 + cx + b & \text{quadratic}\\
h(x) &= ex^3 + dx^2 + cx + b & \text{cubic}
\end{aligned}
$$

In addition to the default \ref command to reference labels, amsmath defines the \eqref command, which puts the equation number in parentheses when referred to in the text (cf. the text in Example 06-11-4). The fancyref package provides further ways of referencing.

The \tag command ignores any \nonumber instructions for individual lines and any starred versions of the environment; the argument of \tag is output regardless, and for this line the environment is not a starred version.

$$\sqrt{a} \times \sqrt{b} = \sqrt{ab}$$

$$\frac{\sqrt{a}}{\sqrt{b}} = \sqrt{\frac{a}{b}} \qquad (\bullet)$$

$$a\sqrt{b} = \sqrt{a^2 b}$$

```
\usepackage{amsmath}

\begin{align*}
  \sqrt{\vphantom{b}a}\times\sqrt{b}
    &= \sqrt{ab} \\
  \frac{\sqrt{a}}{\sqrt{b}}
    &= \sqrt{\frac{a}{b}} \tag{$\bullet$}\\
  a\sqrt{b} & =\sqrt{a^2 b}
\end{align*}
```

06-11-6

*ext mode* \tag typesets its argument in text mode; therefore a math symbol like \bullet is only a valid argument of \tag if it is enclosed in $...$, as seen in the example above. This is independent of \tag itself being used in math mode.

There are two special package options for label which can be given when amsmath is loaded. One is centertags and the other is tbtags. A description of these options can be found at the beginning of this chapter on page 79.

## 6.11.2 Subequations

In standard LaTeX, you can't number subequations. However, there are several packages that will enable you to do so, or a custom way of counting subequations is not too hard to implement either. When using amsmath you can use the subequations environment. The "normal" math environment must be entirely within subequations.

06-11-7

$$y = d \qquad (6.58a)$$
$$y = cx + d \qquad (6.58b)$$
$$y = bx^2 + cx + d \qquad (6.58c)$$
$$y = ax^3 + bx^2 + cx + d \qquad (6.58d)$$

```
\usepackage{amsmath}

\begin{subequations}
\begin{align}
y &= d\\
y &= cx+d\\
y &= bx^{2}+cx+d\\
y &= ax^{3}+bx^{2}+cx+d
\end{align}
\end{subequations}
```

As can be seen from the example, the subequations are counted by default by appending a lowercase letter. The only way to change this is through the \theparentequation command. It saves the last "normal" parent equation number in the parentequation counter because internally the counter equation, which is also valid outside subequations, is still used. It starts at 1 however. To reset the counter after the end of the environment, it has to be saved in parentequation.

Modifications of the equation number with the \theparentequation command must be done *after* the start of the subequations environment. It is only defined within the environment.

06-11-8

$$y = d \qquad (6.59\text{-}1)$$
$$y = cx + d \qquad (6.59\text{-}2)$$
$$y = bx^2 + cx + d \qquad (6.59\text{-}3)$$
$$y = ax^3 + bx^2 + cx + d \qquad (6.59\text{-}4)$$

A reference to Eq. 6.59-2 on page 1.

```
\usepackage{amsmath}

\begin{subequations}
\renewcommand{\theequation}{%
   \theparentequation -\arabic{equation}}
\begin{align}
y &= d\\
y &= cx+d\label{eq:subequation}\\
y &= bx^{2}+cx+d\\
y &= ax^{3}+bx^{2}+cx+d
\end{align}
\end{subequations}
A reference to Eq.~\ref{eq:subequation}
on page~\pageref{eq:subequation}.
```

Example 06-11-8 shows that references in the text to subequations are done in exactly the same way as a reference to a normal equation.

Defining a custom counting of equations is not too difficult even for beginners; counters are always used in the same way. In this example, two equations are placed next to each other and handled as subequations on the same line. The equation counter in amsmath only counts entire lines; therefore this definition is of practical use.

```
\usepackage{amsmath}  \newcounter{mySubCounter}
\newcommand\twocoleqn[2]{\setcounter{mySubCounter}{0}\let\OldTheEq\theequation%
   \renewcommand\theequation{\OldTheEq\alph{mySubCounter}}%
   \noindent\begin{minipage}{.49\textwidth}
     \begin{align}\refstepcounter{mySubCounter}#1\end{align}
   \end{minipage}\hfill\addtocounter{equation}{-1}%
   \begin{minipage}{.49\textwidth}
     \begin{align}\refstepcounter{mySubCounter}#2\end{align}
   \end{minipage}\let\theequation\OldTheEq}
\twocoleqn{y=f(x)}{y=f(z)}
\begin{align} z &= f(x,y) \end{align}
```

$$y = f(x) \tag{6.60a}$$
$$y = f(z) \tag{6.60b}$$

06-11-9

$$z = f(x, y) \tag{6.61}$$

## 6.12 Limits

By default the limits for sum (\sum) and product (\prod) are placed above and below the symbol. For the integral (\int) however they are placed as super- and subscripts. To achieve this behaviour for all operators that have such limits, the amsmath package can be loaded with the sumlimits and intlimits. For the opposite case, too, there are corresponding package options. All of these were described at the beginning of this chapter on page 79. Also of interest are the \underset and \overset commands, which are described in Section 6.18 on page 120.

We discussed multiple or stacked symbols in Section 2.2.2 on page 7 and in more detail in Section 4.4 on page 36. amsmath defines a special \substack command that is especially useful for multiple limits. The syntax of this command is:

```
\substack{ ... \\ ... \\ ... }
```

\substack has replaced the Sb and SP environments described in [7], which are not supported anymore by amsmath.

Example 04-04-2 on page 37 is a previous example with multiple limits. The same result is achieved by typesetting it with \substack:

$$\sum_{\substack{1 \le i \le p \\ 1 \le j \le q \\ 1 \le k \le r}} a_{ij}b_{jk}c_{ki} \tag{6.61}$$

```
\usepackage{amsmath}

\begin{align}
\sum_{\substack{1\le i\le p\\
     1\le j\le q\\
     1\le k\le r}%
   }a_{ij}b_{jk}c_{ki}
\end{align}
```

06-12-1

This command is no better than those covered earlier at handling limits that are wider than the main symbol.

$$X = \sum_{1 \le i \le j \le n} X_{ij} \tag{6.62}$$

```
\usepackage{amsmath}

\begin{align}
 X = \sum_{1\le i\le j\le n}X_{ij}
\end{align}
```

06-12-2

Using the \makebox command with a chosen box width is a way of fixing this problem. It is not ideal, however, as it does not know about the surrounding math font style, so it can make the limits too large because it automatically switches to text mode:

06-12-3

$$X = \sum_{1 \le i \le j \le n} X_{ij} \qquad (6.63)$$

```
\usepackage{amsmath}

\begin{align}
  X = \sum_{\makebox[0pt]{$1\le i\le j\le n$}}X_{ij}
\end{align}
```

To solve this problem, we can define a \mathclap command, which works similarly to the \rlap and \llap commands known from LATEX. It does not require any horizontal space itself, but is centred on the current point anyway. \mathllap and \mathrlap commands can also be defined similarly. In the following example only the first "Jana" takes the correct space, which is then framed by the table. All the others start in the middle of the line (because of the column type c) and then take up space to the left, to the right, or symmetrically to the left and right, depending on the command used. They do not "officially" require the space they need and therefore overwrite any other text or frame.

06-12-4

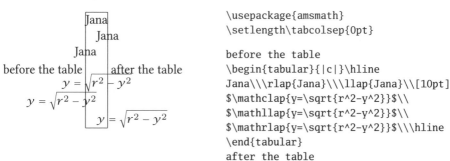

```
\usepackage{amsmath}
\setlength\tabcolsep{0pt}

before the table
\begin{tabular}{|c|}\hline
Jana\\\rlap{Jana}\\\llap{Jana}\\[10pt]
$\mathclap{y=\sqrt{r^2-y^2}}$\\
$\mathllap{y=\sqrt{r^2-y^2}}$\\
$\mathrlap{y=\sqrt{r^2-y^2}}$\\\hline
\end{tabular}
after the table
```

The definitions of these new \mathclap, \mathllap, and \mathrlap commands requires knowledge of some LATEX internals, but are nevertheless given here.

```
\newcommand*\mathllap{\mathstrut\mathpalette\mathllapinternal}
\newcommand*\mathllapinternal[2]{\llap{$\mathsurround=0pt#1{#2}$}}
\newcommand*\clap[1]{\hbox to 0pt{\hss#1\hss}}
\newcommand*\mathclap{\mathpalette\mathclapinternal}
\newcommand*\mathclapinternal[2]{\clap{$\mathsurround=0pt#1{#2}$}}
\newcommand*\mathrlap{\mathpalette\mathrlapinternal}
\newcommand*\mathrlapinternal[2]{\rlap{$\mathsurround=0pt#1{#2}$}}
```

The \mathclap command can now be used to typeset the lower limit, without its width affecting the sum symbol or the following math term.

06-12-5

$$X = \sum_{1 \le i \le j \le n} X_{ij} \qquad (6.64)$$

```
\usepackage{amsmath}

\begin{align}
  X = \sum_{\mathclap{1\le i\le j\le n}}X_{ij}
\end{align}
```

Another problem connected to limits arises if there is only an upper or only a lower limit and you need to put the expression in parentheses:

$$\rule{3cm}{0.4pt}\left[\sum_{\substack{i,j\\i>j}}\frac{1}{2}\cdots\right]\rule{2cm}{0.4pt}$$ (6.65)

06-12-6

```
\usepackage{amsmath}

\begin{align}
\rule[2.5pt]{1cm}{0.2pt} % math axis
  \left[ \sum_{\substack{i,j\\i>j}} \frac{1}{2}
    \dots \right]
\rule[2.5pt]{1cm}{0.2pt} % math axis
\end{align}
```

In math mode, parentheses and large operators are centred vertically on the horizontal "axis" of the equation. This axis is at the height of the math symbols + and − and the fraction bar.

This height of the axis above the base line of a line can be determined with the following trick from [13]:[10]

```
\setbox0=\hbox{$\vcenter{}$}% \ht0 is the axis height
```

For the font used here for example h=2.53528pt, where 1ex=4.31pt.

The correct alignment of the elements can be difficult if the math part outside the parentheses needs to still be on that axis as well. We can't give a general solution because the surrounding context is a major factor in solving this problem. Only one of the possible ways is shown here by the following examples:

$$\rule{3cm}{0.4pt}\left[\sum_{\substack{i,j\\i>j}}\cdots\right]\rule{2cm}{0.4pt}$$ (6.66)

06-12-7

```
\usepackage{amsmath}

\delimitershortfall=-1pt % fits parenthesis
\begin{align}
\rule[2.5pt]{1cm}{0.2pt} % math axis
\left[
 \begin{array}{@{}c@{}}
  \displaystyle\sum_{\substack{i,j\\i>j}}\dots
 \end{array}
\right]
\rule[2.5pt]{1cm}{0.2pt} % math axis
\end{align}
```

The sum was put inside an `array` environment, which automatically adapts the size of the parentheses to its contents. The column definition {@{}c@{}} stands for "no space (@{}) before and after the centred column". This solves the problem of the parentheses, but not that of the vertical alignment. The more stacked limits that are required, the greater is the offset in alignment:

---

[10]To be found in the source code (ftp://dante.ctan.org/tex-archive/system/knuth/tex/texbook.tex) with grep axis $(kpsewhich texbook.tex) | head -1 or the search function of an editor.

06-12-8

$$
\left[ \sum_{\substack{1 \le i \le p \\ 1 \le j \le q \\ 1 \le k \le r}} a_{ij}b_{jk}c_{ki} \right] = A
$$

(6.67)

```
\usepackage{amsmath}

\delimitershortfall=-1pt % fits parenthesis
\begin{align}
\rule[2.5pt]{1cm}{0.2pt} % math axis
\left[ \begin{array}{@{}c@{}}
 \displaystyle\sum_{\substack{%
 1\le i\le p\\1\le j\le q\\1\le k\le r}}
 a_{ij}b_{jk}c_{ki}
\end{array} \right] = A
\rule[2.5pt]{1cm}{0.2pt} % math axis
\end{align}
```

Every solution has advantages and disadvantages with respect to the effort involved and the typographical result. It is up to the user to choose which to use.

The \raisebox command is very effective for translating an object vertically. \raisebox changes into text mode itself, so the argument has to be enclosed in $...$ again. At the same time, display mode must be activated through \displaystyle because $ starts \textstyle, which is not what is wanted here.

06-12-9

$$
\left[ \sum_{\substack{1 \le i \le p \\ 1 \le j \le q \\ 1 \le k \le r}} a_{ij}b_{jk}c_{ki} \right] = A
$$

(6.68)

```
\usepackage{amsmath}

\delimitershortfall=-1pt % fitsparenthesis
\begin{align}
\rule[2.5pt]{1cm}{0.2pt} % math axis
\raisebox{-2.5ex}{$\displaystyle
\left[ \begin{array}{@{}c@{}} \displaystyle
 \sum_{\substack{1\le i\le p\\1\le j \le q\\
 1\le k\le r}} a_{ij}b_{jk}c_{ki}
\end{array} \right]$} = A
\rule[2.5pt]{1cm}{0.2pt} % math axis
\end{align}
```

We should mention again that some of our examples such as the one above do not make math sense. They are constructed to show how arbitrarily complex situations can be solved with LaTeX means in different ways.

The following example of a complex formula from [13] is not perfect; there is horizontal space at the right sum and the parentheses do not extend beyond the lower limits of the left or right sums. However, there is no easy solution for this example like there was for the previous example.

```
\usepackage{amsmath}
\begin{align}
\prod_{j\ge0}\biggl(\sum_{k\ge0}a_{jk}z^k\biggr) = \sum_{n\ge0}z^n\,
 \Biggl(\sum_{\scriptstyle k_0,k_1,\ldots\ge0\atop\scriptstyle
 k_0+k_1+\cdots=n}a_{0k_0}a_{1k_1}\ldots\,\Biggr)
\end{align}
```

06-12-10

$$
\prod_{j \ge 0} \left( \sum_{k \ge 0} a_{jk}z^k \right) = \sum_{n \ge 0} z^n \left( \sum_{\substack{k_0,k_1,\ldots \ge 0 \\ k_0+k_1+\cdots=n}} a_{0k_0}a_{1k_1}\cdots \right)
$$

(6.69)

Just inserting \hspace before and after the second sum immediately improves the expression:

```
\usepackage{amsmath}
\begin{align}
\prod_{j\ge0}\biggl(\sum_{k\ge0}a_{jk}z^k\biggr) = \sum_{n\ge0}z^n\,%
  \Biggl(\hspace{-0.3em}% <======!!!
  \sum_{\scriptstyle k_0,k_1,\ldots\ge0\atop\scriptstyle k_0+k_1+\cdots=n}
  \hspace{-1.2em}% <======!!!
  a_{0k_0}a_{1k_1}\ldots\,\Biggr)
\end{align}
```

$$\prod_{j\ge0}\left(\sum_{k\ge0}a_{jk}z^k\right) = \sum_{n\ge0}z^n\left(\sum_{\substack{k_0,k_1,\ldots\ge0\\k_0+k_1+\cdots=n}}a_{0k_0}a_{1k_1}\cdots\right) \tag{6.70}$$

06-12-11

Next we try a vertical translation of the parentheses beyond the complete limits. However, this does not look good because of the unequal heights of the two sums that result. So the best we can do for this expression is the minimal solution above, not the one below.

```
\usepackage{amsmath}
\begin{align}
\prod_{j\ge0}
\raisebox{-.8ex}{$\displaystyle
  \left(\begin{array}{@{}c@{}}\displaystyle\sum_{k\ge0}a_{jk}z^k\end{array}
  \right)$} = \sum_{n\ge0}z^n\,
\raisebox{-1.75ex}{$\displaystyle
  \left(\begin{array}{@{}c@{}}\displaystyle\sum_{\scriptstyle k_0,k_1,\ldots
    \ge0\atop\scriptstyle k_0+k_1+\cdots=n}
    \kern-.8em a_{0k_0}a_{1k_1}\ldots\end{array}\,\right)$}
\end{align}
```

$$\prod_{j\ge0}\left(\sum_{k\ge0}a_{jk}z^k\right) = \sum_{n\ge0}z^n\left(\sum_{\substack{k_0,k_1,\ldots\ge0\\k_0+k_1+\cdots=n}}a_{0k_0}a_{1k_1}\cdots\right) \tag{6.71}$$

06-12-12

## 6.12.1 \sideset command

The \sideset command is designed for special cases where super- and subscripts are allowed at all four corners of a symbol.

> \sideset{_*lowerLeft*^*upperLeft*}{_*lowerRight*^*upperRight*}\⟨*operator*⟩_*below*^*above*

The command expects three parameters; the first one specifies the limits for the left hand side, the second one for the right hand side, and the third one the operator. The operator must allow limits and may have additional ones for above and below, as can be seen in the example. The assignment is done through the usual symbols _ for below (index) and ^ for above (exponent).

$$\underset{below}{\overset{above}{\underset{LL}{\overset{UL}{\sum}}{}_{LR}^{UR}}} \tag{6.72}$$

```
\usepackage{amsmath}

\begin{align}
  \sideset{_{LL}^{UL}}{_{LR}^{UR}}%
          \sum_{below}^{above}
\end{align}
```

06-12-13

Any of the values can be omitted, as can be seen in the following example, where the second sum uses only the top right position.

06-12-14

$$\sum_{\substack{n<k \\ n \text{ odd}}}^{\prime} nE_n$$

```
\usepackage{amsmath}

\begin{align*}
  \sum^\prime_{n<k\atop n\ \textrm{odd}}nE_{n}
\end{align*}
```

$$\sum_{\substack{n<k \\ n \text{ odd}}}{'} nE_n$$

```
\begin{align*}
  \sideset{}{^\prime}\sum_{n<k\atop n\ \textrm{odd}}nE_{n}
\end{align*}
```

As this example shows, when placing an inverted comma (\prime) at a sum symbol in a displayed equation, it is best to use \sideset, which positions it as an exponent, whereas when the normal notation with ^ was used, it was placed above the sum. Other useful applications include displaying chemical elements or reaction equations:

06-12-15

$${}^{33}_{17}\mathrm{Cl}^{16} \xrightarrow{n,n} {}^{31}_{15}\mathrm{P}^{16} + {}^{4}_{2}\mathrm{He}^{2}$$

```
\usepackage{amsmath}

$\sideset{^{33}_{17}}{^{16}}{\mathop{\mathrm{Cl}}}
 \xrightarrow{n,n}%
 \sideset{^{31}_{15}}{^{16}}{\mathop{\mathrm{P}}}+
 \sideset{^4_2}{^2}{\mathop{\mathrm{He}}}$
```

## 6.13  Operator names

By definition, operator names are typeset with upright characters; variables on the other hand are typeset in italics — $y = \sin(x)$. A list of all command names in standard LaTeX can be found in Section 4.13 on page 62. amsmath defines some more. You can also define custom ones using one of the following commands:

```
\operatorname{operator name}
\operatornamewithlimits{operator name}
```

In the following example, we use \operatorname to declare \Modulo as a math operator. The output is Modulo and the same super- and subscript rules apply that are valid for existing operator names.

06-13-1

$$y = 135 \operatorname{Modulo} 17 = 16$$
$$y = 135\mathrm{Modulo}17 = 16$$

```
\usepackage{amsmath}
\newcommand*\Modulo{\operatorname{Modulo}}

\begin{align*}
  y &= 135\Modulo17=16\\
  y &= 135\mathrm{Modulo}17=16
\end{align*}
```

The differences in the horizontal spacing illustrate that internally \Modulo is an operator.

Names for new operators must, of course, be names that are not already used; however, you can also make a redefinition, with \renewcommand.

*names*

If you want your operator to be able to have limits above and below it, you will need to define it differently, using \operatornamewith limits, as LaTeX distinguishes between operators that may have limits above or below and the rest. If we use \Modulo again, the expression $\displaystyle\Modulo\limits^a {}_b$ outputs $\mathrm{Modulo}_b^a$ where the limits aren't placed above and below Modulo despite having used \limits and \displaystyle. The following example uses both commands to define \Modulo operators, and then shows the difference in the positioning of the limits.

06-13-2

$$\mathrm{Modulo}_b^a \qquad \mathrm{Modulo}_{\;b}^{\;a}$$

```
\usepackage{amsmath}
\newcommand*\ModuloA{\operatorname{Modulo}}
\newcommand*\ModuloB{\operatornamewithlimits{Modulo}}

$\ModuloA\limits^a _b \quad \ModuloB\limits^a _b$
```

Alternatively to defining an operator, you can use the command \mathop to make an expression behave like an operator in a formula. After \mathop has been used, LaTeX will not continue to treat the expression in the argument as an operator when it occurs again. It is therefore a useful command if an expression is to be used only once as an operator.

06-13-3

$$_1\mathrm{B}$$

```
\usepackage{amsmath}

\begin{align*}
  \sideset{_1}{}{\mathop{\mathrm{B}}}
\end{align*}
```

If we had not specified B to be an operator in this example, LaTeX would have given an error message because \sideset must precede an operator.

## 6.14  Text in math mode

For complex text structures within math expressions, see Section 10.7 on page 210.

The \text command, which has already been used in several examples, is superior to \textrm from standard LaTeX (cf. Section 4.6 on page 46).

$$\boxed{\text{\textbackslash text\{}\textit{argument}\text{\}}}$$

It also has an advantage over \mathrm as it typesets its argument using the current text font whereas \textrm\mathrm uses the Roman font. There is also different behaviour with regards to the size of indices and exponents. The following example compares four different ways of inserting text fragments in math mode.

06-14-1

$$\text{\textbackslash mbox} \qquad A_{\text{text}}^{\text{text}}$$
$$\text{\textbackslash text} \qquad A_{\text{text}}^{\text{text}}$$
$$\text{\textbackslash textnormal} \qquad A_{\text{text}}^{\text{text}}$$
$$\text{\textbackslash mathrm} \qquad A_{\text{text}}^{\text{text}}$$

```
\usepackage{amsmath}
\newcommand*\CS[1]{\makebox[4.6em][l]{%
    \normalsize\texttt{\textbackslash#1}}}

\sffamily\Large
\CS{mbox}$A^{\mbox{text}}_{\mbox{text}}$\\
\CS{text}$A^{\text{text}}_{\text{text}}$\\
\CS{textnormal}$A^{\textnormal{text}}_{\textnormal{text}}$\\
\CS{mathrm}$A^{\mathrm{text}}_{\mathrm{text}}$
```

\mbox is the worst of all these ways because it does not take the context into account and sets the font size to that of normal text. \text is the only one that takes both the font family (\sffamily) and the font size of the math context into account. \textnormal uses the right font size, but defaults to the Roman font (as does the following \mathrm) even though \sffamily was specified before.

```
\usepackage{amsmath}
\begin{flalign*}
            && 12(x-1) + 20(y-3) + 14(z-2) &= 0 &&\\
  \text{and} &&         6x + 10y + 7z -50 &= 0   &&
\end{flalign*}
%
\begin{align}
            && 12(x-1) + 20(y-3) + 14(z-2) &= 0\\
  \text{and} &&         6x + 10y + 7z - 50&= 0
\end{align}
```

06-14-2

$$12(x-1) + 20(y-3) + 14(z-2) = 0$$

and

$$6x + 10y + 7z - 50 = 0$$

$$12(x-1) + 20(y-3) + 14(z-2) = 0 \tag{6.73}$$

and

$$6x + 10y + 7z - 50 = 0 \tag{6.74}$$

The \text command is primarily intended for short bits of text to be inserted on a math line. For long passages or separate lines, use the \intertext command. It always puts the text on its own line, and leaves math mode so will wrap text that is longer than a line.

\intertext{*argument*}

This is how it is used within a math environment:

```
\begin{...}% math environment
 ... \\
 ... \\
\intertext{ ... }
 ...
\end{...}  % math environment
```

Note that a line break *may* be inserted in math mode *before* the \intertext command but *not* after it as this would lead to additional vertical space. This example demonstrates the \intertext command:

*line breaks*

```
\usepackage{amsmath}
\newcommand*\diff{\mathop{}\!\mathrm{d}}
\begin{align}
A_{1}
 &= \left|\int_0^1(f(x)-g(x))\diff x\right| +\left| \int _1^2(g(x)-h(x))\diff x
    \right|\nonumber\\
 &= \left|\int_0^1(x^2-3x)\diff x\right| +\left| \int _1^2(x^2-5x+6)\diff x
    \right|\nonumber
\intertext{Now the antiderivative of the two integrals is determined and the
```

```
        values calculated:}
&= \left|\frac{x^3}{3}-\frac{3}{2}x^2\right|_0^1+\left| \frac{x^3}{3}-
   \frac{5}{2}x^2+6x\right|_1^2\nonumber\\
&= \left|\frac{1}{3}-\frac{3}{2}\right| +\left| \frac{8}{3}- \frac{20}{2}+12-
   \left(\frac{1}{3}-\frac{5}{2}+6\right) \right| \nonumber\\
&= \left|-\frac{7}{6}\right| +\left| \frac{28}{6}-\frac{23}{6} \right| =\frac{7}{6}+
   \frac{5}{6}=2\,\text{FE}
\end{align}
```

$$
\begin{aligned}
A_1 &= \left|\int_0^1 (f(x) - g(x))\,\mathrm{d}x\right| + \left|\int_1^2 (g(x) - h(x))\,\mathrm{d}x\right| \\
&= \left|\int_0^1 (x^2 - 3x)\,\mathrm{d}x\right| + \left|\int_1^2 (x^2 - 5x + 6)\,\mathrm{d}x\right|
\end{aligned}
$$

06-14-3

Now the antiderivative of the two integrals is determined and the values calculated:

$$
\begin{aligned}
&= \left|\frac{x^3}{3} - \frac{3}{2}x^2\right|_0^1 + \left|\frac{x^3}{3} - \frac{5}{2}x^2 + 6x\right|_1^2 \\
&= \left|\frac{1}{3} - \frac{3}{2}\right| + \left|\frac{8}{3} - \frac{20}{2} + 12 - \left(\frac{1}{3} - \frac{5}{2} + 6\right)\right| \\
&= \left|-\frac{7}{6}\right| + \left|\frac{28}{6} - \frac{23}{6}\right| = \frac{7}{6} + \frac{5}{6} = 2\,\mathrm{FE}
\end{aligned}
\tag{6.75}
$$

## 6.15 Extensible arrows

A dynamic arrow is a construct of the following kind:

```
\usepackage{amsmath}
$\xrightarrow[\text{below}]{\text{above}}$ $\xrightarrow{x\in [0\ldots\infty)}$
$\xleftarrow[\text{below}]{\text{above}}$  $\xleftarrow{x\in [0\ldots\infty)}$
$\xleftarrow[x\in [0\ldots\infty)]{}$
```

06-15-1

As you can see from the examples, expressions above the arrow go in the required argument and expressions below the arrow go in the optional argument. Also, the arguments are always typeset in math mode. This lets you easily typeset expressions like $x \xrightarrow{n\to\infty} x_0$ (coding: `$x\xrightarrow{n\rightarrow\infty}x_0$`).

Defining custom dynamic (extensible) arrows is fairly straightforward. In the next example we will define a double arrow, \xlongleftrightarrow. (We will follow the conventions that double arrows get a capital letter to distinguish them from single arrows, as with \Rightarrow ($\Rightarrow$), and that command names for extensible arrows begin with an x.) At the same time we will also define \xlongleftrightarrow for a simple arrow.

Tof do this we use the \ext@arrow command, which is defined in amsmath. The coding "0055" is explained below. An arrow always consists of three parts; the start, the extensible part, and the end. In this example these parts are \Leftarrow\Relbar\Rightarrow. Just positioning them next to one another already results in a non-extensible arrow, albeit separated by the usual inter-character spacing — $\Leftarrow=\Rightarrow$. If we locally remove these spaces by inserting the "math space"[11] -3mu, the result is an arrow without spaces.

```
\usepackage{amsmath}
\makeatletter
\newcommand\xLongLeftRightArrow[2][]{\ext@arrow 0055{\LongLeftRightArrowfill@}{#1}{#2}}
\def\LongLeftRightArrowfill@{\arrowfill@\Leftarrow\Relbar\Rightarrow}
\newcommand\xlongleftrightarrow[2][]{\ext@arrow 0055{\longleftrightarrowfill@}{#1}{#2}}
\def\longleftrightarrowfill@{\arrowfill@\leftarrow\relbar\rightarrow}
\makeatother
$\Leftarrow\Relbar\Rightarrow$
$\Leftarrow\mkern-3mu\Relbar\mkern-3mu\Rightarrow$
$\xLongLeftRightArrow[\text{UUU}]{\text{0000000}}$
$\xlongleftrightarrow[y=f(x,v,w)]{x=f(w)}$
```

06-15-2

$$\Leftarrow=\Rightarrow \quad \Longleftrightarrow \quad \xleftrightarrow[\text{UUU}]{\text{0000000}} \quad \xleftrightarrow[y=f(x,v,w)]{x=f(w)}$$

The `\arrowfill@` command ensures that the arrow is extensible, stretching the middle part to the length of the longer parameter, whether it is the upper or the lower, to make sure they are completely above or below the centre part. We can now use the new arrow in exactly the same way as `\xrightarrow`.

The sequence of digits 0055 immediately after the command `\ext@arrow` defines the position of the upper and lower parameters relative to the extended arrow. It is not a number, but four different parameters; they are input without the usual curly braces here because only digits are possible. Each digit stands for a length in the math unit mu (cf. Section 8.1 on page 149).

The meaning of the individual digits is
- 1st digit — space to the left
- 2nd digit — space to the right
- 3rd digit — space to the left and right
- 4th digit — space relative to the arrowhead

The examples below change each parameter in turn to illustrate the effect of each one. The hyphens before and after "below" are inserted just to make this word longer compared to "above" to be able to show the differences more clearly.

```
\usepackage{amsmath,tabularx}
\makeatletter
\def\cs#1{\texttt{\$\textbackslash#1\$}}
\def\mapstofill@{\arrowfill@{\mapstochar\relbar}\relbar\rightarrow}
\newcommand*\xmapsto[6][]{\ext@arrow #3#4#5#6\mapstofill@{#1}{#2}}
\makeatother
\begin{tabularx}{\linewidth}{@{}XX@{}}
\cs{ext@arrow 0000}\quad{\large  $\xmapsto[\fbox{-below-}]{\fbox{above}}0000$}
&\cs{ext@arrow 9000}\quad{\large $\xmapsto[\fbox{-below-}]{\fbox{above}}9000$}\\
%
\cs{ext@arrow 0900}\quad{\large  $\xmapsto[\fbox{-below-}]{\fbox{above}}0900$}
```

---

[11] cf. Section 8.1 on page 149

```
&\cs{ext@arrow 0090}\quad{\large $\xmapsto[\fbox{-below-}]{\fbox{above}}0090$}\\
%
\cs{ext@arrow 0009}\quad{\large  $\xmapsto[\fbox{-below-}]{\fbox{above}}0009$}
&\cs{ext@arrow 0099}\quad{\large $\xmapsto[\fbox{-below-}]{\fbox{above}}0099$}\\
%
\cs{ext@arrow 9999}\quad{\large  $\xmapsto[\fbox{-below-}]{\fbox{above}}9999$}
\end{tabularx}
```

06-15-3

The commands used at the beginning of this section, \xrightarrow and \xleftarrow, are defined internally as:

```
\newcommand{\xrightarrow}[2][]{\ext@arrow 0359\rightarrowfill@{#1}{#2}}
\newcommand{\xleftarrow}[2][]{\ext@arrow 3095\leftarrowfill@{#1}{#2}}
```

## 6.16 Frames

amsmath defines the \boxed command, which can be used to frame either whole or part equations in inline mode or else individual **lines** of equations within an environment. The equation number is never included in the frame.

```
$...\boxed{...}...$
\begin{...} % math environment
\boxed{...}
\end{...} % math environment
```

You can put almost anything inside a \boxed command, but it must not include a column separator & or line break.

```
\usepackage{amsmath}
\newcommand*\diff{\mathop{}\!\mathrm{d}}
$a\boxed{b+c}$
\begin{align}
        f(x) = \int_1^{\infty}\frac{1}{x^2}\diff t = 1\\
  \boxed{f(x) = \int_1^{\infty}\frac{1}{x^4}\diff t = \frac{1}{3}}
\end{align}
```

06-16-1   $a\boxed{b+c}$

$$f(x) = \int_1^\infty \frac{1}{x^2}\,dt = 1 \tag{6.76}$$

$$\boxed{f(x) = \int_1^\infty \frac{1}{x^4}\,dt = \frac{1}{3}} \tag{6.77}$$

To align the equations above horizontally, insert a column separator (outside the box) at the start of each line of the environment. The problem with doing this is that the following parts of the equation are then left-aligned, which means that the equation will still not look aligned because of the box. You must also adjust for the space between the equation and the frame and the line width as well:

```
\usepackage{amsmath}
\newcommand*\diff{\mathop{}\!\mathrm{d}}
\begin{align}
  & \kern\fboxsep\kern\fboxrule f(x) = \int_1^{\infty}\frac{1}{x^2}\diff t = 1\\
  & \boxed{f(x) = \int_1^{\infty}\frac{1}{x^4}\diff t = \frac{1}{3}}
\end{align}
```

06-16-2

$$f(x) = \int_1^\infty \frac{1}{x^2}\,dt = 1 \tag{6.78}$$

$$\boxed{f(x) = \int_1^\infty \frac{1}{x^4}\,dt = \frac{1}{3}} \tag{6.79}$$

You can create coloured boxes with the empheq package. Examples can be found in Section 9.12 on page 176 among others.

## 6.17  Greek letters

We mentioned the ability of amsmath to output lowercase and bold Greek letters in Section 4.14 on page 63. For the bold variant, use the \boldsymbol command. However, if neither \mathbf nor \boldsymbol provide a bold symbol, use the \pmb command (poor man's bold). This just superimposes the letters three times with small horizontal and vertical translations. The effect is only visible when the letter is significantly enlarged, and is practically invisible when printing the document in its original size.

06-17-1

```
\usepackage{amsmath,graphicx}

\scalebox{8}{$\alpha$}
\scalebox{8}{$\pmb{\alpha}$}
$\pmb{\alpha}$
\scalebox{8}{$\boldsymbol{\alpha}$}
```

Table 6.5: Lowercase Greek letters from amsmath with a bold variant

| name | | \boldsymbol | \pmb | name | | \boldsymbol | \pmb |
|---|---|---|---|---|---|---|---|
| \alpha | α | **α** | **α** | \beta | β | **β** | **β** |
| \chi | χ | **χ** | **χ** | \delta | δ | **δ** | **δ** |
| \epsilon | ϵ | **ϵ** | **ϵ** | \eta | η | **η** | **η** |
| \gamma | γ | **γ** | **γ** | \iota | ι | **ι** | **ι** |
| \kappa | κ | **κ** | **κ** | \lambda | λ | **λ** | **λ** |
| \mu | μ | **μ** | **μ** | \nu | ν | **ν** | **ν** |
| \omega | ω | **ω** | **ω** | \phi | ϕ | **ϕ** | **ϕ** |
| \pi | π | **π** | **π** | \psi | ψ | **ψ** | **ψ** |
| \rho | ρ | **ρ** | **ρ** | \sigma | σ | **σ** | **σ** |
| \tau | τ | **τ** | **τ** | \theta | θ | **θ** | **θ** |
| \upsilon | υ | **υ** | **υ** | \varepsilon | ε | **ε** | **ε** |
| \varphi | φ | **φ** | **φ** | \varpi | ϖ | **ϖ** | **ϖ** |
| \varrho | ϱ | **ϱ** | **ϱ** | \varsigma | ς | **ς** | **ς** |
| \vartheta | ϑ | **ϑ** | **ϑ** | \xi | ξ | **ξ** | **ξ** |
| \zeta | ζ | **ζ** | **ζ** | | | | |

## 6.18  Miscellaneous commands

There are some additional commands that can be used in math mode that have not been described so far.

\underset can be very helpful when placing indices (limits) for non-operators (cf. page 169). \overset works similarly.

06-18-1

```
_baseline_
   under

    over
_baseline_
```

```
\usepackage{amsmath}

\_$\underset{under}{baseline}$\_\\
\_$\overset{over}{baseline}$\_
```

In addition to the dot sequences defined in standard LaTeX (cf. Table 4.7 on page 58), amsmath defines more dot commands, which are listed in Table 6.6:

Table 6.6: Dots in math mode with amsmath

\dotsb  ⋯  \dotsc  …  \dotsi  ⋯  \dotsm  ⋯  \dotso  …

## 6.19  Problems with amsmath

amsmath is a powerful package for math typesetting, but has some quirks as well. If you used an align environment inside a gather environment, you would expect the equations to be centred by gather. However, this is only the case if there is an equation number (tag) for the line or an additional column separator. Otherwise the lines are right-aligned.

```
\usepackage{amsmath}
\begin{gather*}
  \begin{align*}
    m_2 &= m_2' + m_2''\\
       &= \frac{V_2'}{v_2'} + \frac{V_2''}{v_2''}
  \end{align*}\\
  \Rightarrow m_2 v_2' = V - V_2'' + V_2''\frac{v_2'}{v_2''}
\end{gather*}
\begin{gather*}
  \begin{align*}
    m_2 &= m_2' + m_2''\\
       &= \frac{V_2'}{v_2'} + \frac{V_2''}{v_2''} & %<<<====
  \end{align*}\\
  \Rightarrow m_2 v_2' = V - V_2'' + V_2''\frac{v_2'}{v_2''}
\end{gather*}
```

06-19-1

$$m_2 = m_2' + m_2''$$
$$= \frac{V_2'}{v_2'} + \frac{V_2''}{v_2''}$$
$$\Rightarrow m_2 v_2' = V - V_2'' + V_2'' \frac{v_2'}{v_2''}$$

$$m_2 = m_2' + m_2''$$
$$= \frac{V_2'}{v_2'} + \frac{V_2''}{v_2''}$$
$$\Rightarrow m_2 v_2' = V - V_2'' + V_2'' \frac{v_2'}{v_2''}$$

This effect is due to the calculation of the horizontal width of the box that the math equation is placed in. This is wrong in the first example because of the missing equation number. The right hand (empty) part of the box is simply cut off while the left hand space stays. Thus the equation is "unsymmetrical" and is not typeset correctly. Adding another column separator to the right hand side, however, prevents it from being cutting off by amsmath.

Another curiosity arises for practically the same reason; again space is cut off the right hand side, only now within a \parbox or a minipage and through a leading \noindent.

```
\usepackage{amsmath}
\fbox{\parbox{10cm}{\begin{align*} a &= b \\ c &= d \end{align*}}}

\fbox{\parbox{10cm}{\noindent\begin{align*} a &= b \\ c &= d\end{align*}}}
```

06-19-2

$$a = b$$
$$c = d$$

$$a = b$$
$$c = d$$

The following example shows a problem with the gathered environment. The first [p] gets lost because gathered takes this as an optional argument.

$$= 100$$
$$[v] = 200$$

```
\usepackage{amsmath}

\[
  \begin{gathered}
    [p] = 100 \\
    [v] = 200
  \end{gathered}
\]
```

06-19-3

Adding a \relax after \begin{gathered} can help.

$$[p] = 100$$
$$[v] = 200$$

```
\usepackage{amsmath}

\[
  \begin{gathered}\relax
    [p] = 100 \\
    [v] = 200
  \end{gathered}
\]
```

06-19-4

## 6.20 Remarks

The xalignat and xxalignat environments can be regarded as obsolete. However, since they are still supported in amsmath we will describe them here. The xalignat environment behaves like the alignat environment, but is "stretched" further apart — the "eXtended" version. It expects a parameter which specifies the number of blocks and thus the number of column separators to expect as well.

```
\begin{xalignat*}{number of blocks}
  left & right & left & right...\\
...
  left & right & left & right...
\end{xalignat*}
```

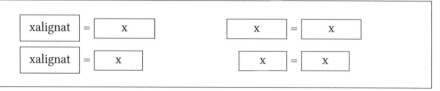

```
\usepackage{amsmath,calc}
\begin{xalignat}{3}
 i_{11} & =0.25 & i_{12} & =i_{21} & i_{13} & =i_{23}\nonumber \\
 i_{21} & =\frac{1}{3}i_{11} & i_{22} & =0.5i_{12} & i_{23} & =i_{31}\\
 i_{31} & =0.33i_{22} & i_{32} & =0.15i_{32} & i_{33} & =i_{11}
\end{xalignat}
```

$$i_{11} = 0.25 \qquad i_{12} = i_{21} \qquad i_{13} = i_{23}$$

$$i_{21} = \frac{1}{3}i_{11} \qquad i_{22} = 0.5i_{12} \qquad i_{23} = i_{31} \qquad (6.80)$$

$$i_{31} = 0.33i_{22} \qquad i_{32} = 0.15i_{32} \qquad i_{33} = i_{11} \qquad (6.81)$$

The same applies to the even more rarely used xxalignat environment, the "eXtremely eXtended" version. It occupies the largest possible space on a line, so never has an equation number for a displayed formula (and therefore there is no starred version of it).

```
\begin{xxalignat}{number of blocks}
  left & right & left & right...\\
...
  left & right & left & right...
\end{xxalignat}
```

```
\usepackage{amsmath,calc}
\begin{xxalignat}{3}
 i_{11} & =0.25        & i_{12} & =i_{21} & i_{13}        &=i_{23}\\
 i_{21} & =\frac{1}{3}i_{11} & i_{22}    & =0.5i_{12} & i_{23} & =i_{31}\\
 i_{31} & =0.33i_{22} & i_{32}           & =0.15i_{32}& i_{33} & =i_{11}
\end{xxalignat}
```

$$i_{11} = 0.25 \qquad\qquad i_{12} = i_{21} \qquad\qquad i_{13} = i_{23}$$

06-20-4

$$i_{21} = \frac{1}{3}i_{11} \qquad\qquad i_{22} = 0.5i_{12} \qquad\qquad i_{23} = i_{31}$$

$$i_{31} = 0.33i_{22} \qquad\qquad i_{32} = 0.15i_{32} \qquad\qquad i_{33} = i_{11}$$

## 6.21 Example[12]

Here are two functions:

$$f(x) = -x^2 + 6x - 5 \tag{6.82}$$

and

$$g(x) = -\frac{1}{3}x^2 + \frac{4}{3}x + \frac{5}{3} \tag{6.83}$$

How large is the area enclosed by the graphs of $f$ and $g$ in the first quadrant?

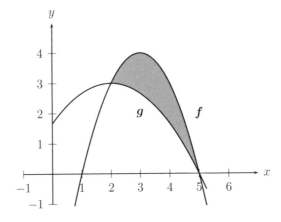

1. Determine the intersection point by equating the two equations:

$$f(x) = g(x) \tag{6.84}$$

$$-x^2 + 6x - 5 = -\frac{1}{3}x^2 + \frac{4}{3}x + \frac{5}{3} \tag{6.85}$$

$$0 = \frac{2}{3}x^2 - \frac{14}{3}x + \frac{20}{3} \tag{6.86}$$

$$0 = x^2 - 7x + 10 \tag{6.87}$$

$$0 = (x - 2)(x - 5) \tag{6.88}$$

$$\mathbb{L} = \{2;\ 5\} \tag{6.89}$$

---

[12] The complete LaTeX source code of this example can be found in 6-21.1tx.

2. There are only two nulls, both of which are positive; therefore the integration limits are fixed:

$$F = \int_{2}^{5} (g(x) - f(x)) \, dx \tag{6.90}$$

The difference $g(x) - f(x)$ is already given by Equation 6.86; it is considered in this form (not as in Equation 6.87). The sketch suggests that a negative value is to be expected because in the integration interval $g(x) < f(x)$:

$$F = \int_{2}^{5} \left( \frac{2}{3}x^2 \frac{14}{3}x + \frac{20}{3} \right) dx \tag{6.91}$$

$$= \left[ \frac{2}{9}x^3 - \frac{14}{6}x^2 + \frac{20}{3}x \right]_{2}^{5} \tag{6.92}$$

$$= \underbrace{\frac{2}{9} \times 125 - \frac{14}{6} \times 25 + \frac{20}{3} \times 5}_{\text{upper limit}} - \underbrace{\left( \frac{2}{9} \times 8 - \frac{14}{6} \times 4 + \frac{20}{3} \times 2 \right)}_{\text{lower limit}} \tag{6.93}$$

$$= \frac{500 - 1050 + 600}{18} - \frac{32 - 168 + 240}{18} \tag{6.94}$$

$$= \frac{50}{18} - \frac{104}{18} \tag{6.95}$$

$$= -\frac{54}{18} = -3 \tag{6.96}$$

The size of the area is 3AU!

3. Some remarks on the initial equations:
   As mentioned before frequently, displaying them in vertex form or in product form can be very helpful.
   For $f(x)$:

$$f(x) = -x^2 + 6x - 5 = -\left( x^2 - 6x + 5 \right) \tag{6.97}$$

$$= -(x - 1)(x - 5) \tag{6.98}$$

This determines the nulls — $\mathbb{L} = \{1; 5\}$. For the vertex form:

$$f(x) = -x^2 + 6x - 5 = -\left( x^2 - 6x + 5 \right) \tag{6.99}$$

$$= -\left( \underbrace{x^2 - 2 \times 3 \times x + 3^2}_{(x-3)^2} \underbrace{- 3^2 + 5}_{-4} \right) \tag{6.100}$$

and therefore $VP(3;4)$. The same can be done with $g(x)$:

$$g(x) = -\frac{1}{3}x^2 + \frac{4}{3}x + \frac{5}{3} \tag{6.101}$$

$$= -\frac{1}{3}\left(x^2 - 4x - 5\right) \tag{6.102}$$

$$= -\frac{1}{3}\left(x + 1\right)\left(x - 5\right) \qquad \boxed{\mathbb{L} = \{-1; 5\}} \tag{6.103}$$

$$g(x) = -\frac{1}{3}x^2 + \frac{4}{3}x + \frac{5}{3} \tag{6.104}$$

$$= -\frac{1}{3}\left(x^2 - 4x - 5\right) \tag{6.105}$$

$$= -\frac{1}{3}\left(x^2 - 2 \times 2 \times x + 2^2 - 2^2 - 5\right) \tag{6.106}$$

$$= -\frac{1}{3}\left((x - 2)^2 - 9\right) \tag{6.107}$$

$$= -\frac{1}{3}(x - 2)^2 + 3 \qquad \boxed{SP(2; 3)} \tag{6.108}$$

# Symbols

In this chapter all relevant math symbols provided by standard LaTeX as well as the numerous other packages are given. An almost complete list of all possible symbols for both math and text mode can be found at [22]. The order of the packages in this chapter is purely based on optimising page breaks.

## 7.1 Standard LaTeX

The symbol class \mathord contains the characters that are not treated separately by LaTeX with respect to their horizontal spacing; they are regarded as "normal" characters. The capital letters are not given in Table 7.1 because order and form depend on the current font.

*\mathord*

### Table 7.1: Small Greek letters of the \mathord class

| | | | | |
|---|---|---|---|---|
| α \alpha | β \beta | χ \chi | δ \delta | ε \epsilon |
| η \eta | γ \gamma | ι \iota | κ \kappa | λ \lambda |
| μ \mu | ν \nu | ω \omega | φ \phi | π \pi |
| ψ \psi | ρ \rho | σ \sigma | τ \tau | θ \theta |
| υ \upsilon | ε \varepsilon | φ \varphi | ϖ \varpi | ϱ \varrho |
| ς \varsigma | ϑ \vartheta | ξ \xi | ζ \zeta | |

07-01-1

### Table 7.2: Letter-like symbols of the \mathord class, several command names exist for individual symbols

| | | | |
|---|---|---|---|
| \$ $ | ℵ \aleph | ℑ \Im | ℜ \Re |
| ℓ \ell | ℏ \hbar | ı \imath | ȷ \jmath |
| $ \mathdollar | ¶ \mathparagraph | § \mathsection | £ \mathsterling |
| ¶ \P | ∂ \partial | £ \pounds | § \S |
| ℘ \wp | | | |

07-01-2

### Table 7.3: Miscellaneous symbols of the \mathord class

| | | | | |
|---|---|---|---|---|
| ! ! | . . | / / | ? ? | @ @ |
| \| \| | # \# | & \& | _ \_ | ‖ \\| |
| ∠ \angle | ‖ \Arrowvert | \| \arrowvert | \ \backslash | ⊥ \bot |
| , \bracevert | ♣ \clubsuit | © \copyright | ♦ \diamondsuit | ∅ \emptyset |
| ∃ \exists | ♭ \flat | ∀ \forall | ♥ \heartsuit | ∞ \infty |
| ¬ \lnot | ∇ \nabla | ♮ \natural | ¬ \neg | ′ \prime |
| ♯ \sharp | ♠ \spadesuit | √ \surd | ⊤ \top | △ \triangle |
| ‖ \Vert | \| \vert | | | |

07-01-3

### Table 7.4: Symbols of the \mathbin class

| | | | |
|---|---|---|---|
| ◯ \bigcirc | ⊙ \odot | ⊖ \ominus | ⊕ \oplus |
| ⊘ \oslash | ⊗ \otimes | • \bullet | · \cdot |
| ∘ \circ | * * | + + | − - |
| ⨿ \amalg | ∗ \ast | ▽ \bigtriangledown | △ \bigtriangleup |
| ∩ \cap | ∪ \cup | † \dag | † \dagger |
| ‡ \ddag | ‡ \ddagger | ◇ \diamond | ÷ \div |
| ∧ \land | ∨ \lor | ∓ \mp | ± \pm |
| \ \setminus | ⊓ \sqcap | ⊔ \sqcup | ⋆ \star |
| × \times | ◁ \triangleleft | ▷ \triangleright | ⊎ \uplus |
| ∨ \vee | ∧ \wedge | ≀ \wr | |

07-01-4

**Table 7.5:** Arrow symbols of the \mathrel class

07-01-5

| | | |
|---|---|---|
| ⇓ \Downarrow | ↓ \downarrow | ← \gets |
| ↩ \hookleftarrow | ↪ \hookrightarrow | ⇐ \Leftarrow |
| ← \leftarrow | ↽ \leftharpoondown | ↼ \leftharpoonup |
| ⇔ \Leftrightarrow | ↔ \leftrightarrow | ⇐ \Longleftarrow |
| ⟵ \longleftarrow | ⟺ \Longleftrightarrow | ⟷ \longleftrightarrow |
| ⟼ \longmapsto | ⟹ \Longrightarrow | ⟶ \longrightarrow |
| ↦ \mapsto | ↗ \nearrow | ↖ \nwarrow |
| ⇒ \Rightarrow | → \rightarrow | ⇁ \rightharpoondown |
| ⇀ \rightharpoonup | ↘ \searrow | ↙ \swarrow |
| → \to | ⇑ \Uparrow | ↑ \uparrow |
| ↕ \Updownarrow | ↕ \updownarrow | ˒ \lhook |
| ˓ \mapstochar | \Relhook | ˒ \rhook |

**Table 7.6:** Comparison symbols of the \mathrel class

07-01-6

| | | | | | |
|---|---|---|---|---|---|
| < < | = = | > > | ≈ \approx | ≍ \asymp | ≅ \cong |
| ≐ \doteq | ≡ \equiv | ≥ \ge | ≥ \geq | ≫ \gg | ≤ \le |
| ≤ \leq | ≪ \ll | ≠ \neq | ≺ \prec | ⪯ \preceq | ~ \sim |
| ≃ \simeq | ≻ \succ | ⪰ \succeq | | | |

**Table 7.7:** Set symbols of the \mathrel class

07-01-7

| | | | | |
|---|---|---|---|---|
| ∈ \in | ∋ \ni | ∉ \notin | ∋ \owns | ⊑ \sqsubseteq |
| ⊒ \sqsupseteq | ⊂ \subset | ⊆ \subseteq | ⊃ \supset | ⊇ \supseteq |

**Table 7.8:** Miscellaneous symbols of the \mathrel class

07-01-8

| | | | | |
|---|---|---|---|---|
| : : | ⋈ \bowtie | ⊣ \dashv | ⌢ \frown | \| \mid | ⊨ \models |
| ‖ \parallel | ⊥ \perp | ∝ \propto | ⌣ \smile | ⊢ \vdash |

The \mathop class encompasses operators of two different sizes. They are either accessed  \mathop
through the font style or else explicitly through \big preceding the command name.

**Table 7.9**: Symbols of the \mathop class

| | | | | |
|---|---|---|---|---|
| $\int$ \int | $\displaystyle\int$ \displaystyle\int | $\oint$ \oint | $\displaystyle\oint$ \displaystyle\oint | |
| $\cap$ \cap | $\bigcap$ \bigcap | $\cup$ \cup | $\bigcup$ \bigcup | |
| $\odot$ \odot | $\bigodot$ \bigodot | $\oplus$ \oplus | $\bigoplus$ \bigoplus | |
| $\otimes$ \otimes | $\bigotimes$ \bigotimes | $\sqcup$ \sqcup | $\bigsqcup$ \bigsqcup | |
| $\uplus$ \uplus | $\biguplus$ \biguplus | $\vee$ \vee | $\bigvee$ \bigvee | |
| $\wedge$ \wedge | $\bigwedge$ \bigwedge | $\coprod$ \coprod | $\displaystyle\coprod$ \displaystyle\coprod | |
| $\prod$ \prod | $\displaystyle\prod$ \displaystyle\prod | $\sum$ \sum | $\displaystyle\sum$ \displaystyle\sum | |
| $\smallint$ \smallint | | | | |

07-01-9

The symbols of the classes \mathopen and \mathclose are arbitrarily stretchable; their height can be adapted to the enclosed expression.

**Table 7.10**: Symbols of the \mathopen and \mathclose classes

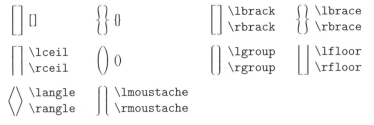

07-01-10

The dot symbols from different classes are all given here purely for systematic reasons.

**Table 7.11**: Dot symbols from several classes

| | | | | | |
|---|---|---|---|---|---|
| , , | $\cdots$ \cdots | ... \ldots | ... \mathellipsis | | |
| ; ; | : \colon | $\ddots$ \ddots | $\vdots$ \vdots | | |

07-01-11

Text symbols can't be used in math mode without first changing to text mode with \mbox or similar commands. For some symbols LaTeX defines corresponding math forms:

**Table 7.12**: Text-like symbols for math mode

| | | | | | |
|---|---|---|---|---|---|
| { \lbrace | } \rbrace | † \dagger | | | |
| ‡ \ddagger | $ \mathdollar | ¶ \mathparagraph | | | |
| £ \mathsterling | ... \mathellipsis | § \mathsection | | | |
| _ \mathunderscore | | | | | |

07-01-12

The following short forms, which are defined in the file `latex.ltx`[1], exist: `\$` instead of `\mathdollar`, `\{` instead of `\lbrace`, `\}` instead of `\rbrace`, `\P` instead of `\mathparagraph`, `\S` instead of `\mathsection`, `\dag` instead of `\dagger`, `\ddag` instead of `\ddagger`, `\pounds` instead of `\mathsterling`, `\dots` instead of `\mathellipsis`.

## 7.2 Symbols of the $\mathcal{AMS}$ packages

The symbols from standard LaTeX listed in Section 7 on page 127 are amended and in some cases modified by the `amssymb` package. Here only the symbols as defined by `amssymb` are given. All further extensions can be found in [22].

**Table 7.13:** amssymb — symbols of the \mathord class

07-02-1

| | | |
|---|---|---|
| ∠ \angle | 𝕜 \Bbbk | ℸ \beth |
| ∎ \blacksquare | Ⓢ \circledS | ∁ \complement |
| ℸ \daleth | ╲ \diagdown | ╱ \diagup |
| Ϝ \digamma | ð \eth | ⅃ \Finv |
| ⅁ \Game | ℷ \gimel | ℏ \hbar |
| ℏ \hslash | ◊ \lozenge | ◆ \blacklozenge |
| ‵ \backprime | ★ \bigstar | ▼ \blacktriangledown |
| ▲ \blacktriangle | ⋖ \lessdot | ∡ \measuredangle |
| ℧ \mho | ∄ \nexists | ∢ \sphericalangle |
| □ \square | ▽ \triangledown | ϰ \varkappa |
| ∅ \varnothing | | |

**Table 7.14:** amssymb — symbols of the \mathbin class

07-02-2

| | | |
|---|---|---|
| ⊡ \boxdot | ⊞ \boxplus | ⊠ \boxtimes |
| ⸳ \centerdot | ⊟ \boxminus | ⊻ \veebar |
| ⊼ \barwedge | ⩞ \doublebarwedge | ⋓ \Cup |
| ⋒ \Cap | ⋏ \curlywedge | ⋎ \curlyvee |
| ⋌ \leftthreetimes | ⋋ \rightthreetimes | ∔ \dotplus |
| ⊺ \intercal | ⊚ \circledcirc | ⊛ \circledast |
| ⊝ \circleddash | ╱ \diagup | ╲ \diagdown |
| ⊛ \divideontimes | ⋖ \lessdot | ⋗ \gtrdot |
| ⋉ \ltimes | ⋊ \rtimes | ∖ \smallsetminus |

---

[1]$TEXMF/tex/latex/base/latex.ltx

**Table 7.15**: amssymb — symbols of the \mathrel class (character set AMSa)

07-02-3

| | | |
|---|---|---|
| ↻ \circlearrowright | ↺ \circlearrowleft | ⇌ \rightleftharpoons |
| ⇋ \leftrightharpoons | ⊩ \Vdash | ⊪ \Vvdash |
| ⊨ \vDash | ↠ \twoheadrightarrow | ↞ \twoheadleftarrow |
| ⇇ \leftleftarrows | ⇉ \rightrightarrows | ⇈ \upuparrows |
| ⇊ \downdownarrows | ↾ \upharpoonright | |
| ⇂ \downharpoonright | ↿ \upharpoonleft | ⇃ \downharpoonleft |
| ↣ \rightarrowtail | ↢ \leftarrowtail | ⇆ \leftrightarrows |
| ⇄ \rightleftarrows | ↰ \Lsh | ↱ \Rsh |
| ⇝ \rightsquigarrow | ↭ \leftrightsquigarrow | ↫ \looparrowleft |
| ↬ \looparrowright | ≗ \circeq | ≿ \succsim |
| ≳ \gtrsim | ⪆ \gtrapprox | ⊸ \multimap |
| ∴ \therefore | ∵ \because | ≑ \doteqdot |
| ≜ \triangleq | ≾ \precsim | ≲ \lesssim |
| ⪅ \lessapprox | ⋖ \eqslantless | ⋗ \eqslantgtr |
| ⋞ \curlyeqprec | ⋟ \curlyeqsucc | ≼ \preccurlyeq |
| ≦ \leqq | ⩽ \leqslant | ⪋ \lessgtr |
| ≓ \risingdotseq | ≒ \fallingdotseq | ≽ \succcurlyeq |
| ≧ \geqq | ⩾ \geqslant | ≷ \gtrless |
| ▷ \vartriangleright | ◁ \vartriangleleft | ⊵ \trianglerighteq |
| ⊴ \trianglelefteq | ≬ \between | ▶ \blacktriangleright |
| ◀ \blacktriangleleft | △ \vartriangle | ≖ \eqcirc |
| ⋚ \lesseqgtr | ⋛ \gtreqless | ⪋ \lesseqqgtr |
| ⪌ \gtreqqless | ⇛ \Rrightarrow | ⇚ \Lleftarrow |
| ∝ \varpropto | ⌣ \smallsmile | ⌢ \smallfrown |
| ⋐ \Subset | ⋑ \Supset | ⊆ \subseteqq |
| ⊇ \supseteqq | ≏ \bumpeq | ≎ \Bumpeq |
| ⋘ \lll | ⋙ \ggg | ⋔ \pitchfork |
| ∽ \backsim | ⋍ \backsimeq | |

**Table 7.16**: amssymb — symbols of the \mathrel class (character set AMSb)

07-02-4

| | | | |
|---|---|---|---|
| $\lvertneqq$ \lvertneqq | $\gvertneqq$ \gvertneqq | $\nleq$ \nleq | $\ngeq$ \ngeq |
| $\nless$ \nless | $\ngtr$ \ngtr | $\nprec$ \nprec | $\nsucc$ \nsucc |
| $\lneqq$ \lneqq | $\gneqq$ \gneqq | $\nleqslant$ \nleqslant | $\ngeqslant$ \ngeqslant |
| $\lneq$ \lneq | $\gneq$ \gneq | $\npreceq$ \npreceq | $\nsucceq$ \nsucceq |
| $\precnsim$ \precnsim | $\succnsim$ \succnsim | $\lnsim$ \lnsim | $\gnsim$ \gnsim |
| $\nleqq$ \nleqq | $\ngeqq$ \ngeqq | $\precneqq$ \precneqq | $\succneqq$ \succneqq |
| $\precnapprox$ \precnapprox | $\succnapprox$ \succnapprox | $\lnapprox$ \lnapprox | $\gnapprox$ \gnapprox |
| $\nsim$ \nsim | $\ncong$ \ncong | $\varsubsetneq$ \varsubsetneq | $\varsupsetneq$ \varsupsetneq |
| $\nsubseteqq$ \nsubseteqq | $\nsupseteqq$ \nsupseteqq | $\subsetneqq$ \subsetneqq | $\supsetneqq$ \supsetneqq |
| $\varsubsetneqq$ \varsubsetneqq | $\varsupsetneqq$ \varsupsetneqq | $\subsetneq$ \subsetneq | $\supsetneq$ \supsetneq |
| $\nsubseteq$ \nsubseteq | $\nsupseteq$ \nsupseteq | $\nparallel$ \nparallel | $\nmid$ \nmid |
| $\nshortmid$ \nshortmid | $\nshortparallel$ \nshortparallel | $\nvdash$ \nvdash | $\nVdash$ \nVdash |
| $\nvDash$ \nvDash | $\nVDash$ \nVDash | $\ntrianglerighteq$ \ntrianglerighteq | $\ntrianglelefteq$ \ntrianglelefteq |
| $\ntriangleleft$ \ntriangleleft | $\ntriangleright$ \ntriangleright | $\nleftarrow$ \nleftarrow | $\nrightarrow$ \nrightarrow |
| $\nLeftarrow$ \nLeftarrow | $\nRightarrow$ \nRightarrow | $\nLeftrightarrow$ \nLeftrightarrow | $\nleftrightarrow$ \nleftrightarrow |
| $\eqsim$ \eqsim | $\shortmid$ \shortmid | $\shortparallel$ \shortparallel | $\thicksim$ \thicksim |
| $\thickapprox$ \thickapprox | $\approxeq$ \approxeq | $\succapprox$ \succapprox | $\precapprox$ \precapprox |
| $\curvearrowleft$ \curvearrowleft | $\curvearrowright$ \curvearrowright | $\backepsilon$ \backepsilon | |

## 7.3 textcomp

In principle the special commands provided by this package are not needed as the same can be achieved more easily with the \oldstylenums command:

07-03-1

| | |
|---|---|
| 0123456789 | `$\oldstylenums{0123456789}$\\` |
| 123.034.445, 32 | `$\oldstylenums{123.034.445,32}$` |

The characters in the following table are strictly speaking not math symbols because the textcomp package provides them outside math mode. This difference is not immediately obvious for figures though, so we won't deal with it here.

07-03-2

| | | |
|---|---|---|
| 0 \textzerooldstyle | 1 \textoneoldstyle | 2 \texttwooldstyle |
| 3 \textthreeoldstyle | 4 \textfouroldstyle | 5 \textfiveoldstyle |
| 6 \textsixoldstyle | 7 \textsevenoldstyle | 8 \texteightoldstyle |
| 9 \textnineoldstyle | | |

## 7.4 `mathcomp`

Similar to the `textcomp` package (cf. Section 7.3 on the previous page), this package by Tilmann Böß defines some useful symbols.

**Table** 7.17: mathcomp — miscellaneous symbols

| | | | |
|---|---|---|---|
| °C `\tccentigrade` | Ω `\tcohm` | ‰ `\tcperthousand` | μ `\tcmu` |
| ‰₀ `\tcpertenthousand` | † `\dagger` | ‡ `\ddagger` | |

`07-04-1`

## 7.5 `mathabx`

All the symbols of this package by Anthony Phan are only available in METAFONT form; vector fonts don't exist yet. This package is sometimes missing in older TeX distributions, but it can be installed manually.[2] The package can be loaded with the options `matha`, `mathb`, or `mathx`. The last letter stands for the character set to load; in here `matha` is the most important one. If no option is specified, all available character sets are loaded.

**Table** 7.18: mathabx — binary operators

| | | | |
|---|---|---|---|
| ∗ `\ast` | ⋏ `\curlywedge` | ⊓ `\sqcap` | ✳ `\Asterisk` |
| ∸ `\divdot` | ⊔ `\sqcup` | ⊼ `\barwedge` | ✴ `\divideontimes` |
| ⊓ `\sqdoublecap` | ★ `\bigstar` | ∸ `\dotdiv` | ⊔ `\sqdoublecup` |
| ★ `\bigvarstar` | ∔ `\dotplus` | □ `\square` | ◆ `\blackdiamond` |
| ẋ `\dottimes` | ⊎ `\squplus` | ∩ `\cap` | ⊼ `\doublebarwedge` |
| · `\udot` | ⚈ `\circplus` | ⋒ `\doublecap` | ⊎ `\uplus` |
| ∗ `\coasterisk` | ⋓ `\doublecup` | ⋆ `\varstar` | ✳ `\coAsterisk` |
| ⋉ `\ltimes` | ∨ `\vee` | ✳ `\convolution` | ⊕ `\pluscirc` |
| ⊻ `\veebar` | ∪ `\cup` | ⋈ `\rtimes` | ⊻ `\veedoublebar` |
| ⋎ `\curlyvee` | ∎ `\sqbullet` | ∧ `\wedge` | |

`07-05-1`

**Table** 7.19: mathabx — geometric binary operators

| | | | |
|---|---|---|---|
| ⊞ `\boxright` | ⊖ `\ominus` | ⧄ `\boxslash` | ⊕ `\oplus` |
| ⊠ `\boxtimes` | ⊕ `\oright` | ⊟ `\boxtop` | ⊘ `\oslash` |
| ⊞ `\boxasterisk` | ⊿ `\boxtriangleup` | ⊗ `\otimes` | ⧅ `\boxbackslash` |
| □ `\boxvoid` | ⊕ `\otop` | ⊟ `\boxbot` | ⊛ `\oasterisk` |
| ⊕ `\otriangleup` | ⊡ `\boxcirc` | ⊘ `\obackslash` | ○ `\ovoid` |
| ⊞ `\boxcoasterisk` | ⊕ `\obot` | ⊟ `\boxdiv` | ⊙ `\ocirc` |
| ⊡ `\boxdot` | ⊛ `\ocoasterisk` | ⊞ `\boxleft` | ⊕ `\odiv` |
| ⊟ `\boxminus` | ⊙ `\odot` | ⊞ `\boxplus` | ⊕ `\oleft` |

`07-05-2`

---

[2] `ftp://dante.ctan.org/tex-archive/fonts/mathabx/`

▾ \blacktriangledown    ◄ \blacktriangleleft
▸ \blacktriangleright   ▲ \blacktriangleup
▽ \smalltriangledown    ◁ \smalltriangleleft
▷ \smalltriangleright   △ \smalltriangleup

**Table 7.20:** mathabx — operators of variable size (\textstyle — \displaystyle)

| | | |
|---|---|---|
| Υ Υ \bigcurlyvee | ◻◻ \bigboxslash | ⊕⊕ \bigoright |
| ⊓⊓ \bigsqcap | ⊠⊠ \bigboxtimes | ⊘⊘ \bigoslash |
| 人人 \bigcurlywedge | ⊞⊞ \bigboxtop | ⊖⊖ \bigotop |
| ✳✳ \bigboxasterisk | △△ \bigboxtriangleup | ⊜⊜ \bigotriangleup |
| ◺◺ \bigboxbackslash | ▢▢ \bigboxvoid | ○○ \bigovoid |
| ⊟⊟ \bigboxbot | ℂℂ \bigcomplementop | ++ \bigplus |
| ⊡⊡ \bigboxcirc | ✳✳ \bigoasterisk | ⊎⊎ \bigsquplus |
| ✳✳ \bigboxcoasterisk | ⊘⊘ \bigobackslash | ×× \bigtimes |
| ⊟⊟ \bigboxdiv | ⊖⊖ \bigobot | ∭ \iiint |
| ⊡⊡ \bigboxdot | ⊚⊚ \bigocirc | ∬ \iint |
| ⊟⊟ \bigboxleft | ✳✳ \bigocoasterisk | ∫ \int |
| ⊟⊟ \bigboxminus | ⊖⊖ \bigodiv | ∯ \oiint |
| ⊞⊞ \bigboxplus | ⊕⊕ \bigoleft | ∮ \oint |
| ⊞⊞ \bigboxright | ⊖⊖ \bigominus | |

**Table 7.21:** mathabx — relation symbols

| | | | |
|---|---|---|---|
| ∮ \between | \| \divides | ≒ \risingdotseq | ≑ \botdoteq |
| ≑ \dotseq | ≳ \succapprox | ≎ \Bumpedeq | ≏ \eqbumped |
| ⪰ \succcurlyeq | ≏ \bumpedeq | ≖ \eqcirc | ⋗ \succdot |
| ≗ \circeq | =: \eqcolon | ≿ \succsim | := \coloneq |
| ≓ \fallingdotseq | ∴ \therefore | ≙ \corresponds | ⋙ \ggcurly |
| ≐ \topdoteq | ⋞ \curlyeqprec | ⋘ \llcurly | ⊨ \vDash |
| ⋟ \curlyeqsucc | ≾ \precapprox | ⊪ \Vdash | ⫤ \DashV |
| ⋞ \preccurlyeq | ⊩ \VDash | ⊣ \Dashv | ⋖ \precdot |
| ⊪ \Vvdash | ⫣ \dashVv | ≾ \precsim | |

| | | | |
|---|---|---|---|
| ≉ \napprox | ⊾ \notperp | ⊭ \nvDash | ≇ \ncong |
| ⊀ \nprec | ⊯ \nVDash | ⋡ \ncurlyeqprec | ⋨ \nprecapprox |
| ⊮ \nVdash | ⋭ \ncurlyeqsucc | ⋠ \npreccurlyeq | ⊬ \nvdash |
| ⊭ \nDashv | ⋠ \npreceq | ⊮ \nVvash | ⊮ \ndashV |
| ⋨ \nprecsim | ⋨ \precnapprox | ⊣ \ndashv | ⊀ \nsim |
| ⋨ \precneq | ⊭ \nDashV | ≄ \nsimeq | ⋨ \precnsim |
| ⊭ \ndashVv | ⊁ \nsucc | ⋩ \succnapprox | ≠ \neq |
| ⋩ \nsuccapprox | ⋡ \succneq | ≭ \notasymp | ⋭ \nsucccurlyeq |
| ⋩ \succnsim | ∤ \notdivides | ⋡ \nsucceq | ≢ \notequiv |
| ⋩ \nsuccsim | | | |

07-05-6

| | | | |
|---|---|---|---|
| ⋢ \nsqsubset | ⊅ \nsupset | ⊒ \sqsupseteq | ⊇ \supseteq |
| ⋢ \nsqSubset | ⊅ \nSupset | ⊒ \sqsupseteqq | ⊇ \supseteqq |
| ⋢ \nsqsubseteq | ⊉ \nsupseteq | ⋣ \sqsupsetneq | ⊋ \supsetneq |
| ⋢ \nsqsubseteqq | ⊉ \nsupseteqq | ⋣ \sqsupsetneqq | ⊋ \supsetneqq |
| ⊐ \nsqsupset | ⊏ \sqsubset | ⊂ \subset | ⋤ \varsqsubsetneq |
| ⊐ \nsqSupset | ⋐ \sqSubset | ⋐ \Subset | ⋤ \varsqsubsetneqq |
| ⊐ \nsqsupseteq | ⊑ \sqsubseteq | ⊆ \subseteq | ⋣ \varsqsupsetneq |
| ⊐ \nsqsupseteqq | ⊑ \sqsubseteqq | ⊆ \subseteqq | ⋣ \varsqsupsetneqq |
| ⊄ \nsubset | ⋤ \sqsubsetneq | ⊊ \subsetneq | ⊊ \varsubsetneq |
| ⋢ \nSubset | ⋤ \sqsubsetneqq | ⊊ \subsetneqq | ⊊ \varsubsetneqq |
| ⊈ \nsubseteq | ⊐ \sqSupset | ⊃ \supset | ⊋ \varsupsetneq |
| ⊈ \nsubseteqq | ⊐ \sqsupset | ⊃ \Supset | ⊋ \varsupsetneqq |

07-05-7

| | | | |
|---|---|---|---|
| ⩾ \eqslantgtr | ⋛ \gtreqless | ≲ \lesssim | ≯ \ngtr |
| ⩽ \eqslantless | ⪌ \gtreqqless | ≪ \ll | ⋧ \ngtrapprox |
| ≥ \geq | ≷ \gtrless | ⋘ \lll | ⋧ \ngtrsim |
| ≧ \geqq | ≳ \gtrsim | ⋦ \lnapprox | ≰ \nleq |
| ≫ \gg | ⪈ \gvertneqq | ≨ \lneq | ≰ \nleqq |
| ⋙ \ggg | ≤ \leq | ⪇ \lneqq | ≮ \nless |
| ⋧ \gnapprox | ≦ \leqq | ⋦ \lnsim | ⋦ \nlessapprox |
| ⪈ \gneq | ≲ \lessapprox | ⪇ \lvertneqq | ⋦ \nlesssim |
| ⪊ \gneqq | ⋖ \lessdot | ⋟ \neqslantgtr | ⋧ \nvargeq |
| ⋧ \gnsim | ⋚ \lesseqgtr | ⋞ \neqslantless | ⋦ \nvarleq |
| ≳ \gtrapprox | ⪋ \lesseqqgtr | ≱ \ngeq | ≥ \vargeq |
| ⋗ \gtrdot | ≶ \lessgtr | ≱ \ngeqq | ≤ \varleq |

07-05-8

| | | |
|---|---|---|
| ⋪ \ntriangleleft | ⋭ \ntrianglerighteq | ▷ \triangleright |
| ⊳ \vartriangleright | ⋬ \ntrianglelefteq | ◁ \triangleleft |
| ⊵ \trianglerighteq | ⋫ \ntriangleright | ⊴ \trianglelefteq |
| ⊲ \vartriangleleft | | |

07-05-9

**Table** 7.22: mathabx — arrows and hooks

| | | |
|---|---|---|
| $\circlearrowleft$ \circlearrowleft | $\leftarrow$ \leftarrow | $\nwarrow$ \nwarrow |
| $\circlearrowright$ \circlearrowright | $\leftleftarrows$ \leftleftarrows | $\restriction$ \restriction |
| $\curvearrowbotleft$ \curvearrowbotleft | $\leftrightarrow$ \leftrightarrow | $\rightarrow$ \rightarrow |
| $\curvearrowbotleftright$ \curvearrowbotleftright | $\leftrightarrows$ \leftrightarrows | $\rightleftarrows$ \rightleftarrows |
| $\curvearrowbotright$ \curvearrowbotright | $\leftrightsquigarrow$ \leftrightsquigarrow | $\rightrightarrows$ \rightrightarrows |
| $\curvearrowleft$ \curvearrowleft | $\leftsquigarrow$ \leftsquigarrow | $\rightsquigarrow$ \rightsquigarrow |
| $\curvearrowleftright$ \curvearrowleftright | $\lefttorightarrow$ \lefttorightarrow | $\righttoleftarrow$ \righttoleftarrow |
| $\curvearrowright$ \curvearrowright | $\looparrowdownleft$ \looparrowdownleft | $\Rsh$ \Rsh |
| $\dlsh$ \dlsh | $\looparrowdownright$ \looparrowdownright | $\searrow$ \searrow |
| $\downdownarrows$ \downdownarrows | $\looparrowleft$ \looparrowleft | $\swarrow$ \swarrow |
| $\downtouparrow$ \downtouparrow | $\looparrowright$ \looparrowright | $\updownarrows$ \updownarrows |
| $\downuparrows$ \downuparrows | $\Lsh$ \Lsh | $\uptodownarrow$ \uptodownarrow |
| $\drsh$ \drsh | $\nearrow$ \nearrow | $\upuparrows$ \upuparrows |
| $\nLeftarrow$ \nLeftarrow | $\nleftrightarrow$ \nleftrightarrow | $\nrightarrow$ \nrightarrow |
| $\nleftarrow$ \nleftarrow | $\nLeftrightarrow$ \nLeftrightarrow | $\nRightarrow$ \nRightarrow |

| | | |
|---|---|---|
| $\barleftharpoon$ \barleftharpoon | $\leftharpoonup$ \leftharpoonup | $\rightleftharpoons$ \rightleftharpoons |
| $\barrightharpoon$ \barrightharpoon | $\leftleftharpoons$ \leftleftharpoons | $\rightrightharpoons$ \rightrightharpoons |
| $\downdownharpoons$ \downdownharpoons | $\leftrightharpoon$ \leftrightharpoon | $\updownharpoons$ \updownharpoons |
| $\downharpoonleft$ \downharpoonleft | $\leftrightharpoons$ \leftrightharpoons | $\upharpoonleft$ \upharpoonleft |
| $\downharpoonright$ \downharpoonright | $\rightbarharpoon$ \rightbarharpoon | $\upharpoonright$ \upharpoonright |
| $\downupharpoons$ \downupharpoons | $\rightharpoondown$ \rightharpoondown | $\upupharpoons$ \upupharpoons |
| $\leftbarharpoon$ \leftbarharpoon | $\rightharpoonup$ \rightharpoonup | $\leftharpoondown$ \leftharpoondown |
| $\rightleftharpoon$ \rightleftharpoon | | |

**Table** 7.23: mathabx — delimiters

| | | | | |
|---|---|---|---|---|
| $\lcorners$ \lcorners | $\rcorners$ \rcorners | $\ulcorner$ \ulcorner | $\urcorner$ \urcorner | $\llcorner$ \llcorner |
| $\lrcorner$ \lrcorner | $\ldbrack$ \ldbrack | $\rdbrack$ \rdbrack | $\lfilet$ \lfilet | $\rfilet$ \rfilet |
| $\thickvert$ \thickvert | $\vvvert$ \vvvert | | | |

**Table** 7.24: mathabx — horizontal braces

| | | |
|---|---|---|
| $\overbrace{abc}$ \overbrace{abc} | $\widebar{abc}$ \widebar{abc} | $\overgroup{abc}$ \overgroup{abc} |
| $\widecheck{abc}$ \widecheck{abc} | $\underbrace{abc}$ \underbrace{abc} | $\wideparen{abc}$ \wideparen{abc} |
| $\undergroup{abc}$ \undergroup{abc} | $\widering{abc}$ \widering{abc} | $\widearrow{abc}$ \widearrow{abc} |

**Table 7.25:** mathabx — miscellaneous symbols

07-05-14

| ∘ \degree | //// \fourth | ∡ \measuredangle | // \second |
|---|---|---|---|
| ╲ \diagdown | # \hash | ⋔ \pitchfork | ∢ \sphericalangle |
| ╱ \diagup | ∞ \infty | ∝ \propto | /// \third |
| ⌀ \diameter | ⋋ \leftthreetimes | ⋌ \rightthreetimes | # \varhash |
| ē \barin | ∈ \in | ⊤̸ \nottop | ∉ \varnotin |
| ∁ \complement | ∄ \nexists | ∋ \owns | ∌ \varnotowner |
| ∃ \exists | ⊥̸ \notbot | ∋̄ \ownsbar | ⅂ \Finv |
| ∉ \notin | ∂ \partial | ↺ \Game | ∌ \notowner |
| ∂̸ \partialslash | | | |

## 7.6 **stmaryrd**

This package's strange name is derived from "St Mary's Road symbol package". The package by Jeremy Gibbons and Alan Jeffrey overwrites several of the standard symbols of LATEX. If you need to preserve them, they must be saved *before* stmaryrd is loaded, for example through:

```
\let\Oldbigtriangleup\bigtriangleup  \let\Oldbigtriangledown\bigtriangledown
\usepackage{stmaryrd}
```

The package lets you specify exactly which symbols to load when loading the package. You select them by using an optional argument of only, and then specifying the required symbol, for example – \usepackage [only,bigparallel] {*stmaryrd*}.

**Table 7.26:** stmaryrd — operators

07-06-1

| ⅄ \Ydown | ≺ \Yleft | ≻ \Yright | ⅄ \Yup |
|---|---|---|---|
| φ \baro | ⫽ \bbslash | & \binampersand | ⅋ \bindnasrepma |
| ⊞ \boxast | ⊟ \boxbar | ▣ \boxbox | ◺ \boxbslash |
| ◎ \boxcircle | ⊡ \boxdot | □ \boxempty | ▨ \boxslash |
| ⅄ \curlyveedownarrow | | ⅄ \curlyveeuparrow | |
| ⅄ \curlywedgedownarrow | | ⅄ \curlywedgeuparrow | |
| ⫽ \fatslash | ⨟ \fatsemi | ⫻ \fatslash | ⫴ \interleave |
| ◁ \leftslice | ⋏ \merge | ⊖ \minuso | ⋇ \moo |
| ⋔ \nplus | ⦶ \obar | □ \oblong | ⦸ \obslash |
| ⧁ \ogreaterthan | ⧀ \olessthan | ⦽ \ovee | ⦼ \owedge |
| ▷ \rightslice | ⫽ \sslash | ⫾ \talloblong | ◯ \varbigcirc |
| ⅄ \varcurlyvee | ⋏ \varcurlywedge | ⊛ \varoast | ⦶ \varobar |
| ⦸ \varobslash | ⊚ \varocircle | ⊙ \varodot | ⧁ \varogreaterthan |
| ⦵ \varolessthan | ⊖ \varominus | ⊕ \varoplus | ⊘ \varoslash |
| ⊗ \varotimes | ⦽ \varovee | ⦼ \varowedge | Ⅹ \vartimes |

**Table 7.27:** stmaryrd — large operators

07-06-2

| | | |
|---|---|---|
| ☐ \bigbox | Υ \bigcurlyvee | ⋏ \bigcurlywedge |
| ⦀ \biginterleave | ⊞ \bignplus | ‖ \bigparallel |
| ⊓ \bigsqcap | ▽ \bigtriangledown | △ \bigtriangleup |

**Table 7.28:** stmaryrd — relations

07-06-3

| | | |
|---|---|---|
| ⋵ \inplus | ⋶ \niplus | ⋬ \ntrianglelefteqslant |
| ⋭ \ntrianglerighteqslant | ⋐ \subsetplus | ⫅ \subsetpluseq |
| ⋑ \supsetplus | ⫆ \supsetpluseq | ⊲ \trianglelefteqslant |
| ▷ \trianglerighteqslant | | |

All arrows can be negated symbolically with a line by preceding them with \Arrownot, but this only works properly for horizontal arrows.

**Table 7.29:** stmaryrd — arrows

07-06-4

| | | |
|---|---|---|
| ↦̸ \Arrownot\Longmapsto | ↛ \Arrownot\shortrightarrow | |
| ↭ \leftrightarrowtriangle | | ↯ \lightning |
| ↤ \longmapsfrom | ⟸ \Longmapsfrom | ⟹ \Longmapsto |
| ↦ \Mapsto | ⇷ \Mapsfrom | ← \leftarrowtriangle |
| ⇆ \leftrightarroweq | ↪ \mapsfrom | ↗ \nnearrow |
| ↖ \nnwarrow | →> \rightarrowtriangle | ⦆ \rrparenthesis |
| ↓ \shortdownarrow | ← \shortleftarrow | → \shortrightarrow |
| ↑ \shortuparrow | ↘ \ssearrow | ↙ \sswarrow |

**Table 7.30:** stmaryrd — delimiters

07-06-5

| | | | |
|---|---|---|---|
| ⁅ \Lbag | ⁆ \Rbag | ⁅ \lbag | ⟦ \llbracket |
| ⌈ \llceil | ⌊ \llfloor | ⦇ \llparenthesis | ⁆ \rbag |
| ⟧ \rrbracket | ⌉ \rrceil | ⌋ \rrfloor | |

Delimiters can be used as usual in conjunction with the \left and \right commands:

07-06-6

$$\left\llbracket \bigbox_{i \in I}^{a \oplus b} P_i \right\rrbracket$$

```
\usepackage{stmaryrd,lucidabr}

\[ \left\llbracket
   \bigbox_{i\inplus I}^{a \varoplus b} P_i
\right\rrbracket \]
```

## 7.7 `trfsigns`

This package by Kai Rascher contains transformation symbols as they are used in systems theory and electronics. They can be used in both math and text mode.

**Table 7.31:** `trfsigns` — transformation symbols

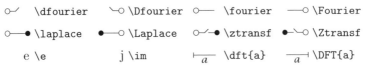

| ○⌄ `\dfourier` | ⌄○ `\Dfourier` | ○— `\fourier` | —○ `\Fourier` |
| ○—● `\laplace` | ●—○ `\Laplace` | ○⌄● `\ztransf` | ●⌄○ `\Ztransf` |
| e `\e` | j `\im` | $\vdash_{\overline{a}}$ `\dft{a}` | $\overline{a}\dashv$ `\DFT{a}` |

07-07-1

## 7.8 `MnSymbol`

The package `MnSymbol` by Achim Blumensath makes it possible to select the calligraphic characters from the fonts `mathabx`, `lmodern`, or `MnSymbolS` through the package options abx, cmsy, or mnsy. The additional packages `textcomp` and `amsmath` are loaded automatically; options for those two packages should be passed through `\PassOptionsToPackage{`*options*`}{`*package*`}`; otherwise an *option clash* may result. Note that the `MnSymbolS` package is incompatible with the `amsfonts` and `amssymb` packages.

**Table 7.32:** `MnSymbol` — binary symbols

| ⊔ `\amalg` | ⊔⊔ `\doublesqcup` | ∴ `\righttherefore` | * `\ast` |
| W `\doublevee` | ⋋ `\backslashdiv` | ⋀ `\doublewedge` | ⊢ `\rightY` |
| ⋈ `\bowtie` | ∵ `\downtherefore` | ⋊ `\rtimes` | • `\bullet` |
| ⋎ `\downY` | ⋌ `\slashdiv` | ∩ `\cap` | ⋉ `\dtimes` |
| ∏ `\smallprod` | ⩍ `\capdot` | ⋰ `\fivedots` | ⊓ `\sqcap` |
| ⩒ `\capplus` | ∝∝ `\hbipropto` | ⊡ `\sqcapdot` | · `\cdot` |
| ⋯ `\hdotdot` | ⊡ `\sqcapplus` | ○ `\circ` | ⊔ `\sqcup` |
| ⊔ `\sqcupdot` | ∴ `\lefttherefore` | ⊔ `\sqcupplus` | ∪ `\cup` |
| ∷ `\squaredots` | ⊍ `\cupdot` | ⊰ `\leftY` | × `\times` |
| ⊎ `\cupplus` | ⋉ `\ltimes` | ∴ `\udotdot` | Y `\curlyvee` |
| ╲ `\medbackslash` | ∴ `\uptherefore` | Ẏ `\curlyveedot` | ○ `\medcircle` |
| ⊥ `\upY` | ⋏ `\curlywedge` | ╱ `\medslash` | ⋈ `\utimes` |
| ⋏ `\curlywedgedot` | ∣ `\medvert` | ⅄ `\vbipropto` | ∴ `\ddotdot` |
| ⊩ `\medvertdot` | ∶ `\vdotdot` | ∴ `\diamonddots` | − `\minus` |
| ∨ `\vee` | ÷ `\div` | ⊤ `\minusdot` | V̇ `\veedot` |
| ⊣ `\dotmedvert` | ∓ `\mp` | ⋈ `\vertbowtie` | ⊥ `\dotminus` |
| ♂ `\neswbipropto` | ⊩ `\vertdiv` | ⋒ `\doublecap` | ♋ `\nwsebipropto` |
| ∧ `\wedge` | ⋓ `\doublecup` | + `\plus` | ∧̇ `\wedgedot` |
| W `\doublecurlyvee` | ± `\pm` | ≀ `\wreath` | ⋏⋏ `\doublecurlywedge` |
| ⋒ `\doublesqcap` | | | |

07-08-1

**Table** 7.33: MnSymbol — geometric binary symbols

07-08-2

| | | |
|---|---|---|
| ◩ \boxbackslash | ◎ \ocirc | ▣ \boxbox |
| ⊙ \odot | ⊡ \boxdot | ⊖ \ominus |
| ⊟ \boxminus | ⊕ \oplus | ⊞ \boxplus |
| ∎ \filledsquare | ⊘ \oslash | ☑ \boxslash |
| ★ \filledstar | ⊛ \ostar | ⊠ \boxtimes |
| ▼ \filledtriangledown | ⊗ \otimes | ⊡ \boxvert |
| ◀ \filledtriangleleft | ⊲ \otriangle | ◇ \diamondbackslash |
| ▶ \filledtriangleright | ⓘ \overt | ◈ \diamonddiamond |
| ▲ \filledtriangleup | ☆ \pentagram | ◇ \diamonddot |
| ◇ \meddiamond | ◇ \smalldiamond | ◇ \diamondminus |
| □ \medsquare | ▫ \smallsquare | ◈ \diamondplus |
| ☆ \medstar | ⋆ \smallstar | ◇ \diamondslash |
| ▽ \medtriangledown | ▿ \smalltriangledown | ◈ \diamondtimes |
| ◁ \medtriangleleft | ◁ \smalltriangleleft | ◇ \diamondvert |
| ▷ \medtriangleright | ▷ \smalltriangleright | △ \medtriangleup |
| △ \smalltriangleup | ◆ \filleddiamond | ⊛ \oast |
| ⋆ \thinstar | ⊘ \obackslash | |

**Table** 7.34: MnSymbol — operators

07-08-3

| | | |
|---|---|---|
| ⋂ \bigcap | ⊖ \bigominus | \complement |
| ⋂̇ \bigcapdot | ⊕ \bigoplus | ⨿ \coprod |
| ⋂₊ \bigcapplus | ⊘ \bigoslash | ∫⋯∫ \idotsint |
| ◯ \bigcircle | ⊛ \bigostar | ⨌ \iiiint |
| ⋃ \bigcup | ⊗ \bigotimes | ∭ \iiint |
| ⋃̇ \bigcupdot | ⊴ \bigotriangle | ∬ \iint |
| ⋃₊ \bigcupplus | ⓘ \bigovert | ∫ \int |
| ⅄ \bigcurlyvee | ＋ \bigplus | ∮ \landdownint |
| ⅄̇ \bigcurlyveedot | ⊓ \bigsqcap | ∮ \landupint |
| ⋏ \bigcurlywedge | ⊓̇ \bigsqcapdot | ∮ \lcircleleftint |
| ⋏̇ \bigcurlywedgedot | ⊞ \bigsqcapplus | ∮ \lcirclerightint |
| ⋁ \bigdoublecurlyvee | ⊔ \bigsqcup | ∯ \oiint |
| ⋀ \bigdoublecurlywedge | ⊔̇ \bigsqcupdot | ∮ \oint |
| ⋁ \bigdoublevee | ⊎ \bigsqcupplus | ∏ \prod |
| ⋀ \bigdoublewedge | ⨯ \bigtimes | ∮ \rcircleleftint |
| ⊛ \bigoast | ⋁ \bigvee | ∮ \rcirclerightint |
| ⊘ \bigobackslash | ⅋ \bigveedot | ∫ \strokedint |
| ◎ \bigocirc | ⋀ \bigwedge | ∑ \sum |
| ⊙ \bigodot | ⋀ \bigwedgedot | ∮ \sumint |

## 7.9 `wasysym`

The name of this package by Axel Kielhorn derives from the author of the fonts Roland Waldi. It provides only a few math symbols; the integral symbols are not recommended stylistically. The package option `nointegrals` prevents loading those symbols and overwriting any existing *no integrals* ones. *option*

**Table 7.35:** wasysym — integrals

| | | | |
|---|---|---|---|
| ∫ \int | ∬ \iint | ∭ \iiint | ∫ \varint |
| ∮ \varoint | ∯ \oiint | | |

07-09-1

**Table 7.36:** wasysym — miscellaneous mathematical symbols

| | | | |
|---|---|---|---|
| ⪈ \apprge | ⪉ \apprle | □ \Box | ◇ \Diamond | ⌐ \invneg |
| ⋈ \Join | ⤳ \leadsto | ◁ \lhd | ◀ \LHD | ⊗ \logof |
| ℧ \mho | ○ \ocircle | ▷ \rhd | ▶ \RHD | ⊏ \sqsubset |
| ⊐ \sqsupset | ⊴ \unlhd | ⊵ \unrhd | ∢ \varangle | ∝ \wasypropto |
| ∴ \wasytherefore | | | | |

07-09-2

## 7.10 `mathdesign`

*three* The `mathdesign` package by Paket Pichaureau supports the three font types Utopia, Garamond *different* and Charter. They can be selected through the optional argument. If no argument is given, no *fonts* symbols are loaded.

**Table 7.37:** mathdesign — operators of different sizes for Utopia

| | | |
|---|---|---|
| ∮∮ \intclockwise | ∮∮ \ointclockwise | ∰∰ \oiiint |
| ∮∮ \ointctrclockwise | ∯∯ \oiint | |

07-10-1

**Table 7.38:** mathdesign — operators of different sizes for Garamond

| | | |
|---|---|---|
| ∮∮ \intclockwise | ∮∮ \ointclockwise | ∰∰ \oiiint |
| ∮∮ \ointctrclockwise | ∯∯ \oiint | |

07-10-2

Table 7.39: mathdesign — operators of different sizes for Charter

07-10-3

$$\text{\intclockwise} \qquad \text{\ointclockwise} \qquad \text{\oiiint}$$

$$\text{\ointctrclockwise} \qquad \text{\oiint}$$

The following delimiters do not need \left or \right commands; they are added internally. "Missing" delimiters still have to be marked in the usual manner with \left. or \right..

Table 7.40: mathdesign — delimiters of different sizes, here for \textstyle — \displaystyle and with \left and \right.

07-10-4

$$\text{\leftwave} \quad \text{\rightwave} \quad \text{\leftevaw} \quad \text{\rightevaw}$$

## 7.11 **esint**

Some characters vary significantly in different fonts. "ß" is one such character, and in math mode the integral sign is another. The esint package by Eddie Saudrais defines a series of balanced integral symbols that are also available as Type-1-Font. The following symbols are provided:

Table 7.41: esint — integrals

07-11-1

| | | |
|---|---|---|
| \int | \iint | \iiintop |
| \iiiintop | \dotsintop | \ointop |
| \oiint | \sqint | \sqiint |
| \ointctrclockwise | \ointclockwise | \varointclockwise |
| \varointctrclockwise | \fint | \varoiint |
| \landupint | \landdownint | |

## 7.12 `upgreek` and `fixmath`

By default, the lowercase Greek letters are typeset in italics (cf. Section 4.14 on page 63). However, the upgreek package by Walter Schmidt lets you typeset them upright. Depending on the option, the package takes the characters from the default PostScript character set (option Symbol) or the Euler character set (option Euler). The difference is quite noticeable; the characters from the Euler font appear much thinner. The fixmath package by the same author provides bold versions of the italic Greek uppercase letters.

Table 7.42: upgreek — upright Greek letters from the Euler font

| | | | | |
|---|---|---|---|---|
| α \upalpha | θ \uptheta | π \uppi | φ \upphi | β \upbeta |
| ϑ \upvartheta | ϖ \upvarpi | φ \upvarphi | γ \upgamma | ι \upiota |
| ρ \uprho | χ \upchi | δ \updelta | κ \upkappa | ρ \upvarrho |
| ψ \uppsi | ϵ \upepsilon | λ \uplambda | σ \upsigma | ω \upomega |
| ε \upvarepsilon | μ \upmu | σ \upvarsigma | ζ \upzeta | ν \upnu |
| τ \uptau | η \upeta | ξ \upxi | υ \upupsilon | Γ \Upgamma |
| Λ \Uplambda | Σ \Upsigma | Ψ \Uppsi | Δ \Updelta | Ξ \Upxi |
| Υ \Upupsilon | Ω \Upomega | Θ \Uptheta | Π \Uppi | Φ \Upphi |

07-12-1

Table 7.43: upgreek — upright Greek letters from the PostScript font

| | | | | |
|---|---|---|---|---|
| α \upalpha | θ \uptheta | π \uppi | φ \upphi | β \upbeta |
| ϑ \upvartheta | ϖ \upvarpi | φ \upvarphi | γ \upgamma | ι \upiota |
| ρ \uprho | χ \upchi | δ \updelta | κ \upkappa | ρ \upvarrho |
| ψ \uppsi | ε \upepsilon | λ \uplambda | σ \upsigma | ω \upomega |
| ε \upvarepsilon | μ \upmu | σ \upvarsigma | ζ \upzeta | ν \upnu |
| τ \uptau | η \upeta | ξ \upxi | υ \upupsilon | Γ \Upgamma |
| Λ \Uplambda | Σ \Upsigma | Ψ \Uppsi | Δ \Updelta | Ξ \Upxi |
| Υ \Upupsilon | Ω \Upomega | Θ \Uptheta | Π \Uppi | Φ \Upphi |

07-12-2

Table 7.44: fixmath — bold and italic Greek uppercase letters from the CM font

| | | |
|---|---|---|
| Γ \Gamma | Γ \boldmath\Gamma | |
| Δ \Delta | Δ \boldmath\Delta | Δ \upDelta |
| Θ \Theta | Θ \boldmath\Theta | |
| Λ \Lambda | Λ \boldmath\Lambda | |
| Ξ \Xi | Ξ \boldmath\Xi | |
| Π \Pi | Π \boldmath\Pi | |
| Σ \Sigma | Σ \boldmath\Sigma | |
| Υ \Upsilon | Υ \boldmath\Upsilon | |
| Φ \Phi | Φ \boldmath\Phi | |
| Ψ \Psi | Ψ \boldmath\Psi | |
| Ω \Omega | Ω \boldmath\Omega | Ω \upOmega |

07-12-3

# 7.13  **gensymb**

The gensymb package by Walter Schmidt defines frequently needed symbols that can be used in both math and text mode in the same way. It is advisable to always load the textcomp package before loading this one.

**Table** 7.45: gensymb — frequently used symbols for text and math mode

07-13-1

° \degree  °C \celsius  ‰ \perthousand  Ω \ohm  μ \micro

# 7.14  **pxfonts/txfonts**

Both packages are by Young Ryu and provide symbols for Palatino (pxfonts) and Times-Roman (txfonts). There is no difference in terms of the symbols; therefore they are not distinguished here.

**Table** 7.46: pxfonts/txfonts — operators of variable size

07-14-1

| | | | |
|---|---|---|---|
| ⊞ ⊞ \bigsqcapplus | | ∮ ∮ \ointclockwise |
| ⊎ ⊎ \bigsqcupplus | | ∮ ∮ \ointctrclockwise |
| ∱ ∱ \fint | | ∰ ∰ \sqiiint |
| ∫···∫ ∫···∫ \idotsint | | ∯ ∯ \sqiint |
| ⨌ ⨌ \iiiint | | ∮ ∮ \sqint |
| ∭ ∭ \iiint | | ∰ ∰ \varoiiintclockwise |
| ∬ ∬ \iint | | ∰ ∰ \varoiiintctrclockwise |
| ∰ ∰ \oiiintclockwise | | ∯ ∯ \varoiintclockwise |
| ∰ ∰ \oiiintctrclockwise | | ∯ ∯ \varoiintctrclockwise |
| ∰ ∰ \oiiint | | ∮ ∮ \varointclockwise |
| ∯ ∯ \oiintclockwise | | ∮ ∮ \varointctrclockwise |
| ∯ ∯ \oiintctrclockwise | | × × \varprod |
| ∯ ∯ \oiint | | |

Table 7.47: pxfonts/txfonts — binary operators

| | | | | |
|---|---|---|---|---|
| ① \circledbar | ⊘ \circledwedge | ○ \medcirc | ⊘ \circledbslash | ⅋ \invamp |
| ⊞ \sqcapplus | ⊘ \circledvee | ● \medbullet | ⊎ \sqcupplus | |

07-14-2

Table 7.48: pxfonts/txfonts — binary relations

| | | |
|---|---|---|
| ⊘ \circledgtr | ⋈ \lJoin | ⨯ \opentimes |
| ⊘ \circledless | ⋈ \lrtimes | ⊥⊥ \Perp |
| :≈ \colonapprox | ⊸ \multimap | ≦ \preceqq |
| ::≈ \Colonapprox | ∘─∘ \multimapboth | ≨ \precneqq |
| :− \coloneq | ⅃ \multimapbothvert | ⋈ \rJoin |
| ::− \Coloneq | ─• \multimapdot | ⊱ \strictfi |
| ::= \Coloneqq | •─• \multimapdotboth | ⊰ \strictif |
| \coloneqq* | ∘─• \multimapdotbothA | ⊰⊰ \strictiff |
| ::∼ \Colonsim | ⅃ \multimapdotbothAvert | ≧ \succeqq |
| :∼ \colonsim | •─∘ \multimapdotbothB | ≩ \succneqq |
| −:: \Eqcolon | ⅃ \multimapdotbothBvert | // \varparallel |
| −: \eqcolon | ⅃ \multimapdotbothvert | \\ \varparallelinv |
| =: \eqqcolon | •─ \multimapdotinv | ⊩ \VvDash |
| =:: \Eqqcolon | ∘─ \multimapinv | |
| ≈ \eqsim | ⨯ \openJoin | |

07-14-3

Table 7.49: pxfonts/txfonts — negated binary operators

| | | |
|---|---|---|
| ≉ \napproxeq | ⋠ \npreccurlyeq | ≉ \nthickapprox |
| ≭ \nasymp | ≨ \npreceqq | ↞ \ntwoheadleftarrow |
| ⋠ \nbacksim | ⋨ \nprecsim | ↠ \ntwoheadrightarrow |
| ≇ \nbacksimeq | ≉ \nsimeq | ∦ \nvarparallel |
| ≠ \nbumpeq | ⋩ \nsuccapprox | ∦ \nvarparallelinv |
| ≠ \nBumpeq | ⋡ \nsucccurlyeq | ⊮ \nVdash |
| ≢ \nequiv | ≩ \nsucceqq | |
| ⋨ \nprecapprox | ⋩ \nsuccsim | |

07-14-4

Table 7.50: pxfonts/txfonts — set symbols

| | | | |
|---|---|---|---|
| ⊄ \nsqsubset | ⊉ \nsqsupseteq | ⊉ \nSupset | ⊈ \nsqsubseteq |
| ⋢ \nSubset | ⊅ \nsqsupset | ⊈ \nsubseteqq | |

07-14-5

Table 7.51: pxfonts/txfonts — negated comparison symbols

| | | | |
|---|---|---|---|
| ≫̸ \ngg | ≵ \ngtrsim | ⋦ \nlesssim | ⋧ \ngtrapprox |
| ⋦ \nlessapprox | ≪̸ \nll | ≹ \ngtrless | ≸ \nlessgtr |

07-14-6

**Table 7.52**: pxfonts/txfonts — arrows

07-14-7

| | | |
|---|---|---|
| ⇐⊡ \boxdotLeft | ⊙→ \circleddotright | ↔◇ \Diamondleft |
| ←⊡ \boxdotleft | ←○ \circleleft | ◇→ \Diamondright |
| ⊡→ \boxdotright | ○→ \circleright | ◇⇒ \DiamondRight |
| ⊡⇒ \boxdotRight | ↔ \dashleftrightarrow | ↜ \leftsquigarrow |
| ⇐□ \boxLeft | ↔◇ \DiamonddotLeft | ↗ \Nearrow |
| ←□ \boxleft | ←◇ \Diamonddotleft | ↖ \Nwarrow |
| □→ \boxright | ◇→ \Diamonddotright | ⇛ \Rrightarrow |
| □⇒ \boxRight | ◇⇒ \DiamonddotRight | ↘ \Searrow |
| ←⊙ \circleddotleft | ↔◇ \DiamondLeft | ↙ \Swarrow |

**Table 7.53**: pxfonts/txfonts — upright Greek letters

07-14-8

| | | | | |
|---|---|---|---|---|
| $\alpha$ \alphaup | $\theta$ \thetaup | $\pi$ \piup | $\phi$ \phiup | $\beta$ \betaup |
| $\vartheta$ \varthetaup | $\varpi$ \varpiup | $\varphi$ \varphiup | $\gamma$ \gammaup | $\iota$ \iotaup |
| $\rho$ \rhoup | $\chi$ \chiup | $\delta$ \deltaup | $\kappa$ \kappaup | $\varrho$ \varrhoup |
| $\psi$ \psiup | $\epsilon$ \epsilonup | $\lambda$ \lambdaup | $\sigma$ \sigmaup | $\omega$ \omegaup |
| $\varepsilon$ \varepsilonup | $\mu$ \muup | $\varsigma$ \varsigmaup | $\zeta$ \zetaup | $\nu$ \nuup |
| $\tau$ \tauup | $\eta$ \etaup | $\xi$ \xiup | $\upsilon$ \upsilonup | |

The following four characters are only available when the txfonts package is loaded with the option varg. Only the characters *g*, *v*, *w* and *y* are replaced.

**Table 7.54**: pxfonts/txfonts — alternative characters

07-14-9

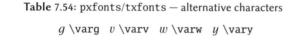

*g* \varg   *v* \varv   *w* \varw   *y* \vary

**Table 7.55**: pxfonts/txfonts — letter-like symbols

07-14-10

| | | | |
|---|---|---|---|
| ¢ \mathcent | £ \mathsterling | ∉ \notin | ∌ \notni |
| ◆ \Diamondblack | ƛ \lambdaslash | ♥ \varheartsuit | ◇ \Diamonddot |
| ♣ \varclubsuit | ♠ \varspadesuit | ƛ \lambdabar | ◆ \vardiamondsuit |

## 7.15 **nath**

The nath (natural math) package by Michal Marvan redefines all delimiters such that they do not require the usual \left and \right commands. For vertical delimiters \lVert and \rVert have to be used, however, because it cannot be determined unambiguously on which side of a math expression the command occurs.

Table 7.56: nath — delimiters

⌞ \niv    ⌟ \vin

07-15-1

Table 7.57: nath — delimiters of variable size (\normalsize — \huge)

《⟪ \lAngle ⟫》 \rAngle    ⟦⟦ \lBrack ⟧⟧ \rBrack ⌈⌈ \lCeil

07-15-2

⌉⌉ \rCeil ⌊⌊ \lFloor ⌋⌋ \rFloor ‖‖ \lVert ‖‖ \rVert

‖‖ \ldouble| ‖‖ \rdouble| 《⟪ \double< ⟫》 \double>

⟦⟦ \double[ ⟧⟧ \double] ‖‖ \ltriple| ‖‖ \rtriple|

《⟪ \triple< ⟫》 \triple> ⟦⟦ \triple[ ⟧⟧ \triple]

## 7.16 `mathtools`

This package by Morten Høgholm defines a few symbols.

Table 7.58: mathtools — brackets, arrows and hooks

07-16-1

$\overbrace{abc}$ \overbrace{abc}    $\underbrace{abc}$ \underbrace{abc}

$\overbracket{abc}$ \overbracket{abc}    $\underbracket{abc}$ \underbracket{abc}

$\xhookleftarrow{abc}$ \xhookleftarrow{abc}    $\xleftrightharpoons{abc}$ \xleftrightharpoons{abc}

$\xhookrightarrow{abc}$ \xhookrightarrow{abc}    $\xmapsto{abc}$ \xmapsto{abc}

$\xLeftarrow{abc}$ \xLeftarrow{abc}    $\xRightarrow{abc}$ \xRightarrow{abc}

$\xleftharpoondown{abc}$ \xleftharpoondown{abc}    $\xrightharpoondown{abc}$ \xrightharpoondown{abc}

$\xleftharpoonup{abc}$ \xleftharpoonup{abc}    $\xrightharpoonup{abc}$ \xrightharpoonup{abc}

$\xleftrightarrow{abc}$ \xleftrightarrow{abc}    $\xrightleftharpoons{abc}$ \xrightleftharpoons{abc}

$\xLeftrightarrow{abc}$ \xLeftrightarrow{abc}

Table 7.59: mathtools — binary relations

07-16-2

::≈ \Colonapprox    :− \coloneq    −:: \Eqcolon    :≈ \colonapprox

:∼ \colonsim    : \eqqcolon    : \coloneqq    ::∼ \Colonsim

:: \Eqqcolon    :: \Coloneqq    :: \dblcolon    ::− \Coloneq

−: \eqcolon

# TEX and math

In theory there is no need to use the commands described in this chapter; LATEX as the superordinate command package contains usually improved versions. Nevertheless people who develop their own packages or want to extend existing commands do use the TEX commands, either out of choice or necessity. For this reason therefore, as well as for completeness, this chapter lists all the TEX length registers and TEX commands relevant to math typesetting.

Watch out for the commands given here not working in the way described; LATEX and many packages redefine a lot of them or define completely new versions. When problems arise, the first thing to check is whether the command has been redefined.

*redefine command*

## 8.1 Length registers

These are commands in TEX and LATEX that contain and save a value in one of the valid units. Lengths can be added, subtracted, multiplied and divided. They can be output in the running text by prefixing them with a `\the` (cf. Example 08-01-1 on the next page). A length can be fitted with glue, which allows TEX to stretch the length up to a value of `plus`... or compress it by up to `minus`..., e.g. `\the\medmuskip=4.0mu plus 2.0mu minus 4.0mu`, is by default 4mu, but TEX can stretch it by adding up to 2mu and schrink it by substract up to 4mu: 0mu ≤ `\medmuskip` ≤ 6mu. Another example is `\the\parfillskip=0.0pt plus 1.0fil`, which is by default 0pt, but TEX can stretch it with any possible value.

| | |
|---|---|
| \parskip: | 0.0pt plus 1.0pt |
| \thinmuskip: | 3.0mu |
| \medmuskip: | 4.0mu plus 2.0mu minus 4.0mu |
| \thickmuskip: | 5.0mu plus 5.0mu |

08-01-1

```
\usepackage{array}
\setlength\arraycolsep{2pt}

\begin{tabular}{@{}>{%
  \ttfamily\textbackslash}ll@{}}
parskip:   & \the\parskip\\
thinmuskip: & \the\thinmuskip\\
medmuskip:  & \the\medmuskip \\
thickmuskip:& \the\thickmuskip
\end{tabular}
```

### \abovedisplayshortskip, \abovedisplayskip

This is a length, with glue, that is used above displayed equations (cf. Section 4.8.4 for an example). In amsmath these lengths default to \abovedisplayshortskip=0.0pt plus 3.0pt and \abovedisplayskip=10.0pt plus 2.0pt minus 5.0pt.

### \belowdisplayshortskip, \belowdisplayskip

This is a length, with glue, that is used below displayed equations (cf. Section 4.8.4 for an example). In amsmath these lengths default to \belowdisplayshortskip=6.0pt plus 3.0pt minus 3.0pt and \belowdisplayskip=10.0pt plus 2.0pt minus 5.0pt.

### \delimiterfactor

The height of a brace is frequently suboptimal, and in some cases significantly too small. The factor \delimiterfactor can be used to influence the height. The height of a delimiter is ⟨calculated height⟩ × ⟨\delimiterfactor⟩/1000. The default is \delimiterfactor=901.

$$y = \begin{cases} x^2 + 2x & \text{if } x < 0, \\ x^3 & \text{if } 0 \le x < 1, \\ x^2 + x & \text{if } 1 \le x < 2, \\ x^3 - x^2 & \text{if } 2 \le x. \end{cases}$$

08-01-2

```
\[
y = \left\{%
  \begin{array}{ll}
    x^2+2x  &\textrm{if }x<0,\\
    x^3     &\textrm{if }0\le x<1,\\
    x^2+x   &\textrm{if }1\le x<2,\\
    x^3-x^2 &\textrm{if }2\le x.
  \end{array}%
\right.
\]
```

$$y = \begin{cases} x^2 + 2x & \text{if } x < 0, \\ x^3 & \text{if } 0 \le x < 1, \\ x^2 + x & \text{if } 1 \le x < 2, \\ x^3 - x^2 & \text{if } 2 \le x. \end{cases}$$

08-01-3

```
\delimiterfactor=1500
\[
y = \left\{%
  \begin{array}{ll}
    x^2+2x  &\textrm{if }x<0,\\
    x^3     &\textrm{if }0\le x<1,\\
    x^2+x   &\textrm{if }1\le x<2,\\
    x^3-x^2 &\textrm{if }2\le x.
  \end{array}%
\right.
\]
```

### \delimitershortfall

In addition to the \delimiterfactor explained above, the height of a brace can be influenced with \delimitershortfall. TEX makes the brace not higher than ⟨*calculated height*⟩ × ⟨*delimiterfactor*⟩ / 1000 and ⟨*calculated height*⟩ − ⟨*delimitershortfall*⟩. This makes it possible to determine the optimal height for every brace, especially for nested ones.

08-01-4

$$x \cdot \left( \left( x^2 - y^2 \right) - 3 \right)$$

$$x \cdot \left( \left( x^2 - y^2 \right) - 3 \right)$$

```
$x\cdot\left(\left(x^2-y^2\right)-3\right)$ \\[7pt]
\setlength\delimitershortfall{-1pt}
$x\cdot\left(\left(x^2-y^2\right)-3\right)$
```

08-01-5

$$(((A)))$$

$$\left( \left( (A) \right) \right)$$

```
$\left(\left(\left(A\right)\right)\right)$\\[7pt]
\setlength\delimitershortfall{-1pt}
$\left(\left(\left(A\right)\right)\right)$
```

### \displayindent

This length denotes the additional space on the left side of a displayed equation. It is 0pt by default, but can be set to an arbitrary value within display environments.

```
\newcommand*\diff{\mathop{}\!\mathrm{d}}
\[ f(x) = \int\frac{\sin x}{x}\diff x\  \fbox{\the\displayindent} \]
\[\displayindent=50pt
  f(x) = \int\frac{\sin x}{x}\diff x\  \fbox{\the\displayindent} \]
\[\displayindent=100pt
  f(x) = \int\frac{\sin x}{x}\diff x\  \fbox{\the\displayindent}\]
```

08-01-6

$$f(x) = \int \frac{\sin x}{x} \, \mathrm{d}x \boxed{0.0\text{pt}}$$

$$f(x) = \int \frac{\sin x}{x} \, \mathrm{d}x \boxed{50.0\text{pt}}$$

$$f(x) = \int \frac{\sin x}{x} \, \mathrm{d}x \boxed{100.0\text{pt}}$$

### \displaywidth

The width of a displayed equation is usually \linewidth. It can also be centred in the first half, as seen in the second example, or the second half, as seen in the last example.

```
\newcommand*\diff{\mathop{}\!\mathrm{d}}
\[ f(x) = \int \frac{\sin x}{x}\,\diff x \]
\[ \displaywidth=0.5\linewidth%
   f(x) = \int \frac{\sin x}{x}\,\diff x \]
\[ \displaywidth=1.5\linewidth%
  f(x) = \int \frac{\sin x}{x}\,\diff x \]
```

$$f(x) = \int \frac{\sin x}{x}\, dx$$

08-01-7

$$f(x) = \int \frac{\sin x}{x}\, dx$$

$$f(x) = \int \frac{\sin x}{x}\, dx$$

### \mathsurround

This length was already described in Section 2.2.6 on page 11.

### \medmuskip

See Section 4.8 on page 48 for an example.

### \mkern

\mkern works similarly to \kern, but expects its parameter to be in the math unit mu (cf. Section 4.8 on page 48).

### \mskip

This command is the math version of \skip and may only be used in math mode. The parameter has to be a length of unit mu.

### \muskip

There are 256 \muskip length registers altogether — \muskip0 to \muskip255. Each one may contain a length of unit mu, which may be fitted with glue. The \advance, \multiply, and \divide commands can be used for a restricted set of arithmetic operations. The registers are local to their math environment; they can only be used globally if they have been assigned with \global.

```
\global\muskip1=30mu plus20mu minus10mu
\global\muskip3=40mu plus 1fil% 1fil -> dynamic length
$A\mskip\muskip1 B\mskip\muskip3 C$
```

$$A \quad B \qquad\qquad\qquad\qquad\qquad C$$

08-01-8

In [13] Knuth writes –

> All assignments to the scratch registers whose numbers are 1, 3, 5, 7, and 9 should be \global; all assignments to the other scratch registers (0, 2, 4, 6, 8, 255) should be non-\global.

### \muskipdef

Instead of the predefined 256 register names \muskip<[0...255]>, \muskipdef can be used to assign a symbolic name and use it accordingly. To achieve this, one of the 256 registers is assigned to the name.

<table>
<tr><td>08-01-9</td><td>

`\muLength:30.0mu plus 20.0mu minus 10.0mu`

</td><td>

```
$\muskipdef\muLength=100 %reg. 100
\muLength=30mu plus20mu minus10mu
\mbox{\ttfamily\textbackslash
      muLength:\the\muLength}$
```

</td></tr>
</table>

## \nonscript

For the math styles `\scriptstyle` and `\scriptscriptstyle`, `\nonscript` can be used to suppress space immediately following them. The command has no effect for the other styles.

<table>
<tr><td>08-01-10</td><td>

$A \xrightarrow{C} B$

$A \xrightarrow{C} B$

$A \qquad \xrightarrow{C} B$

$A \qquad \xrightarrow{C} B$

$A \qquad \xrightarrow{C} B$

</td><td>

```
\usepackage{amsmath}

$\scriptscriptstyle A\nonscript\mkern40mu\xrightarrow{C} B$\\
$\scriptstyle      A\nonscript\mkern40mu\xrightarrow{C} B$\\
$\displaystyle     A\nonscript\mkern40mu\xrightarrow{C} B$\\
$                  A\nonscript\mkern40mu\xrightarrow{C} B$\\
$\textstyle        A\nonscript\mkern40mu\xrightarrow{C} B$
```

</td></tr>
</table>

## \nulldelimiterspace

For "empty" delimiters in expressions like "`\left( ... \right.`", where "`\right.`" is "empty", the length `\nulldelimiterspace` determines the width of the right (empty) box that replaces the non-existent delimiter.

```
default: \verb+\the\nulldelimiterspace+: \the\nulldelimiterspace
$$ X \left( x=x \right. X $$
changed: \nulldelimiterspace=10\nulldelimiterspace
\verb+\the\nulldelimiterspace+: \the\nulldelimiterspace
$$ X \left( x=x \right. X $$
```

<table>
<tr><td>08-01-11</td><td>

default: `\the\nulldelimiterspace`: 1.2pt

$$X\,(x = x\ X$$

changed: `\the\nulldelimiterspace`: 11.99997pt

$$X\,(x = x \quad X$$

</td></tr>
</table>

## \predisplaysize

The effective width of a line of text that precedes a displayed formula is saved in `\predisplaysize`; in addition to the text, 2em of the current font are added. Based on this length, TeX decides whether to use the length `\abovedisplayskip` or alternatively `\abovedisplayshortskip` for the vertical spacing between text and formula.

```
\newcommand*\diff{\mathop{}\!\mathrm{d}}
\newlength\PREsave
The effective \ldots
\[
  \fbox{\the\predisplaysize}\quad F(x)=\int f(x)\diff x
  \global\PREsave=\predisplaysize
\]
\advance\PREsave by -2em\rule{\PREsave}{1pt}
```

The effective …

$\boxed{\text{82.96986pt}}$    $F(x) = \int f(x)\,\mathrm{d}x$

08-01-12

---

```
\newcommand*\diff{\mathop{}\!\mathrm{d}}
\newlength\PREsave
The effective width of a line of text preceding a displayed formula \ldots
\[
  \fbox{\the\predisplaysize}\quad F(x)=\int f(x)\diff x
  \global\PREsave=\predisplaysize
\]
\advance\PREsave by -2em\rule{\PREsave}{1pt}
```

The effective width of a line of text preceding a displayed formula …

08-01-13

$\boxed{\text{294.2795pt}}$    $F(x) = \int f(x)\,\mathrm{d}x$

---

If there is no line before the formula, the value is set to $-$\maxdimen=-16383.99998pt. If the preceding line does not extend to its natural width because of the limited stretchability of \parfillskip, the value is set to $+$\maxdimen.

### \scriptspace

After an index or exponent, the space \scriptspace is inserted; it defaults to 0.5pt.

$y = x^4 - x^3 + x$     $y = x^4 \quad - x^3 \quad + x$

```
$  y=x^4-x^3+x $ \qquad
\scriptspace=10pt $ y=x^4-x^3+x $
```

08-01-14

### \thickmuskip

See Section 4.8 on page 48 for an example.

### \thinmuskip

The frequently used short forms for positive and negative spacing are:

```
\def\,{\mskip\thinmuskip}
\def\!{\mskip-\thinmuskip}
```

This spacing is used especially with integrals to improve the typesetting (cf. Section 4.8 on page 48).

$\sqrt{2}x - \sqrt{2}\,x$

$\sqrt{\log x} - \sqrt{\log x}$

$P\left(1/\sqrt{n}\right) - P\left(1/\sqrt{n}\,\right)$

$[0, 1) - [\,0, 1)$

$x^2/2 - x^2\!/2$

```
$\sqrt 2 x$ --- $\sqrt 2\,x$       \\[6pt]
$\sqrt{\log x}$ --- $\sqrt{\,\log x}$\\[6pt]
$P\left({1/\sqrt n}\right)$ ---
$P\left({1/ \sqrt n}\,\right)$  \\[10pt]
$[0,1)$ --- $[\,0,1)$             \\[6pt]
$x^2/2$ --- $x^2\!/2$
```

08-01-15

08-01-16

$$\int\!\!\int_D \mathrm{d}x\mathrm{d}y \qquad \int\!\!\int_D \mathrm{d}x\,\mathrm{d}y$$

```
\newcommand*\dx{\mathrm{d}x}
\newcommand*\dy{\mathrm{d}y}
```

$$\int\!\!\int_D \mathrm{d}x\,\mathrm{d}y \qquad \int\!\!\int_D \mathrm{d}x\,\mathrm{d}y$$

$$\int\!\!\int_D \mathrm{d}x\,\mathrm{d}y \qquad \int_D \mathrm{d}x\,\mathrm{d}y$$

$$\int\!\!\int_D \mathrm{d}x\mathrm{d}y$$

```
\[\int\int_D \dx\dy \quad
   \int\!\int_D \dx\,\dy \]
\[\int\!\!\int_D \dx\,\,\dy  \quad
   \int\!\!\!\int_D \dx\,\,\,\dy \]
\[\int\!\!\!\!\int_D \dx\,\,\,\,\dy \quad
   \int\!\!\!\!\!\int_D \dx\,\,\,\,\,\dy \]
\[\int\!\!\!\int_D \dx\dy\]
```

### \medmuskip

See Section 4.8 on page 48 for an example.

## 8.2  Math font commands

### \delcode

Every character is assigned to a category in TeX and has besides this \catcode and \mathcode a \delcode. This code describes how an individual character behaves as math delimiter in math mode. This command is used when defining a math delimiter:

```
\def\set@@mathdelimiter#1#2#3#4#5{%
  \global\delcode'#3="\hexnumber@#1#4\hexnumber@#2#5\relax}
```

Also see the following example for \delimiter.

### \delimiter

Every character can be used as a delimiter for math expressions. TeX only needs to know which character to use for the standard font and which one for the large one. For LaTeX the command \DeclareMathDelimiter is used (cf. Section 4.5.2).

In the following example, \tdela is the character $0x22$ (↑) of font number 2 (csmy) and character $0x78$ of font number 3 (cmex) for the large version. \tdelb is the same, only upside down (↓).

08-02-1

$$\uparrow x - y \downarrow (x + y) = x^2 - y^2$$

$$\uparrow \sum_{n=0}^{\infty} \frac{1}{2^n} \downarrow^2 = 4$$

$$\left| \sum_{n=0}^{\infty} \frac{1}{2^n} \right|^2 = 4$$

```
\def\tdela{\delimiter"4222378\relax}
\def\tdelb{\delimiter"5223379\relax}

$\tdela x-y\tdelb(x+y)=x^2-y^2$
\[ \tdela\sum_{n=0}^\infty
        {1\over2^n}\tdelb^2 = 4  \]
\[ \left\tdela\sum_{n=0}^\infty
        {1\over2^n}\right\tdelb^2 = 4 \]
```

The meaning of the parameters for \delimiter is explained in Table 8.1 and above for the characters mentioned for \delcode. 4222378 is decomposed into the character sequence $4 - 2 - 22 - 3 - 78$:

4    opening parenthesis according to Table 8.1;
2    font family 2;
22   hexadecimal character number;
3    font family 3;
78   hexadecimal character number.

**Table 8.1**: Meaning of the first character in the parameter of \delcode

| type | meaning | type | meaning |
|------|---------|------|---------|
| 0 | normal character | 1 | large operator |
| 2 | binary operator | 3 | relation operator |
| 4 | opening parenthesis | 5 | closing parenthesis |
| 6 | type character | 7 | variable character |

**\displaystyle**

See Section 4.9 on page 55 for an example.

**\fam**

When T<sub>E</sub>X changes from text into math mode, the fonts for \textfont, \scriptfont, and \scriptscriptfont must be known. This selection is usually called a "family" and internally determined by a number $[0\ldots15]$; usually only some of them are used. The T<sub>E</sub>X default is \fam=-1.

| \fam | example | \fam | example |
|------|---------|------|---------|
| \fam=-1 | 123aicAICαιγ | \fam=0 | 123aicAICαιγ |
| \fam=1 | 123aicAICαιγ | \fam=2 | ∞∈∋⊣⟩⌋𝒜𝒥𝒞αιγ |
| \fam=3 | | \fam=4 | 123aicAICαιγ |
| \fam=5 | ↨↩↪⇔⇎⇗AIICαιγ | \fam=6 | 123aicAICαιγ |

08-02-2

123aicAICαιγ8                \usepackage{array,mathptmx}

08-02-3

123aicAICαιγ1               $\mathrm{123aicAIC\alpha\iota\gamma(\the\fam)}$\\[5pt]
                           $\mathnormal{123aicAIC\alpha\iota\gamma(\the\fam)}$\\[5pt]
∞∈∋⊣⟩⌋𝒜𝒥𝒞αιγ∈            $\mathcal{123aicAIC\alpha\iota\gamma(\the\fam)}$\\[5pt]
**123aicAICαιγ9**           $\mathbf{123aicAIC\alpha\iota\gamma(\the\fam)}$\\[5pt]
123aicAICαιγ10             $\mathit{123aicAIC\alpha\iota\gamma(\the\fam)}$\\[5pt]
                           $\mathtt{123aicAIC\alpha\iota\gamma(\the\fam)}$\\[5pt]
123aicAICαιγ11             $\mathsf{123aicAIC\alpha\iota\gamma(\the\fam)}$\\[5pt]

123aicAICαιγ12

**\mathaccent**

It expects three parameters which are given encoded in one hexadecimal number:

08-02-4

$$\acute{A}$$

```
\def\dA{\mathaccent"7015\relax}

\Large $\dA{A}$
```

The first digit (7) denotes the class according to Table 8.1 on the preceding page. The next digit (0) selects one of the 16 font families [0..f]. The last two digits (15) denote the number of the character in the font ($15_{16} = 21_{10} = 25_8$) (cf. Table 11.1 on page 245).

## \mathbin

\mathbin defines a character (or a math expression enclosed in curly braces) as a binary symbol, thereby giving it the appropriate spacing before and after it:

08-02-5

$$a|b \quad a \mid b$$

```
\Large$a|b \quad a\mathbin| b$
```

## \mathchar

The definition of a character as a math one is done with \mathchar, which expects the same type of four-digit parameter as \mathaccent. In the example, these are 1 for the class (big operators), 3 for the font family (math extension font) and $58_{16}$ for the character (big sum) (from Table 11.5 on page 247). The largest possible value for the argument is $7FFF_{16} = 32767_{10}$.

08-02-6

$$a\sum_{i=1}^{\infty} b \quad a\sum_{i=1}^{\infty} b$$

```
\Large
$a\sum\limits_{i=1}^{\infty} b \quad
a\mathchar"1358\limits_{i=1}^{\infty} b$
```

The definition of a character as "mathematical" means that it can only be used in math mode; otherwise an error message is given:

```
\def\bMath{\mathchar"0042} % definition of B as \bMath
\bMath                     % error, not in math mode

 Missing $ inserted.
<inserted text>
                $
<to be read again>
                   \mathchar
\bMath ->\mathchar
              "0042
l.2 \bMath
```

## \mathchardef

The only difference to \mathchar is that the preceding \def or the \newcommand in LaTeX is not required when working with \mathchardef.

08-02-7

$$a\sum_{i=1}^{\infty} \sqrt{i+1}$$

$$a\sum_{i=1}^{\infty} \sqrt{i+1}$$

```
\mathchardef\Sum="1358 % large sum character
$a\Sum\limits_{i=1}^{\infty}\sqrt{i+1}$\\[10pt]
$a\sum\limits_{i=1}^{\infty}\sqrt{i+1}$
```

### \mathchoice

A definition can be given specific definitions for the four math styles \displaystyle −
\textstyle − \scriptstyle – \scriptscriptstyle through \mathchoice. In the following example, a black box is drawn depending on the underlying math style.

```
\def\myRule{%
   \mathchoice{\rule{20pt}{20pt}}{\rule{10pt}{10pt}}
   {\rule{5pt}{5pt}}{\rule{2.5pt}{2.5pt}}\mkern2mu}
\def\Stil{\mathchoice{D}{T}{S}{SS}\mkern2mu}
\large
$\myRule\sum\limits_{\myRule i=1}^{\myRule\infty} \qquad
 \myRule\frac{\myRule\sqrt{\myRule i+1}}{\myRule i^{\myRule2}}$
%
$\Stil\sum\limits_{\Stil i=1}^{\Stil\infty} \qquad
 \Stil\frac{\Stil\sqrt{\Stil i+1}}{\Stil i^{\Stil2}}$
```

$$\blacksquare\sum_{\blacksquare i=1}^{\blacksquare\infty} \qquad \blacksquare\frac{\blacksquare\sqrt{\blacksquare i+1}}{\blacksquare i\blacksquare 2} \; T \sum_{S i=1}^{S \infty} \qquad T \frac{S \sqrt{S i+1}}{S i^{SS 2}}$$

08-02-8

### \mathclose

\mathclose assigns the parameter to class 5 (closing character). The parameter may be an
individual character or a math expression.

$$A : \frac{B}{C} : D \qquad A : \frac{B}{C} : D$$

```
\large $A:\frac{B}{C}:D$ \quad
$A\mathopen:\frac{B}{C}\mathclose: D $
```

08-02-9

### \mathcode

A math font is very different to a text font. Every character has to be assigned to a class and a
font family. The majority of the characters are defined through \mathcode, which expects three
parameters — class, font family, and character number. For example \mathcode'\<="313C
defines the character "<" as a symbol of class 3 (a relation symbol), font family 1, and the assigned
character 0x3C, which corresponds to the same character here. Any other character can be
assigned as well, as can be seen in the following example. Although the / character is output
each time, this happens as a result of the "<" character being defined as a relation symbol in the
first case. This also causes the differences in spacing.

$$a / b \qquad a/b$$

```
\mathcode'\<="313D % corresponds to the / character
$a<b$ \qquad $a/b$
```

08-02-10

### \mathop

Assigns type large operator (class 1) to the parameter, which may be an individual character of a
math expression. Also see Section 6.13 on page 113 for further examples.

$$A_{i=1}^{\infty}$$

```
\[ A_{i=1}^{\infty} \]
\[ \mathop{A}_{i=1}^{\infty} \]
```

08-02-11

$$\overset{\infty}{\underset{i=1}{A}}$$

### \mathopen

This command is simply the opposite of \mathclose.

### \mathord

Assigns the type ordinary character (class 0) to the parameter, which may be an individual character or a math expression.

08-02-12

$$y = f(x)$$
$$y = f(x)$$

```
\large
$y = f(x)$\\[5pt]
$y \mathord= f(x)$
```

### \mathpunct

Assigns the type punctuation (class 6) to the parameter, which may be an individual character or a math expression (also see Section 4.8.3 on page 51 for an example).

### \mathrel

Assigns the type relation symbol (class 3) to the parameter, which may be an individual character or a math expression.

08-02-13

$$x_1 o x_2 o x_3$$
$$x_1 o x_2 o x_3$$

```
\large
$x_1 o x_2 o x_3$\\[5pt]
$x_1\mathrel o x_2\mathrel o x_3$
```

### \scriptfont

The difference to \scriptstyle is that \scriptfont only determines the font that is usually used for super- and subscripts.

08-02-14

$$A_1 \; A_1$$

```
$A_1$
\font\tenxii=cmr12
\scriptfont0=\tenxii
$A_1$
```

### \scriptscriptfont

Ditto for the \scriptscriptstyle.

### \scriptscriptstyle

Selects the math style \scriptscriptstyle for the characters that follow it.

### \scriptstyle

Selects the math style \scriptstyle for the characters that follow it.

### \skew

In math mode all variables are typeset in italics; in some cases this makes it difficult to fit an accent immediately above a character. The \skew command can help here. It expects three parameters:

horizontal shift    The horizontal shift of the accent in mu (math unit).
the accent    As the name suggests.
the character    The base character, may have an accent already.

$$\tilde{\imath} \quad \tilde{\hat{A}} \quad \tilde{\imath} \quad \tilde{\hat{A}}$$

```
\large
$\tilde i$ \quad $\tilde{\hat{A}}$ \quad
$\skew{3}{\tilde}{i}$ \quad
$\skew{7}{\tilde}{\hat{A}}$
```

08-02-15

The characters above were all created *without* amsmath. amsmath redefines how to set accents; when using this package, there is little need to use \skew. The same accents created *with* amsmath, but *without* \skew are: $\tilde{\imath}$ and $\tilde{\hat{A}}$ ($\tilde{i}$ and $\tilde{\hat{A}}$). The advantage of loading the package is clearly visible.

### \skewchar

This command expects two parameters; the character set and the number of the comparison symbol. The latter is used to place the accent as optimally as possible. If no comparison character is defined, \skewchar defaults to $-1$.

### \textfont

This defines the text font for the math style \textstyle. Here "text" also means math expressions on a line of text (cf. Chapter 2 on page 5).

### \textstyle

This selects the math style \textstyle for the characters that follow it. "Text" means the same as above.

## 8.3 Commands for math mode

Apart from a few exceptions, the commands given here are not of interest to users of L<sup>A</sup>T<sub>E</sub>X. Nevertheless they show the philosophy behind the concept of T<sub>E</sub>X and what to keep in mind when developing new L<sup>A</sup>T<sub>E</sub>X commands. In cases where the examples for the commands are self-explanatory, no further comment is given.

### \above

The \above command corresponds to the command for creating fractions; the fraction bar is optional through a default of 0pt and therefore also suitable for binomials.

$$\frac{a}{b} \quad \binom{a}{b} \quad \frac{a}{b} \quad \frac{a}{b} \quad \frac{a}{b} \quad \binom{a}{b}$$

```
$a\above0pt b$ \quad
$\left(a\above0pt b\right)$ \quad
${a\above1pt b}$\quad
${a\above2.5pt b}$\quad
$\displaystyle{a\above0pt b}$\quad
$\displaystyle\left({a\above0pt b}\right)$
```

08-03-1

## \abovewithdelims

08-03-2

$$\binom{a}{b}$$

```
$a\abovewithdelims()0pt b$   \medskip
```

$$\left\{\frac{a}{b}\right\}$$

```
\def\fdelimA{\abovewithdelims\{)1.0pt}
${a\fdelimA b}$                \medskip
```

$$\left[\frac{a}{b}\right]$$

$$\left\{\begin{matrix}a\\b\end{matrix}\right.$$

```
\def\fdelimB{\abovewithdelims[]2.0pt}
${a\fdelimB b}$                \medskip

\def\fdelimC{\abovewithdelims\{.0pt}
$\displaystyle{a\fdelimC b}$
```

## \atop

08-03-3

$$\frac{a}{b} \binom{n}{k} = \frac{n!}{k!(n-k)!} \begin{matrix}a\\b\end{matrix}$$

```
$a\atop b$
$({n \atop k}) = {n!\above1pt k!(n-k)!}$
$\displaystyle{a\atop b}$
```

## \atopwithdelims

08-03-4

$$\binom{a}{b}$$

```
$a\atopwithdelims() b$         \medskip
```

$$\binom{n}{k} = \frac{n!}{k!(n-k)!}$$

```
${n \atopwithdelims() k} =
  {n!\above1pt k!(n-k)!}$      \medskip
```

$$\left\{\begin{matrix}a\\b\end{matrix}\right.$$

```
$\displaystyle{a\atopwithdelims\{. b}$
```

## \displaylimits

This command can be used to reset the handling of limits to the default of the current style. This means the the integration limits are usually put next to the integral.

## \eqno

This command expects a "parameter" and puts it as the equation number next to the formula on the right. The corresponding equation counter is not affected however. The parameter is everything to the right of \eqno until the end of the equation.

08-03-5

$$y = f(x) \qquad\qquad (A12)$$

```
$$ y=f(x) \eqno(A12) $$
```

## \everydisplay

This puts the parameter, which may contain arbitrary code, before every displayed formula after opening math mode.

08-03-6

$$f(x) = \int \frac{\sin x}{x}\,\mathrm{d}x$$

$$g(x) = \int \frac{\sin^2 x}{x^2}\,\mathrm{d}x$$

```
\usepackage{color}
\newcommand*\diff{\mathop{}\!\mathrm{d}}

\everydisplay{\color{red}}
\[ f(x) = \int\frac{\sin x}{x}\diff x \]
\[ g(x) = \int\frac{\sin^2 x}{x^2}\diff x \]
```

## \everymath

Puts the parameter, which may contain arbitrary code, before every math expression in inline mode after opening math mode.

$$f(x) = \int \frac{\sin x}{x} dx$$

instead of $\frac{\sin x}{x}$ now with $\frac{\cos x}{x}$:

$$g(x) = \int \frac{\cos x}{x} dx$$

08-03-7

```
\usepackage{color}
\everymath{\color{red}\displaystyle}

\[ f(x) = \int \frac{\sin x}{x}dx \]
instead of $\frac{\sin x}{x}$
  now with $\frac{\cos x}{x}$:
\[ g(x) = \int \frac{\cos x}{x}dx \]
```

*footnotes* Footnote numbers as superscripts are also typeset in math mode; bear this in mind if you want to use \everymath.

## \left

TEX determines the height of the delimiter given as a parameter for the left hand side of a formula. A \right is required.

## \leqno

This command is identical to \eqno except that it puts the equation number on the left hand side.

## \limits

This puts limits always above and below an operator instead of next to it (cf. Section 4.4 on page 36).

## \mathinner

This defines the math subexpression that follows it as "subformula".

## \nolimits

The opposite of \limits; limits are put next to the operators.

## \over

\over corresponds to the \frac command and is the same as \overwithdelims.

$\frac{a}{b}$    $\frac{\frac{m}{n}}{a+b}$     $\dfrac{\frac{m}{n}}{a+b}$

08-03-8

```
$ {a\over b} \qquad {{m\over n}\over{a+b}} $
\[ {m\over n}\over{a+b} \]
```

## \overline

This places a line above the parameter, which may be an individual character or a math expression; it suffers from the same problems as \underline.

08-03-9

$$\overline{x} + \overline{y} = \overline{z}$$
$$\overline{x} + \overline{A} = \overline{z}$$

$$\overline{x} + \overline{A} = \overline{z}$$

```
\def\yPh{\vphantom{A}}
\let\ol\overline

$\overline{x}+\overline{y}=\overline{z}$\\
$ \ol{x} + \ol{A} = \ol{z} $\\[5pt]
$ \ol{x\yPh} + \ol{A} = \ol{z\yPh} $
```

## \overwithdelims

This is a command for "generalised fractions" with predefined width of the fraction bar.

08-03-10

$$\left[ \frac{\left(\frac{a}{b}\right)}{a+b} \quad \frac{m}{n} \right]$$

$$\left\{ \frac{\frac{m}{n}}{a+b} \right.$$

```
$ {a\overwithdelims() b} \qquad
   {m\over n}\overwithdelims[]{a+b} $
\[ {m\over n}\overwithdelims\{.{a+b} \]
```

## \radical

This creates a root symbol (atom) from the parameter encoded as a 24 bit number. The individual character sets usually contain several root symbols. Their use can be controlled through \radical. The parameter contains six specifications altogether and corresponds to a six-digit number when input in hexadecimal. 270371 is interpreted as $2 - 70 - 3 - 71$ and corresponds to font family (normal) − character number − font family (big) − character number.

08-03-11

$$\sqrt{\frac{1}{7}}$$

```
\def\Sqrt#1#2{\radical"270371 {#1 \over #2}}
$ \Sqrt{1}{7} $\\[5pt]
```

$$\sqrt{\frac{1}{7}}$$

```
\def\Sqrt#1#2{\radical"270372 {#1 \over #2}}
$ \Sqrt{1}{7} $\\[5pt]
```

$$\sqrt{\frac{1}{7}}$$

```
\def\Sqrt#1#2{\radical"270373 {#1 \over #2}}
$ \Sqrt{1}{7} $\\[5pt]
```

$$\sqrt{\frac{1}{7}}$$

```
\def\Sqrt#1#2{\radical"270374 {#1 \over #2}}
$ \Sqrt{1}{7} $\\[5pt]
```

## \right

This command is the opposite of \left and may only occur pairwise with it.

## \underline

\underline (like \overline) suffers from problems when individual characters with different depths (or height) are marked by lines below (or above) them. In all of these cases, the \vphantom command helps.

08-03-12

$$\underline{x} + \underline{y} = \underline{z}$$

$$\underline{x} + \underline{y} = \underline{z}$$

$$\underline{x_1} + \underline{y_2} = \underline{z_3}$$

```
\def\yPh{\vphantom{y}}
\let\ul\underline

$\underline{x}+\underline{y}=\underline{z}$\\[8pt]
$ \ul{x\yPh} + \ul{y} = \ul{z\yPh} $\\[8pt]
$ \ul{x_1} + \ul{y_2} = \ul{z_3} $
```

**\vcenter**

This centres vertical material relative to its axis.

## 8.4 Math penalties

The penalties are a significant criterion for TₑX to be able to perform its calculations quickly while leaving lots of ways for the user to interact.

**\binoppenalty**

In Section 2.2.5 on page 10 we discussed ways of having a line break in math inline mode. TₑX only inserts a line break if at least one binary symbol occurs and additionally \binoppenalty is smaller than 10 000.

**\displaywidowpenalty**

Penalty that is added immediately after the last but one line of a displayed math environment. This is to prevent single lines at the beginning of a page before a display environment.

**\postdisplaypenalty**

Penalty that is added immediately after the end of a displayed math environment.

**\predisplaypenalty**

Penalty that is added immediately before the start of a displayed math environment.

**\relpenalty**

Penalty for a line break after a relation symbol if a line break is possible.

C h a p t e r $9$

# Other packages

The following examples are not meant to be a substitute for studying the documentation of the packages. The list of packages, which are more or less related to math, is simply too long to give an example for every package. The selection in this chapter is more or less arbitrary, but common problems are addressed.

Table 9.1: List of mathematical packages

| package | description |
|---|---|
| accents | Multiple accents. |
| alltt | Verbatim-like input where some macros are expanded. |
| alphalph | Macros \alphalph and \AlphAlph to convert from numbers to letters – 1 to a, 26 to z, 27 to aa, 52 to zz etc. |
| amsbsy | Bold mathematical symbols. |
| amslatex | A collection of files by the $\mathcal{A}_{\mathcal{M}}\mathcal{S}$. |
| amscd | Commutative diagrams in LaTeX. |
| amsfonts | TeX fonts by the $\mathcal{A}_{\mathcal{M}}\mathcal{S}$. |
| amsmath | The main package of the $\mathcal{A}_{\mathcal{M}}\mathcal{S}$ distribution. |
| amstext | Defines the \text macro to be used within equations. |
| amsthm | LaTeX package for theorems, created by the $\mathcal{A}_{\mathcal{M}}\mathcal{S}$. |
| bez123 | Linear, quadratic, and cubic Bezier curves. |
| brclc | Package for 16 bit operations "+, −, *, /, ˆ, exp, log, ln, sin, cos, tan, asin, acos, atan". The numerical expressions of the \clc macro are calculated by a preprocessor and read by LaTeX. |
| breqn | Automatic line break in displayed formulae. |
| cancel | Macros to cancel (strike through) mathematical or textual objects. |
| cases | Defines macros to create case differentiations while numbering the equation. |
| comma | Allows to use the comma to separate thousands. |
| datenumber | Macros to convert a date to a number and vice versa and incrementing dates. |
| diagxy | Draws commutative diagrams. |
| dsfont | Based in the CMR font, mathematical symbols for N, Z, Q, R, and C are provided. |
| easy | Several packages for equations, matrices, etc. |
| empheq | Emphasising equations or parts through frames etc. |
| esint | Special (better) integrals. |
| esvect | Special (better) vector arrows. |
| eucal | Euler script symbols in math mode. |
| eufrak | Euler Gothic symbols in math mode. |
| exscale | Automatically scaled integral and sum symbols. |
| fixmath | LaTeX's style for maths made compatible to ISO31-0:1992 to ISO31-13:1992. |
| fltpoint | Supports simple floating point operations (addition, subtraction, multiplication, division, and rounding). |
| gauss | Support for the Gaussian elimination. |
| icomma | Intelligent placement of a comma as a decimal separator. |
| leftidx | Left and right hand side super- and subscript. |
| mathdots | Macros to create "dots" which take the current font size into account. |
| mathematica | Fonts similar to the Mathematica software package. |
| mathtools | Very useful collection of macros to optimise amsmath. |
| nath | Package for inserting equations in "natural form". |
| numprint | Prints columns of numbers with thousands separator. |

continued . . .

... continued

| package | description |
| --- | --- |
| random | Generation of random numbers in TEX. |
| romannum | Conversion of Arabic numbers into Roman numerals. |
| xlop | Visualisation of the basic arithmetic operations. |

The packages described here refer to the following versions of TEX Live 2009.

| | |
| --- | --- |
| accents.sty | 2006/05/12 v1.3 Math Accent Tools |
| alltt.sty | 1997/06/16 v2.0g defines alltt environment |
| amscd.sty | 1999/11/29 v1.2d |
| amsopn.sty | 1999/12/14 v2.01 operator names |
| bigdelim.sty | 1999/11/05 v1.0 |
| bm.sty | 2004/02/26 v1.1c Bold Symbol Support (DPC/FMi) |
| braket.sty | 2006/09/12 |
| cancel.sty | 2000/03/12 v2.1 Cancel math terms |
| cases.sty | 2002/05/01 v2.5 |
| delarray.sty | 1994/03/14 v1.01 array delimiter package (DPC) |
| dsfont.sty | 1995/08/01 v0.1 Double stroke roman fonts |
| empheq.sty | 2007/12/03 v2.12 Emphasising equations (MH) |
| esvect.sty | 2000/07/09 v1.2 Typesetting vectors (es) |
| eucal.sty | 2001/10/01 v2.2d Euler Script fonts |
| eufrak.sty | 2001/10/01 v2.2e Euler Gothic fonts |
| framed.sty | 2007/10/04 v0.95 framed or shaded text with page breaks |
| gauss.sty | 2002/10/11 v2.14 A Package for Typesetting Matrix Operations |
| mathtools.sty | 2008/08/01 v1.06 Extension to amsmath |
| pstricks.sty | 2008/11/26 v0.40 LaTeX wrapper for 'PSTricks' (RN,HV) |
| pstricks.tex | 2009/11/17 v1.31 'PSTricks' (tvz,hv) |
| pst-node.tex | 2008/11/26 v1.01 Nodes and connections |
| delarray.sty | 1994/03/14 v1.01 array delimiter package (DPC) |
| xypic.sty | 1999/02/16 Xy-pic version 3.7 |
| xlop.sty | 2006/01/04 v0.22 |

## 9.1 accents

LaTeX knows \underline and \underbar; both yield the same result however. The accents package by Javier Bezos provides an \underaccent command, which allows "accents" as an underscore.

09-01-1

```
\underline{$M$} M    $\underbar{M}$ M    \underbar{$M$} M
$\underaccent{\bar}{M}$ M    $\underaccent{--}{M}$ M
```

## 9.2 `alltt` — verbatim-like environment

This package by Johannes Brahms provides the `alltt` environment of the same name. It should be used in preference to the usual `verbatim` environment if certain text or math expressions should be shown as the result, not as the source code.

```
\usepackage{alltt}
\begin{alltt}
Similar to verbatim, but \textbf{commands} are \emph{executed}.
Some special characters --- # $ % ^ & ~ _; can be input without backslash.
And a math expression --- \( \sqrt{a^2+b^2} \). It \emph{must} be enclosed in
\verb+\(+ ... \verb+\)+ for inline mode, the usual $ ... $ method does not work
here.
\end{alltt}
```

Similar to verbatim, but **commands** are *executed*.
Some special characters --- # $ % ^ & ~ _; can be input without backslash.
And a math expression ---  $\sqrt{a^2+b^2}$ . It *must* be enclosed in
\( ... \) for inline mode, the usual $ ... $ method does not work
here.

09-02-1

## 9.3 `amscd` — commutative diagrams

The `amscd` package is part of `amsmath`, but has to be loaded explicitly. It doesn't support diagonal arrows, but is much easier to use than most of the other packages that provide support for commutative diagrams. Further information can be found in [26], albeit outdated, and in the very good, new and current summary [19].

```
\usepackage{amsmath,amscd}
\begin{align}
\begin{CD}
  R\times S\times T @>\text{restriction}>> S\times T \\
  @V\text{proj}VV                          @VV\text{proj}V \\
  R\times S            @<<\text{inclusion}<         S
\end{CD}
\end{align}
```

$$R \times S \times T \xrightarrow{\text{restriction}} S \times T$$
$$\text{proj} \downarrow \qquad\qquad \downarrow \text{proj} \qquad\qquad\qquad (9.1)$$
$$R \times S \xleftarrow{\text{inclusion}} S$$

09-03-1

```
\usepackage{amsmath,amscd}
\newcommand\cov{\operatorname{cov}}\newcommand\non{\operatorname{non}}
\newcommand\cf{\operatorname{cf}}  \newcommand\add{\operatorname{add}}
\begin{equation}
\begin{CD}
```

```
\cov(\mathcal{L}) @>>> \non(\mathcal{K}) @>>> \cf(\mathcal{K}) @>>>
\cf(\mathcal{L})\\
@VVV @AAA @AAA @VVV\\
\add(\mathcal{L}) @>>> \add(\mathcal{K}) @>>> \cov(\mathcal{K}) @>>>
\non(\mathcal{L})
\end{CD}
\end{equation}
```

09-03-2

$$\begin{CD}
\mathrm{cov}(\mathcal{L}) @>>> \mathrm{non}(\mathcal{K}) @>>> \mathrm{cf}(\mathcal{K}) @>>> \mathrm{cf}(\mathcal{L})\\
@VVV @AAA @AAA @VVV\\
\mathrm{add}(\mathcal{L}) @>>> \mathrm{add}(\mathcal{K}) @>>> \mathrm{cov}(\mathcal{K}) @>>> \mathrm{non}(\mathcal{L})
\end{CD} \tag{9.2}$$

## 9.4 amsopn — new operators

The amsopn package is automatically loaded with the amsmath package. It can also be loaded separately though, and makes it very easy for you to define new operators.

Table 9.3 shows a summary of the operators defined by the amsopn package. The operator names defined by standard LATEX can be found in Table 4.10 on page 63. amsmath redefines some of them.

**Table 9.3**: amsopn and the predefined operator names

| | | | | | | | |
|---|---|---|---|---|---|---|---|
| \arccos | arccos | \arcsin | arcsin | \arctan | arctan | \arg | arg |
| \cos | cos | \cosh | cosh | \cot | cot | \coth | coth |
| \csc | csc | \deg | deg | \det | det | \dim | dim |
| \exp | exp | \gcd | gcd | \hom | hom | \inf | inf |
| \injlim | inj lim | \ker | ker | \lg | lg | \lim | lim |
| \liminf | lim inf | \limsup | lim sup | \ln | ln | \log | log |
| \max | max | \min | min | \Pr | Pr | \projlim | proj lim |
| \sec | sec | \sin | sin | \sinh | sinh | \sup | sup |
| \tan | tan | \tanh | tanh | | | | |

09-04-1

$$\underset{s=p}{Res} - \underset{s=p}{\mathrm{Res}}$$
$$Res_{s=p} - \mathrm{Res}_{s=p} - \underset{s=p}{\mathrm{Res}}$$

```
\usepackage{amsmath,amsopn}
\DeclareMathOperator*{\Res}{Res}

$\underset{s=p}{Res} \text{ --- }
 \underset{s=p}{\Res}$\\[5pt]
$\mathop{Res}_{s=p} \text{ --- }
 \Res_{s=p} \text{ --- } \Res\limits_{s=p}$
```

## 9.5 bigdelim — delimiters for arrays

bigdelim by Piet van Oostrum is interesting in conjunction with the package multirow — it provides a simple method of adding short braces (that do not span all the rows) to a matrix (array).

The syntax of the two commands \ldelim and \rdelim is:

```
\ldelim⟨delimiter⟩{n columns}{space} [text]
\rdelim⟨delimiter⟩{n columns}{space} [text]
```

The following example would be possible with just the normal array environment, but the braces would need to be nested, which would take much more effort than using bigdelim. The trick is to reserve an extra column for the braces; the cells remain empty, but are later spanned by a brace.

```
\usepackage{multirow,bigdelim}
\usepackage{amsmath}% Wegen pmatrix
\delimitershortfall=-4pt
\[ \begin{pmatrix}
    & x_{11} & x_{12} & \dots & x_{1p} & \rdelim\}{4}{3cm}[some text]\\
    \ldelim[{5}{1cm}[Text] & x_{21} & x_{22} & \dots & x_{2p} \\
    & \vdots\\
    & x_{n_1 1}& x_{n_1 2} & \dots & x_{n_1 p}\\
    & x_{n_1+1,1}&x_{n_1+1,2} & \dots & x_{n_1+1, p} &
        \rdelim\}{3}{3cm}[some more text]\\
    & \vdots\\
    & x_{n_1+n_2, 1} & x_{n_1+n_2,2} & \dots & x_{n_1+n_2,p}\\
    & \vdots \\
X & - & - & -& &- & X
\end{pmatrix} \]
```

09-05-1

In the example above, the last row is intentionally marked X – – – – X to emphasise the "empty" columns. The example also shows that cells that are not spanned by the brace may contain content as usual.

All the characters that can be used as brace characters in math mode can also be used as delimiters; i.e. ( ) [ ] { } | etc. (cf. Section 4.5 on page 38). The additional text is always typeset in text mode.

The package `bigdelim` requires the package `multirow`, but does not load it automatically — you must remember to take care of that.

⚠ *Load* *multiro*

09-05-2

$$
\left(
\begin{array}{cccc}
x & x & x & x \\
x & x & x & x \\
x & x & x & x \\
x & x & x & x
\end{array}
\right)
\begin{array}{c}
1 \\ i \\ j \\
\end{array}
\quad (9.3)
$$

$$
\begin{array}{ccc}
y & & u \quad v
\end{array}
$$

```
\usepackage{multirow,bigdelim}

\begin{equation}
\begin{array}{*{7}{c}}
 \ldelim({4}{4mm}& x & x & x & x
   &\rdelim){4}{4mm} & 1 \\
 & x & x & x & x &  & i\\
 & x & x & x & x &  & j\\
 & x & x & x & x &  \\
y &  & u & v &  &  &
\end{array}
\end{equation}
```

## 9.6  bm — bold math

It has been mentioned several times already that bold font is not consistent in math mode. `\mathbf` always typesets its argument in upright characters. The package bm (bold math) by David Carlisle and Frank Mittelbach provides the `\bm` command of the same name, which leaves normal characters in italics. If a whole formula is to be typeset in bold font, better ways are shown in Section 4.18 on page 65.

```
\usepackage{bm}
$f(x)=\sin(x)$ $\mathbf{f(x)=\sin(x)}$ $\bm{f(x)=\sin(x)}$ \\
$f(\alpha)=\sin(\alpha)$ $\mathbf{f(\alpha)=\sin(\alpha)}$
                   $\bm{f(\alpha)=\sin(\alpha)}$
```

09-06-1

$$f(x) = \sin(x) \; \mathbf{f(x) = \sin(x)} \; \boldsymbol{f(x) = \sin(x)}$$
$$f(\alpha) = \sin(\alpha) \; \mathbf{f(\alpha) = \sin(\alpha)} \; \boldsymbol{f(\alpha) = \sin(\alpha)}$$

If amsmath is loaded at the same time, negative interactions will occur. More details can be found in the documentation of bm. For bold sequences which are frequently used the package offers a permanent definition as well:

⚠ *amsmath*

```
\usepackage{bm,amsmath}
\DeclareBoldMathCommand[bold]{\Sin}{\mathrm{sin}}
\DeclareBoldMathCommand[bold]{\Alpha}{\alpha}
$ f(x)=\sin(x) $ $ \mathbf{f(x)=\sin(x)} $ $ \bm{f(x)=\sin(x)} $ \\
$ f(x)=\Sin(x) $ $ \mathbf{f(x)=\Sin(x)} $ $ \bm{f(x)=\Sin(x)} $ \\
$ f(\Alpha)=\Sin(\Alpha)$ $\mathbf{f(\Alpha)=\Sin(\Alpha)}$
                   $\bm{f(\Alpha)=\Sin(\Alpha)}$
```

09-06-2

$$f(x) = \sin(x) \; \mathbf{f(x) = \sin(x)} \; \boldsymbol{f(x) = \sin(x)}$$
$$f(x) = \mathbf{\sin}(x) \; \mathbf{f(x) = \sin(x)} \; \boldsymbol{f(x) = \sin(x)}$$
$$f(\alpha) = \mathbf{\sin(\alpha)} \; \mathbf{f(\alpha) = \sin(\alpha)} \; \boldsymbol{f(\alpha) = \sin(\alpha)}$$

## 9.7 **braket** — delimiters in math expressions

Brackets can be inserted in a number of ways within a math expression. It is more difficult, however, to add a single separator between them, for example:

$$\left\{ x\in\mathbf{R}|0<|x|<\frac{5}{3}\right\}$$

```
\[ \left\{ x\in\mathbf{R} | 0<{|x|}<\frac{5}{3}
                                    \right\} \]
```

09-07-1

The display is simply wrong; the first vertical bar should be much larger and the same height as the outer braces. This can be fixed by using \vphantom:

$$\left\{ x\in\mathbf{R}\,\left|\,0<|x|<\frac{5}{3}\right.\right\}$$

```
\[ \left\{\vphantom{\frac{5}{3}}x\in\mathbf{R}
   \right|\left. 0<{|x|}<\frac{5}{3}\right\} \]
```

09-07-2

The point of the braket package by Donald Arseneau is to automate the manual adjustment of the height of the brace.[1] It provides the following commands:

```
\Bra{math expression}
\Ket{math expression}
\Braket{math expression}
\Set{math expression}
```

Applying the four commands above to the example yields:

```
\usepackage{braket,tabularx}
\begin{tabularx}{\linewidth}{@{}XXX@{}}
\[ \Bra{x\in\mathbf{R} | 0<|x|<\frac{5}{3}} \]
  & \[ \Ket{x\in\mathbf{R} | 0<|x|<\frac{5}{3}} \]
  & \[ \Braket{x\in\mathbf{R} | 0<|x|<\frac{5}{3}} \] \\[-4ex]
\[ \Braket{x\in\mathbf{R} | 0<\vert x\vert <\frac{5}{3}} \]
  & \[ \Set{x\in\mathbf{R} | 0<|x|<\frac{5}{3}} \]
\end{tabularx}
```

$$\left\langle x\in\mathbf{R}|0<|x|<\frac{5}{3}\right| \qquad \left| x\in\mathbf{R}|0<|x|<\frac{5}{3}\right\rangle \qquad \left\langle x\in\mathbf{R}\,\left|\,0<\left|x\right|<\frac{5}{3}\right\rangle\right.$$

09-07-3

$$\left\langle x\in\mathbf{R}\,\left|\,0<|x|<\frac{5}{3}\right\rangle\right. \qquad \left\{ x\in\mathbf{R}\,\left|\,0<|x|<\frac{5}{3}\right.\right\}$$

The difference between the \Set and \Braket commands is in how vertical lines are handled. In \Set only the first vertical line is made the same size as the surrounding braces. The \Braket command, however, makes all the vertical lines the same height. If the vertical lines should not be changed, use \Bra or \Ket.

$$\left\langle \phi\,\left|\,\frac{\partial^2}{\partial t^2}\,\right|\,\psi\right\rangle$$

$$\left\{ \phi\,\left|\,\frac{\partial^2}{\partial t^2}|\psi\right.\right\}$$

```
\usepackage{braket}

\[ \Braket{ \phi | \frac{\partial^2}%
            {\partial t^2} | \psi }\]
\[ \Set{ {\phi} | \frac{\partial^2}%
            {\partial t^2} | \psi }\]
```

09-07-4

---

[1]Additionally there are similarly named commands that start with lowercase letters; they are not of interest here, however.

## 9.8  cancel — cancelling

The cancel package by Donald Arseneau provides frequently requested support for cancelling math expressions, especially for cancelling fractions. The cancelling of an expression may be done with a slash, backslash or X. A horizontal line is not supported by the package, but a custom definition can be added easily (cf. Example 09-08-1). Alternatively, individual characters can be cancelled with the accents package (cf. Section 4.11 on page 59).

```
\cancel{expression}
\bcancel{expression}
\xcancel{expression}
\renewcommand{\CancelColor}{colour}
```

The cancellation character used by each command is denoted by the first letter of their names; \cancel as the standard command uses the variable slash, \bcancel the variable backslash, \xcancel the capital X.

In the next example, we will also define \hcancel to cancel characters with a horizontal line. The definition will allow choice of colour for the horizontal line; it therefore assumes that either the color or xcolor package has been loaded. The predefined cancel commands can be extended to support colours as well, by renewing the command of the package; the syntax for this is shown in the box above.

09-08-1

$$\cancel{3} \qquad f(x) = \frac{\left(x^2+1\right)\cancel{(x-1)}}{\cancel{(x-1)}(x+1)}$$

```
\usepackage{amsmath,color,cancel}
\newcommand\hcancel[2][black]{%
  \setbox0=\hbox{#2}%
    \rlap{\raisebox{.45\ht0}{%
      \textcolor{#1}{\rule{\wd0}{1pt}}}}#2}

$\cancel{3}\qquad
    f(x)=\dfrac{\left(x^2+1\right)
      \cancel{(x-1)}}{\cancel{(x-1)}(x+1)}$\\[0.5cm]
$\bcancel{3}\qquad\bcancel{1234567}$\\[0.5cm]
\renewcommand*{\CancelColor}{\color{cyan}}
$\xcancel{3}\qquad\xcancel{1234567}$\\[0.5cm]
$\hcancel{3}\qquad\hcancel[red]{1234567}$
```

Another variant is \cancelto, which turns the slash into an arrow pointing to a substitute value.

```
\cancelto{expression}{expression}
```

No space is reserved; you have to take care of that manually. In the following example, an additional space of 1em was inserted.

09-08-2

$$f(x) = \frac{\left(x^2+1\right)\cancelto{1}{(x-1)}}{\cancelto{1}{(x-1)}(x+1)}$$

```
\usepackage{amsmath,cancel}

$f(x)=\dfrac{\left(x^2+1\right)
    \cancelto{1}{(x-1)}}{%
      \cancelto{1}{(x-1)}\kern1em(x+1)}$
```

The font size of the first argument of \cancelto can be controlled through a package option. It is handled like an exponent by default. Table 9.4 on the following page shows the

Table 9.4: Control of the math style of the \cancelto command

| current style | samesize | package options smaller | Smaller |
|---|---|---|---|
| \displaystyle | \displaystyle | \textstyle | \scriptstyle |
| \textstyle | \textstyle | \scriptstyle | \scriptstyle |
| \scriptstyle | \scriptstyle | \scriptscriptstyle | \scriptscriptstyle |
| \scriptscriptstyle | \scriptscriptstyle | \scriptscriptstyle | \scriptscriptstyle |

possible package options and the respective effects. The default are the settings for smaller, Example 09-08-3 shows it for the option samesize.

$$f(x) = \frac{\left(x^2+1\right)\cancel{(x-1)}^1}{\cancel{(x-1)}^1(x+1)}$$

```
\usepackage{amsmath}
\usepackage[samesize]{cancel}

$f(x)=\dfrac{\left(x^2+1\right)
  \cancelto{1}{(x-1)}}{%
  \cancelto{1}{(x-1)}\kern1em(x+1)}$
```
09-08-3

## 9.9 cases

In Section 10.2 on page 200 we will show another way to label individual lines of a case split with equation numbers. However, the cases package by Donald Arseneau is much simpler.

$$\operatorname{sgn} x = \begin{cases} -1, & \text{for } x < 0 \quad (9.1) \\ 0, & \text{for } x = 0 \quad (9.2) \\ 1, & \text{for } x > 0 \quad (9.3) \end{cases}$$

```
\usepackage{amsmath,cases}
\DeclareMathOperator{\sgn}{sgn}

\begin{numcases}{\sgn{x}=}
  -1, & for $x < 0$\\
   0, & for $x = 0$\\
   1, & for $x > 0$
\end{numcases}
```
09-09-1

The same is possible for subequations:

```
\usepackage{cases}
\begin{subnumcases}{\label{w} w\equiv}
  0 & $c = d = 0$\label{wzero}\\
  \sqrt{|c|}\,\sqrt{\frac{1 + \sqrt{1+(d/c)^2}}{2}} & $|c| \geq |d|$ \\
  \sqrt{|d|}\,\sqrt{\frac{|c/d| + \sqrt{1+(c/d)^2}}{2}} & $|c| < |d|$
\end{subnumcases}
\begin{subnumcases}{\sqrt{c+id}=}
  0                       & $w=0$ (case \ref{wzero})          \\
  w+i\frac{d}{2w}         & $w \neq 0$, $c \geq 0$            \\
  \frac{|d|}{2w} + iw \qquad & $w \neq 0$, $c < 0$, $d \geq 0$\\
  \frac{|d|}{2w} - iw \qquad & $w \neq 0$, $c < 0$, $d < 0$
\end{subnumcases}
```

09-09-2

$$w \equiv \begin{cases} 0 & c = d = 0 \quad\quad (9.4a) \\[2mm] \sqrt{|c|}\,\sqrt{\dfrac{1 + \sqrt{1 + (d/c)^2}}{2}} & |c| \geq |d| \quad\quad (9.4b) \\[2mm] \sqrt{|d|}\,\sqrt{\dfrac{|c/d| + \sqrt{1 + (c/d)^2}}{2}} & |c| < |d| \quad\quad (9.4c) \end{cases}$$

$$\sqrt{c + id} = \begin{cases} 0 & w = 0 \text{ (case 9.4a)} \quad\quad (9.5a) \\[2mm] w + i\dfrac{d}{2w} & w \neq 0, c \geq 0 \quad\quad (9.5b) \\[2mm] \dfrac{|d|}{2w} + iw & w \neq 0, c < 0, d \geq 0 \quad\quad (9.5c) \\[2mm] \dfrac{|d|}{2w} - iw & w \neq 0, c < 0, d < 0 \quad\quad (9.5d) \end{cases}$$

## 9.10 delarray — delimiters for arrays

The delarray package by David Carlisle extends the syntax of the array environment by direct support for the delimiters (braces and similar symbols, cf. Section 4.5 on page 38). The two delimiters are simply given on either side of the preamble of the matrix; the argument handling here is different from the usual LaTeX one.

```
\begin{array} [position] ⟨delLeft⟩{array preamble}⟨delRight⟩
...
\end{array}
```

The delimiter arguments *must* be input without curly braces. To use square brackets [] as delimiters, they must be specified as \lbrack and \rbrack.

09-10-1

$$A = \left\| \begin{array}{cc} a & b \\ c & d \end{array} \right\|$$

$$B = \left[ \begin{array}{cc} a & b \\ c & d \end{array} \right]$$

```
\usepackage{delarray}

\[ A=\begin{array}\|{cc}\|
   a & b\\
   c & d
\end{array} \]
\[ B=\begin{array}\lbrack{cc}\rbrack
   a & b\\
   c & d
\end{array} \]
```

The package expects both delimiters to be specified;[2] it is, however, possible to input a dot, similar to the usual \left ... \right notation, to denote an empty delimiter.

09-10-2

$$A = \left\{ \begin{array}{cc} a & b \\ c & d \end{array} \right.$$

```
\usepackage{delarray}

\[ A=\begin{array}\{{cc}.
   a & b\\
   c & d
\end{array} \]
```

---

[2]It is an *extension* of the array environment; therefore *both* may be missing as well.

This version is useful when amsmath cannot or should not be used.

## 9.11 dsfont — double stroke font for set symbols

The font amsb can be used for the construction of the standard set symbols (cf. Figure 11.7 on page 248):

```
\usepackage{amsfonts}
$\mathbb{A\,B\,C\,D\,E\,F\,G\,H\,I\,J\,K\,L\,M\,N\,O\,P\,Q\,R\,S\,T\,U\,V\,W\,X\,Y\,Z}$
```

$$\mathbb{A\,B\,C\,D\,E\,F\,G\,H\,I\,J\,K\,L\,M\,N\,O\,P\,Q\,R\,S\,T\,U\,V\,W\,X\,Y\,Z}$$

09-11-1

Alternatively, the dsfont package by Olaf Kummer can be used, which provides a corresponding alphabet for capital letters and some other characters. They can be accessed through the \mathds command:

```
\usepackage{dsfont}
$\mathds{A\,B\,C\,D\,E\,F\,G\,H\,I\,J\,K\,L\,M\,N\,O\,P\,Q\,R\,S\,T\,U\,V\,W\,X\,Y\,Z}$\\
$\mathds{a,h,k,1}$
```

$$\mathds{A\,B\,C\,D\,E\,F\,G\,H\,I\,J\,K\,L\,M\,N\,O\,P\,Q\,R\,S\,T\,U\,V\,W\,X\,Y\,Z}$$
$$\mathds{a,h,k,1}$$

09-11-2

```
\usepackage[sans]{dsfont}
$\mathds{A\,B\,C\,D\,E\,F\,G\,H\,I\,J\,K\,L\,M\,N\,O\,P\,Q\,R\,S\,T\,U\,V\,W\,X\,Y\,Z}$\\
$\mathds{a,h,k,1}$
```

$$\mathds{A\,B\,C\,D\,E\,F\,G\,H\,I\,J\,K\,L\,M\,N\,O\,P\,Q\,R\,S\,T\,U\,V\,W\,X\,Y\,Z}$$
$$\mathds{a,h,k,1}$$

09-11-3

## 9.12 empheq — emphasis of parts of an equation

The empheq package ("emphasise equation") by Morten Høgholm offers many useful things for math mode. The following example is taken from the documentation:

```
\usepackage{amsmath,empheq,xcolor,graphicx}
\newcommand*\mycolbox[1]{%
  \colorbox{blue!30!yellow!40}{\hspace{1em}#1\hspace{1em}}}
\begin{subequations}
\begin{empheq}[% start of options
    box=\mycolbox,
    right={\;\makebox[.9em]{$\raisebox{-.5\totalheight+\fontdimen22\textfont2}
          {\resizebox{!}{\EmphEqdisplayheight+\EmphEqdisplaydepth}{!}}$}},
    left={X=Y\Rightarrow\empheqlbrace}]{alignat=3}
%
  A_1&=b_1 & \qquad c_1&=d_1 & \qquad e_1&= f_1
      \tag*{\_A\raisebox{1ex}{silly}\raisebox{-1ex}{tag}\_}\\
```

```
    A_2&=b_2 & \qquad c_2&=d_2 & \qquad  e_2&= f_2 \\
    A_3&=b_3 & \qquad c_3&=d_3 & \qquad  e_3&= f_3
\end{empheq}
\end{subequations}
```

09-12-1

$$X = Y \Rightarrow \begin{cases} A_1 = b_1 & c_1 = d_1 & e_1 = f_1 \\ A_2 = b_2 & c_2 = d_2 & e_2 = f_2 \\ A_3 = b_3 & c_3 = d_3 & e_3 = f_3 \end{cases}$$

$_A{}^{\text{silly}}{}_{\text{tag}-}$

(9.6a)

(9.6b)

Usually the package is used for simpler tasks like the framing of equations. The empheq environment is given the type of frame as option and the type of equation as parameter. For optimal results, only the equation environments of amsmath should be used, as described in Chapter 6 on page 77. The eqnarray environment is not supported at all by empheq; the amsmath package is loaded automatically.

*amsmath*

```
\begin{empheq} [options] {environment}
...% content of the amsmath environment
...
\end{empheq}
```

The simplest box is \fbox, which only frames the equation — the equation number stays outside:

09-12-2

$$f(x) = \int_1^{\infty} \frac{1}{x^2}\,dt = 1$$

(9.7)

$$f(x) = \int_1^{\infty} \frac{1}{x^2}\,dt = 1$$

(9.8)

```
\usepackage{empheq}
\newcommand*\diff{\mathop{}\!\mathrm{d}}

\begin{empheq}[box=\fbox]{align}
    f(x)=\int\limits_1^{\infty}
            \frac{1}{x^2}\diff t=1
\end{empheq}
\begin{empheq}[box={\fboxsep=10pt\fbox}]{align}
    f(x)=\int\limits_1^{\infty}
            \frac{1}{x^2}\diff t=1
\end{empheq}
```

The same application, but with \colorbox:

09-12-3

$$f(x) = \int_1^{\infty} \frac{1}{x^2}\,dt = 1$$

(9.9)

```
\usepackage{empheq,color}
\newcommand*\diff{\mathop{}\!\mathrm{d}}

\begin{empheq}%
    [box={\fboxsep=10pt\colorbox{yellow}}]{align}
    f(x)=\int\limits_1^{\infty}
            \frac{1}{x^2}\diff t=1
\end{empheq}
```

The option box takes any valid LaTeX code as parameters. The framing of subequations has already been shown in the first example. The following example does this for only one of the two:

$$f(x) = \int\limits_1^{\infty} \frac{1}{x^1}\, dt = 1 \qquad (9.10a)$$

$$f(x) = \int\limits_2^{\infty} \frac{1}{x^2}\, dt = 0.25 \qquad (9.10b)$$

```
\usepackage{amsmath,empheq,color}
\newcommand*\diff{\mathop{}\!\mathrm{d}}

\begin{subequations}
\begin{align}
   f(x) & =\int\limits_1^{\infty}\frac{1}{x^1}\diff t=1
\end{align}
\begin{empheq}[box={\fboxsep=10pt\colorbox{cyan}}]{align}
   f(x) & =\int\limits_2^{\infty}\frac{1}{x^2}\diff t=0.25
\end{empheq}
\end{subequations}
```

`09-12-4`

Any displayed equation can be labelled with a comment on the left or right, which will be centred vertically:

```
\usepackage{empheq}
\begin{empheq}[right=\kern1em\empheqrbrace\quad\Leftarrow\text{subtract}]{align}
   3x_1 + 4x_2 -3x_3 &= 4\\   3x_1 - 4x_2 - x_3 &= 1
\end{empheq}
```

$$
\left.\begin{aligned}
3x_1 + 4x_2 - 3x_3 = 4 \\
3x_1 - 4x_2 - x_3 = 1
\end{aligned}\right\} \quad \Leftarrow \text{subtract}
$$

$$(9.11)$$
$$(9.12)$$

`09-12-5`

*PSTricks*  If the package fancybox or pstricks is loaded, the box types of these packages may be used as well. Note that when using PSTricks, PDF documents have to be created through the sequence latex→dvips→ps2pdf.

$$f(x) = \int\limits_1^{\infty} \frac{1}{x^2}\, dt = 1 \qquad (9.13)$$

```
\usepackage{empheq,fancybox}
\newcommand*\diff{\mathop{}\!\mathrm{d}}

\begin{empheq}[box={\fboxsep=10pt\shadowbox*}]{align}
   f(x) &= \int\limits_1^{\infty}\frac{1}{x^2}\diff t=1
\end{empheq}
```

`09-12-6`

$$f(x) = \int\limits_1^{\infty} \frac{1}{x^2}\, dt = 1 \qquad (9.14)$$

```
\usepackage{empheq,pstricks}
\newcommand*\diff{\mathop{}\!\mathrm{d}}

\begin{empheq}[box={\psshadowbox[shadowcolor=black!50,
   framesep=10pt,framearc=0.2]}]{align}
   f(x) &= \int\limits_1^{\infty}\frac{1}{x^2}\diff t=1
\end{empheq}
```

`09-12-7`

## 9.13 esvect — vector arrows

Especially designed for vectors, the esvect package by Eddie Saudrais improves their display significantly. It provides different display types for the arrows through package options (a, b, ... or h):

| package option | a | b | c | d | e | f | g | h |
|---|---|---|---|---|---|---|---|---|
| arrow type | $\rightarrow$ | $\rightarrow$ | $\rightarrow$ | $\rightarrow$ | $\rightarrow$ | $\rightarrow$ | $\rightarrow$ | $\rightarrow$ |

`09-13-1`

Without an explicit specification, the package is loaded with the option d. The advantage of the package becomes obvious in the following example where \vv is juxtaposed with the standard LATEX version with \overrightarrow or \vec. The last example is particularly unsatisfactory.

```
\usepackage{esvect}
\newcommand*\CMD[1]{\texttt{\textbackslash#1}}
\begin{tabular}{@{}c c c@{}}\hline
\texttt{esvect} & \multicolumn{2}{c}{\LaTeX}\\
\CMD{vv}       & \CMD{overrightarrow}      & \CMD{vec}      \\\hline
$\vv{a}$       & $\overrightarrow{a}$      & $\vec{a}$      \\
$\vv{abc}$     & $\overrightarrow{abc}$    & $\vec{abc}$    \\
$\vv{\imath}$  & $\overrightarrow{\imath}$ & $\vec{\imath}$\\
$\vv*{A}{x}$   & $\overrightarrow{A}_{x}$  & $\vec{A_x}$    \\
$\vv{E}_{\vv{u}_{\vv{w}}}$
               & $\overrightarrow{E}_{\overrightarrow{u}_{\overrightarrow{w}}}$
                                           & $\vec{E}_{\vec{u}_{\vec{w}}}$$\\\hline
\end{tabular}
```

09-13-2

| esvect | LATEX | |
|--------|-------|--|
| \vv | \overrightarrow | \vec |
| $\vec{a}$ | $\vec{a}$ | $\vec{a}$ |
| $\vec{abc}$ | $\vec{abc}$ | $\vec{abc}$ |
| $\vec{\imath}$ | $\vec{\imath}$ | $\vec{\imath}$ |
| $\vec{A}_x$ | $\vec{A}_x$ | $\vec{A}_x$ |
| $\vec{E}_{\vec{u}_{\vec{w}}}$ | $\vec{E}_{\vec{u}_{\vec{w}}}$ | $\vec{E}_{\vec{u}_{\vec{w}}}$ |

# 9.14  eucal and eufrak — Script and Gothic characters

The eucal package supports "scriptwriting" of the capital letters. It is actually part of the amsmath bundle and can be found in the amsfonts directory.

The eufrak package also supports Gothic letters in math mode and defines the \mathfrak command for this.

09-14-1

$\mathscr{ABCDEFGHIJKLMNOPQRSTUVWXYZ}$
$\mathscr{MATH\ WITH\ LATEX}$
$\mathcal{ABCDEFGHIJKLMNOPQRSTUVWXYZ}$
$\mathcal{MATH\ WITH\ LATEX}$
$\mathfrak{ABCDEFGHIJKLMNOPQRSTUVWXYZ}$
$\mathfrak{MATH\ WITH\ LATEX}$

```
\usepackage[mathscr]{eucal}
\usepackage{eufrak}

$\mathscr{ABCDEFGHIJKLMNOPQRSTUVWXYZ}$\\
$\mathscr{MATH WITH LATEX}$\\
$\mathcal{ABCDEFGHIJKLMNOPQRSTUVWXYZ}$\\
$\mathcal{MATH WITH LATEX}$\\
$\mathfrak{ABCDEFGHIJKLMNOPQRSTUVWXYZ}$\\
$\mathfrak{MATH WITH LATEX}$
```

This package can cause problems when loaded at the same time as other font packages. For the example above the eucal package with the option mathscr was loaded. Thus the old \mathcal command is still available. The difference to the new \mathscr is obvious.

## 9.15 exscale — large symbols

The following formula was typeset with the default font sizes of this book and therefore looks as expected:

```
\newcommand*\diff{\mathop{}\!\mathrm{d}}
\[\int_{-1}^{+1}\frac{f(x)}{\sqrt{1-x^{2}}}\diff x\approx\frac{\pi}{n}\sum_{i=1}^{n}f
  \left(\cos\left(\frac{2i-1}{2n}\right)\right) \]
```

$$\int_{-1}^{+1} \frac{f(x)}{\sqrt{1-x^2}}\,\mathrm{d}x \approx \frac{\pi}{n}\sum_{i=1}^{n} f\left(\cos\left(\frac{2i-1}{2n}\right)\right)$$

09-15-1

If the equation is created with \huge prepended, the result is surprising because everything but the integral symbol and the sum symbol is enlarged. This is for historic reasons; \int and \sum symbols are not enlarged through the version contained in the font cmex (cf. Figure 11.5 on page 247). At the time, this problem was not of interest, but nowadays symbols as large as this are often used for presentations.

```
\newcommand*\diff{\mathop{}\!\mathrm{d}}
\huge
\[\int_{-1}^{+1}\frac{f(x)}{\sqrt{1-x^{2}}}\diff x\approx\frac{\pi}{n}\sum_{i=1}^{n}f
  \left(\cos\left(\frac{2i-1}{2n}\right)\right) \]
```

$$\int_{-1}^{+1} \frac{f(x)}{\sqrt{1-x^2}}\,\mathrm{d}x \approx \frac{\pi}{n}\sum_{i=1}^{n} f\left(\cos\left(\frac{2i-1}{2n}\right)\right)$$

09-15-2

Using the exscale package gives a better result; it is enough to simply load the package in the preamble of the document:

```
\usepackage{exscale}
\newcommand*\diff{\mathop{}\!\mathrm{d}}
\huge
\[\int_{-1}^{+1}\frac{f(x)}{\sqrt{1-x^{2}}}\diff x\approx\frac{\pi}{n}\sum_{i=1}^{n}f
  \left(\cos\left(\frac{2i-1}{2n}\right)\right) \]
```

$$\int_{-1}^{+1} \frac{f(x)}{\sqrt{1-x^2}}\,\mathrm{d}x \approx \frac{\pi}{n}\sum_{i=1}^{n} f\left(\cos\left(\frac{2i-1}{2n}\right)\right)$$

09-15-3

The problem above does not occur with every combination of math fonts. Some newer font packages solve it themselves. You will only encounter this problem when using the default CM fonts of LaTeX.

# 9.16 **gauss.sty** – Visualisation of the Gaussian elimination

The Gaussian elimination for solving systems of linear equations is regarded as a standard method in natural sciences. The gauss package offers special support for this. It only uses the picture environment to create the polylines; therefore pdflatex can be used without problems to directly create PDFs.

## 9.16.1 **gmatrix**

Similar to the amsmath package, a matrix environment is provided. This is nothing special, apart from the fact that usually no parentheses are used. This can be set through an optional parameter that determines which one of the matrix environments of amsmath will be used in the end. Without parameter, the simple matrix environment of amsmath is used.

```
\usepackage{gauss}
$\begin{gmatrix}    1 & 2\\  3 & 4   \end{gmatrix}$
$\begin{gmatrix}[v] 1 & 2\\  3 & 4   \end{gmatrix}$
$\begin{gmatrix}[V] 1 & 2\\  3 & 4   \end{gmatrix}$
$\begin{gmatrix}[p] 1 & 2\\  3 & 4   \end{gmatrix}$
$\begin{gmatrix}[b] 1 & 2\\  3 & 4   \end{gmatrix}$
$\begin{gmatrix}[B] 1 & 2\\  3 & 4   \end{gmatrix}$
```

09-16-1

$$
\begin{matrix} 1 & 2 \\ 3 & 4 \end{matrix} \quad
\begin{vmatrix} 1 & 2 \\ 3 & 4 \end{vmatrix} \quad
\begin{Vmatrix} 1 & 2 \\ 3 & 4 \end{Vmatrix} \quad
\begin{pmatrix} 1 & 2 \\ 3 & 4 \end{pmatrix} \quad
\begin{bmatrix} 1 & 2 \\ 3 & 4 \end{bmatrix} \quad
\begin{Bmatrix} 1 & 2 \\ 3 & 4 \end{Bmatrix}
$$

In addition to these, the \newmatrix command is supplied, which makes it possible to define new environments with arbitrary parentheses, for example:

09-16-2

$$
\left]\begin{matrix} 1 & 2 \\ 3 & 4 \end{matrix}\right[
$$

```
\usepackage{gauss} \newmatrix{]}{[}{q}

$\begin{gmatrix}[q]
   1 & 2\\  3 & 4
  \end{gmatrix}$
```

The first two parameters determine the parentheses and the third one the shorthand for the option. Obviously v, V, p, b, and B do not make sense as shorthand for custom matrix environments and g is not available.

## 9.16.2 Row and column markers

Counting rows and columns starts in the upper left corner at 0; the lower right corner therefore has (*number of rows* – 1, *number of columns* – 1). Row markers are introduced by the \rowops command and column operations by \colops. There are three commands available for the special markers (operators):

| | |
|---|---|
| \add [summand] {*Z/S*}{*Z/S*} | % addition marker for row/column |
| \mult{*Z/S*}{*factor*} | % multiplication marker for row/column |
| \swap{*Z/S*}{*Z/S*} | % marks two rows/columns with a double arrow |

### 9.16.3 Examples

In a system of linear equations, two rows and columns which are linearly dependent should be marked in the matrix of coefficients:

$$\begin{bmatrix} 1 & 2 & -3 \\ 0.1 & 0.2 & -1 \\ -0.5 & -1 & \frac{3}{2} \end{bmatrix}$$

```
\usepackage{gauss}

$\begin{gmatrix}[b]
    1   & 2   & -3\\
    0.1 & 0.2 & -1\\
    -0.5 &-1   & \frac{3}{2}
\rowops\swap{0}{2} \colops\swap{0}{1}
\end{gmatrix}$
```

09-16-3

In principle the \swap command is meant to mark two rows/columns which can be swapped. To transform this system of equations into a triangular form in the usual manner, for example with the addition method, the Gaussian elimination is the preferred way. For the first two rounds:

$$\begin{bmatrix} 1 & 2 & -3 \\ 0.1 & 0.2 & -1 \\ -0.5 & -1 & \frac{3}{2} \end{bmatrix} \quad |\times(-0.1)$$

$$\begin{bmatrix} 1 & 2 & -3 \\ -0.5 & -1 & \frac{3}{2} \\ 0 & 0 & -0.7 \end{bmatrix} \quad |\times 0.5$$

```
\usepackage{gauss}

$\begin{gmatrix}[b]
    1   & 2 & -3\\   0.1 & 0.2 & -1\\
    -0.5 &-1 & \frac{3}{2}
\rowops\mult{0}{\times(-0.1)}
\add{0}{1}\swap{1}{2}
\end{gmatrix}$\\[2ex]
$\begin{gmatrix}[b]
    1 & 2 & -3\\ -0.5 &-1 & \frac{3}{2}\\
    0 & 0 & -0.7
\rowops\mult{0}{\times0.5}\add{0}{1}
\end{gmatrix}$
```

09-16-4

A complete application is shown in the following example, which solves the system of linear equations $\vec{A} \times \vec{x} = \vec{b}$.

```
\usepackage{gauss}
\begin{equation}
\begin{gmatrix}[b]
  1 &-2 & 1 & 2\\ 2 & 3 &-2 & 3\\   4 &-1 & 3 &-1\\ 3 & 2 &-4 & 5
\end{gmatrix} \times
\begin{gmatrix}[b]   x\\ y \\ z\\ t \end{gmatrix} =
\begin{gmatrix}[b]   8\\ 14\\ 7\\ 5 \end{gmatrix}
\end{equation}
```

$$\begin{bmatrix} 1 & -2 & 1 & 2 \\ 2 & 3 & -2 & 3 \\ 4 & -1 & 3 & -1 \\ 3 & 2 & -4 & 5 \end{bmatrix} \times \begin{bmatrix} x \\ y \\ z \\ t \end{bmatrix} = \begin{bmatrix} 8 \\ 14 \\ 7 \\ 5 \end{bmatrix} \tag{9.15}$$

09-16-5

Vertical alignment can be a problem; on one hand everything has to be left-aligned, on the other hand the individual cells have different horizontal widths. In these cases it is advisable not to create four different matrices, but to put everything into a single gmatrix like in example 09-16-6 on the next page and mark the individual cells.

```
\usepackage{gauss}  \renewcommand\rowmultlabel[1]{|\,\times#1}
\[\begin{gmatrix}
1 &-2 & 1 & 2 && & 8 \\   2 & 3 &-2 & 3 && & 14\\ % rows 0 and 1
4 &-1 & 3 &-1 && & 7 \\   3 & 2 &-4 & 5 && & 5 \\ % 2 and 3
                      \\
1 &-2 & 1 & 2 && & 8 \\   0 & 7 &-4 &-1 & & &-2 \\ % 5 and 6
0 & 7 &-1 &-9 & &-25\\    0 & 8 &-7 &-1 & & &-9 \\ % 7 and 8
                      \\
1 &-2 & 1 & 2 && & 8 \\   0 & 7 &-4 &-1 & & &-2 \\ % 10 and 11
0 & 0 & 3 &-8 & &-23\\    0 & 0 &-17& 1 & &-47\\ % 12 and 13
                      \\
1 &-2 & 1 & 2 && & 8 \\   0 & 7 &-4 &-1 & & &-2 \\ % 15 and 16
0 & 0 & 3 &-8 & &-23\\    0 & 0 & 0 &-\frac{133}{3}& &\frac{-532}{3} % 17 and 18
\rowops
\mult{0}{(-2)}\add{0}{1}\mult{0}{(-4)}\add{0}{2}\mult{0}{(-3)}\add{0}{3}% matrix 1
\mult{6}{(-1)}\add{6}{7}\mult{6}{(-\frac{8}{7})}\add{6}{8}% matrix 2
\mult{12}{\frac{17}{3}}\add{12}{13} % matrix 3
\end{gmatrix} \]
```

If you want to parenthesise and number the matrix parts individually, the problem becomes more difficult. In this case the flegn class option or the package of the same name can be used:

```
\usepackage{gauss,fleqn}\renewcommand\rowmultlabel[1]{|\,\times#1}
\begin{equation}
  \begin{gmatrix}[p]
    1 &-2 & 1 & 2 \\ 2 & 3 &-2 & 3 \\ % rows 0 and 1
    4 &-1 & 3 &-1 \\ 3 & 2 &-4 & 5    % 2 and 3
  \rowops
  \mult{0}{(-2)} \add{0}{1} \mult{0}{(-4)}\add{0}{2} \mult{0}{(-3)} \add{0}{3}
```

```
  \end{gmatrix}
\end{equation}      % end matrix 1
\begin{equation}
  \begin{gmatrix}[p]
   1 &-2 & 1 & 2 \\ 0 & 7 &-4 &-1 \\% rows 0 and 1
   0 & 7 &-1 &-9 \\ 0 & 8 &-7 &-1   % 2 and 3
   \rowops
   \mult{1}{(-1)} \add{1}{2} \mult{1}{(-\frac{8}{7})} \add{1}{3}
   \end{gmatrix}
\end{equation}
```

$$
\begin{pmatrix} 1 & -2 & 1 & 2 \\ 2 & 3 & -2 & 3 \\ 4 & -1 & 3 & -1 \\ 3 & 2 & -4 & 5 \end{pmatrix}
\left|\times(-2)\right. \quad \left|\times(-4)\right. \quad \left|\times(-3)\right.
\tag{9.16}
$$

09-16-7

$$
\begin{pmatrix} 1 & -2 & 1 & 2 \\ 0 & 7 & -4 & -1 \\ 0 & 7 & -1 & -9 \\ 0 & 8 & -7 & -1 \end{pmatrix}
\left|\times(-1)\right. \quad \left|\times(-\tfrac{8}{7})\right.
\tag{9.17}
$$

The \mult function can also be used for normal equation transformations if they are done within the gmatrix environment.

```
\usepackage{amsmath,gauss}
\newcommand*\diff{\mathop{}\!\mathrm{d}}
\setlength\arraycolsep{1.4pt}
\renewcommand\rowmultlabel[1]{\left|\,#1\right.}
\newcommand*\xstrut{\rule{0pt}{5ex}}
\begin{equation}
\renewcommand\arraystretch{1.8}
\begin{gmatrix}
 \dfrac{1}{C}\int i\times\diff t+L\times\dfrac{\diff i}{\diff t}
           +R\times i\hfill &= & 0\\
 \xstrut\dfrac{1}{C}\times i+L\times\dfrac{\diff ^2i}{\diff t^2}+
    R\times\dfrac{\diff i}{\diff t}\hfill &= &0\\
 \xstrut\dfrac{\diff ^2i}{\diff t^2}+\dfrac{R}{L}\times\dfrac{\diff i}{\diff t}+
    \dfrac{1}{LC}\times i\hfill &= & 0\\
 \xstrut\ddot{i}+\delta\dot{i}+\omega_0^2i\hfill & = &0
 \rowops
 \mult{0}{\mbox{differentiation}}  \mult{1}{:L\mbox{ and sorting}}
 \mult{2}{\dfrac{R}{L}=\delta;\ \dfrac{1}{LC}=\omega_0^2}
\end{gmatrix}
\end{equation}
```

09-16-8

$$\frac{1}{C} \int i \times \mathrm{d}t + L \times \frac{\mathrm{d}i}{\mathrm{d}t} + R \times i \; = 0 \;\mid \text{differentiation}$$

$$\frac{1}{C} \times i + L \times \frac{\mathrm{d}^2 i}{\mathrm{d}t^2} + R \times \frac{\mathrm{d}i}{\mathrm{d}t} \; = 0 \;\mid : L \text{ and sorting}$$

$$\frac{\mathrm{d}^2 i}{\mathrm{d}t^2} + \frac{R}{L} \times \frac{\mathrm{d}i}{\mathrm{d}t} + \frac{1}{LC} \times i \; = 0 \;\left|\; \frac{R}{L} = \delta; \; \frac{1}{LC} = \omega_0^2 \right.$$

$$\ddot{i} + \delta \dot{i} + \omega_0^2 i \qquad\qquad = 0$$

(9.18)

## 9.17  **mathtools** — extended and new environments

This package by Morten Høgholm provides many modified and new environments. It can be loaded with the options given in Table 9.5. All unknown options are passed on to the amsmath package, which is loaded implicitly because mathtools is based on it.

```
\usepackage[fleqn,centertags]{mathtools}
```

is therefore identical to

```
\usepackage[fleqn,centertags]{amsmath}
\usepackage{mathtools}
```

**Table 9.5:** Package options for mathtools

| *name* | *meaning* |
|---|---|
| donotfixamsmathbugs | Do not change the kern of amsmath. |
| fixamsmath | (default) Fixes two errors in amsmath. For further information see the bug reports 3591 and 3614 in the official LaTeX bug database (http://www.latex-project.org/cgi-bin/ltxbugs2html) |
| allowspaces | Allows spaces between \begin{...} and a possible optional argument[...]. This can lead to errors if there is no optional argument but the following line of the equation starts with [. In this case the content is interpreted as an optional argument belonging to the environment. |
| disallowspaces | (default) After the start of an environment, an opening square bracket is only interpreted as an optional argument if is preceded immediately by the start of the environment. Otherwise it will be treated as normal text. |

All parameters of the package can be set or changed at any time with the \mathtoolsset command:

```
\mathtoolsset{parameters}
```

### 9.17.1 Limits

The ways of placing limits optimally already described in Section 6.12 on page 108 are supported by mathtools through the same or similar commands.

---

\mathllap [style] {*argument*}

\mathclap [style] {*argument*}

\mathrlap [style] {*argument*}

\clap{*text*}

\mathmbox{*argument*}     \mathmakebox [width] [position] {*argument*}

---

$$x = \sum_{1 \le i \le j \le n} x_{ij}$$

$$x = \sum_{1 \le i \le j \le n} x_{ij}$$

```
\usepackage{mathtools}

\[ x = \sum_{1\le i\le j\le n}x_{ij} \]
\[ x = \sum_{\mathclap{1\le i\le j\le n}}x_{ij} \]
```

09-17-1

The package provides further commands to optimise limits, for example for a horizontal translation.

### 9.17.2 Equation numbers (tags)

The following commands greatly simplify changing the equation numbering or the labelling. The optional *format* can be used for pass arbitrary formatting commands; for example \textsf.

---

\newtagform{*name*} [format] {*left*}{*right*}

\renewtagform{*name*} [format] {*left*}{*right*}

\usetagform{*name*}

---

$$E = m \cdot c^2 \qquad <9.19>$$

$$E \ne mc^3 \qquad [9.20]$$

$$E \ne mc^3 \qquad <9.21>$$

$$E = m \cdot c^2 \qquad (9.22)$$

```
\usepackage{mathtools} \newtagform{angels}{<}{>}
\newtagform{brackets}[\textsf]{[}{]}

\usetagform{angels}
\begin{equation} E = m\cdot c^2 \end{equation}
\usetagform{brackets}
\begin{equation} E \neq m c^3  \end{equation}
\renewtagform{brackets}{<}{>}
\begin{equation} E \neq m c^3  \end{equation}
\usetagform{default}
\begin{equation} E = m\cdot c^2 \end{equation}
```

09-17-2

References to equations are usually typeset in the current text font style. Special formatting of a label is lost; this is usually not important for pure equation numbers, but can be a nuisance when using the \tag* command. mathtools therefore defines \refeq, which outputs the reference with the original formatting.

<table>
<tr><td>09-17-3</td></tr>
</table>

$$E = m \cdot c^2 \qquad \textbf{Einstein}$$

*Either* **Einstein** *or* (**Einstein**) *or the special* **Einstein**.

```
\usepackage{mathtools}

\begin{equation}
E = m\cdot c^2 \label{foo}
        \tag*{\textbf{Einstein}}
\end{equation}
\emph{Either~\ref{foo} or~\eqref{foo} or
the special~\refeq{foo}.}
```

The showonlyrefs and showmanualtags switches for numbering equations behave similarly to the \cite command for literature references. Only the equations that are referenced in the text with \eqref or \refeq are fitted with a number or label. \noeqref can be used similarly to the \nocite command to force a number or label for non-referenced equations.

```
showonlyrefs=true|false
showmanualtags=true|false
\refeq{label}        \noeqref{label1,label2,...}
```

The following example would normally have six labelled equations. However, because of \mathtoolsset{*showonlyrefs, showmanualtags*} only the ones that are referenced are labelled. Additionally, the labels created manually are displayed in any case. This setting is changed after the first equation through showmanualtags=falsesuch that the following label (!!) is not shown. The last equation is numbered despite not being referenced because the standard was restored before it through showonlyrefs=false.

<table>
<tr><td>09-17-4</td></tr>
</table>

$$a = a \qquad <9.23>$$
$$b = b \qquad <**>$$

A reference to Equation 9.23 (a).

$$c = c$$
$$d = d$$

A reference to Equation (e). Label (!!) is not shown now!

$$e = e \qquad (9.23)$$

Reset the default.

$$f = f \qquad (9.24)$$

```
\usepackage{mathtools}
\newtagform{angels}{<}{>}

\mathtoolsset{showonlyrefs,showmanualtags}
\usetagform{angels}
\begin{gather}
  a = a \label{eq:a}\\
  b = b\label{eq:b}\tag{**}
\end{gather}
\mathtoolsset{showmanualtags=false}
A reference to Equation~\refeq{eq:a} (a).
\usetagform{default}% change the labeling
\begin{align}
  c &= c \label{eq:c}\\
  d &= d\label{eq:d}\tag{!!}
\end{align}
A reference to Equation~\refeq{eq:e} (e).
Label (!!) is not shown now!
\begin{align} e = e \label{eq:e}\end{align}
Reset the default.
\mathtoolsset{showonlyrefs=false}
\begin{align} f = f \label{eq:f}\end{align}
```

### 9.17.3 Matrices and cases

The starred versions of the matrix environments (cf. Section 6.3.5 on page 96) allow options for the column alignment – l, c and r.

```
\begin{matrix*} [position] ... \end{matrix*}
\begin{pmatrix*} [position] ... \end{pmatrix*}
\begin{bmatrix*} [position] ... \end{bmatrix*}
\begin{Bmatrix*} [position] ... \end{Bmatrix*}
\begin{vmatrix*} [position] ... \end{vmatrix*}
\begin{Vmatrix*} [position] ... \end{Vmatrix*}
```

$$\begin{pmatrix} -1 & 3 \\ 2 & -4 \end{pmatrix}$$

```
\usepackage{mathtools}

\[ \begin{pmatrix*}[r]
   -1 & 3 \\
   2  & -4
   \end{pmatrix*} \]
```

09-17-5

The cases environment from the amsmath package corresponds in principle to an array environment; this means that fractions and operators are typeset in \textstyle. mathtools defines the dcases environment, which uses the \displaystyle for the individual cells. Now the operator symbols are the right size:

```
\usepackage{mathtools}
\newcommand*\diff{\mathop{}\!\mathrm{d}}
\[ s =
  \begin{dcases}
    \int\limits_a^b\sqrt{1+y\prime^2}\diff x                & y=f(x)\\
    \int\limits_{t_1}^{t_2}\sqrt{\dot{x}^2+\dot{y}^2}\diff t & x=f(t); y=f(t) \\
    \int\limits_{\varphi_1}^{\varphi_2}
         \sqrt{\left(\frac{\diff r}{\diff\varphi}\right)^2+r^2}\,\diff\varphi
                                                            & r=f(\varphi)
  \end{dcases}
\]
```

$$s = \begin{cases} \displaystyle\int_a^b \sqrt{1 + y'^2}\, dx & y = f(x) \\[2ex] \displaystyle\int_{t_1}^{t_2} \sqrt{\dot{x}^2 + \dot{y}^2}\, dt & x = f(t); y = f(t) \\[2ex] \displaystyle\int_{\varphi_1}^{\varphi_2} \sqrt{\left(\dfrac{dr}{d\varphi}\right)^2 + r^2}\, d\varphi & r = f(\varphi) \end{cases}$$

09-17-6

The starred version of the environment automatically typesets the second column in text mode.

```
\begin{dcases}... &... \end{dcases}
\begin{dcases*}... & ⟨text column⟩ \end{dcases*}
```

09-17-7

$$\operatorname{sgn} x = \begin{cases} -1 & x \text{ negative} \\ 0 & x = 0 \\ +1 & x \text{ positive} \end{cases}$$

```
\usepackage{mathtools}

\[ \mathop{\mathrm{sgn}}x =
    \begin{dcases*}
                -1  & $x$ negative \\
  \phantom{+} 0  & $x=0$ \\
                +1  & $x$ positive
    \end{dcases*} \]
```

### 9.17.4  Environments

The gather environment, which is already defined by amsmath, is extended by two new variants in mathtools:

```
\begin{lgathered} [position]
... \\
...
\end{lgathered}
\begin{rgathered} [position]
... \\
...
\end{rgathered}
\newgathered{name}{start of line}{end of line}{end}
\renewgathered{name}{start of line}{end of line}{end}
```

The lgathered and rgathered environments behave similarly to the gather environment demonstrated in Example 06-03-3 on page 89. They are centred as blocks, but left- or right-aligned within, however.

```
\usepackage{mathtools}
\begin{gather}
\begin{lgathered}
 i_{11}+i_{21}=0.25\\ i_{21}=\frac{1}{3}i_{11}\\ i_{31}=0.33i_{22}+i_{11}-0.1i_{21}
\end{lgathered}\\[5pt]
\begin{rgathered}
 i_{11}+i_{21}=0.25\\ i_{21}=\frac{1}{3}i_{11}\\ i_{31}=0.33i_{22}+i_{11}-0.1i_{21}
\end{rgathered}\\[5pt]
\begin{gathered}
 i_{11}+i_{21}=0.25\\ i_{21}=\frac{1}{3}i_{11}\\ i_{31}=0.33i_{22}+i_{11}-0.1i_{21}
\end{gathered}
\end{gather}
```

$$i_{11} + i_{21} = 0.25$$

09-17-8

$$i_{21} = \frac{1}{3}i_{11} \tag{9.26}$$

$$i_{31} = 0.33i_{22} + i_{11} - 0.1i_{21}$$

$$i_{11} + i_{21} = 0.25$$

$$i_{21} = \frac{1}{3}i_{11} \tag{9.27}$$

$$i_{31} = 0.33i_{22} + i_{11} - 0.1i_{21}$$

$$i_{11} + i_{21} = 0.25$$

$$i_{21} = \frac{1}{3}i_{11} \tag{9.28}$$

$$i_{31} = 0.33i_{22} + i_{11} - 0.1i_{21}$$

The \newgathered command can be used to define a new gather environment with extended properties. The parameter *start of line* may contain arbitrary material that will be evaluated at the beginning of each line within the environment. Similarly the next parameter *end of line* at the end of each line. The last parameter *end* is only evaluated once the end of the respective environment is reached. The following example defines an agathered environment with an alpha counter.

```
\usepackage{mathtools}
\newcounter{linecnt}\renewcommand\thelinecnt{(\alph{linecnt})}
\newcommand\stepline{\stepcounter{linecnt}\thelinecnt}
\newgathered{agathered}{\llap{\stepline}\quad\hfil}{\hfil}{\setcounter{linecnt}{0}}
\begin{align}
\begin{agathered}
 i_{11}+i_{21}=0.25\\ i_{21}=\frac{1}{3}i_{11}\\ i_{31}=0.33i_{22}+i_{11}-0.1i_{21}
\end{agathered}\\[7pt]
\begin{agathered}
 f(x)=3x+4\\ g(x)=\sqrt{x^2-1}\\ h(x)=\frac{1}{x}
\end{agathered}
\end{align}
```

$$\text{(a)} \qquad i_{11} + i_{21} = 0.25$$

09-17-9

$$\text{(b)} \qquad i_{21} = \frac{1}{3}i_{11} \tag{9.29}$$

$$\text{(c)} \quad i_{31} = 0.33i_{22} + i_{11} - 0.1i_{21}$$

$$\text{(a)} \quad f(x) = 3x + 4$$

$$\text{(b)} \quad g(x) = \sqrt{x^2 - 1} \tag{9.30}$$

$$\text{(c)} \qquad h(x) = \frac{1}{x}$$

amsmath distinguishes between outer and inner application for some environments. It does not have to be an outer environment itself; it can be part of an including environment. The only important thing is that the outermost environment is an outer environment, for example align. For some of the environments there are "outer-inner-pairs" in amsmath, as for gather and gathered. Through the definition of multlined, mathtools makes this pairing also available for multline. Furthermore additional commands are defined to optimise the typesetting:

```
\begin{multlined} [position] [width]
... \\
\end{multlined}
\shoveleft [length] {argument}
\shoveright [length] {argument}
```

```
\usepackage{mathtools}
\[ A = \begin{multlined}[t]
        \framebox[3cm]{line 1} \\ \vdots \\ \framebox[3cm]{line n}
      \end{multlined} =
  B = \begin{multlined}[c]
        \framebox[3cm]{line 1} \\ \vdots \\ \framebox[3cm]{line n}
      \end{multlined} =
  C = \begin{multlined}[b]
        \framebox[3cm]{line 1} \\ \vdots \\ \framebox[3cm]{line n}
      \end{multlined} =
  D \]
```

The two commands \shoveleft and \shoveright have an extended syntax compared to the ones from amsmath; you are able to specify the translation length through an optional argument (cf. Example ?? on page ??). Even negative indentations are possible, as can be seen in the following example. The first case, with no additional indents, provides a comparison.

```
\usepackage{mathtools}
\newcommand*\CMD[1]{\texttt{\textbackslash#1}}
\[ \begin{multlined}[c][0.75\linewidth]
    \framebox[0.5\columnwidth]{x}\\ \framebox[0.5\columnwidth]{x}\\
    \shoveleft{\framebox[0.5\columnwidth]{\CMD{shoveleft}}}\\
    \shoveright{\framebox[0.5\columnwidth]{\CMD{shoveright}}}\\
    \framebox[0.5\columnwidth]{x}\\ \framebox[0.5\columnwidth]{x}
  \end{multlined} \]
\[ \begin{multlined}[c][0.75\linewidth]
    \framebox[0.5\columnwidth]{x}\\ \framebox[0.5\columnwidth]{x}\\
    \shoveleft[1cm]{\framebox[0.5\columnwidth]{\CMD{shoveleft}}}\\
```

```
    \shoveright[-5mm]{\framebox[0.5\columnwidth]{\CMD{shoveright}}}\\
    \framebox[0.5\columnwidth]{x}\\ \framebox[0.5\columnwidth]{x}
\end{multlined} \]
```

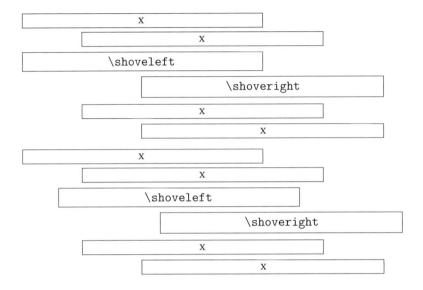

09-17-11

The \MoveEqLeft command is useful when the column separator should not be taken into account. The optional argument changes the translation from the default of 2mm to a part or a multiple of the unit em.

\MoveEqLeft [factor]

Using this command saves using other tricks to disregard the column separator.

```
\usepackage{mathtools}
\newcommand*\diff{\mathop{}\!\mathrm{d}}
\begin{align*}
\MoveEqLeft[6]
  \frac{A}{x-x_1},\ \frac{A}{\left(x-x_1\right)^k},\ \frac{Ax+b}{x^2+px+q},\
    \frac{Ax+b}{\left(x^2+px+q\right)^k} \text{ with } p^2<4q \text{ and } A\ne 0\\
  \int\frac{A}{x-x_1}\diff x &= A\ln\left|x-x_1\right|+C\\
  \int\frac{A}{\left(x-x_1\right)^k}\diff x
    &= -\frac{A}{(k-1)\left(x-x_1\right)^{k-1}}+C\\
  \int\frac{\diff x}{x^2+px+q}
    &= \frac{2}{\sqrt{4q-p^2}}\cdot\arctan\frac{2x+p}{\sqrt{4q-p^2}}+C\\
  \int\frac{Ax+B}{x^2+px+q}\diff x
    &= \frac{A}{2}\ln\left|x^2+px+q\right| +\frac{2B-Ap}{\sqrt{4q-p^2}}
    \cdot\arctan\frac{2x+p}{\sqrt{4q-p^2}}+C
\end{align*}
```

09-17-12

$$\frac{A}{x - x_1}, \quad \frac{A}{(x - x_1)^k}, \quad \frac{Ax + b}{x^2 + px + q}, \quad \frac{Ax + b}{(x^2 + px + q)^k} \quad \text{with } p^2 < 4q \text{ and } A \neq 0$$

$$\int \frac{A}{x - x_1} \, dx = A \ln |x - x_1| + C$$

$$\int \frac{A}{(x - x_1)^k} \, dx = -\frac{A}{(k - 1)(x - x_1)^{k-1}} + C$$

$$\int \frac{dx}{x^2 + px + q} = \frac{2}{\sqrt{4q - p^2}} \cdot \arctan \frac{2x + p}{\sqrt{4q - p^2}} + C$$

$$\int \frac{Ax + B}{x^2 + px + q} \, dx = \frac{A}{2} \ln \left| x^2 + px + q \right| + \frac{2B - Ap}{\sqrt{4q - p^2}} \cdot \arctan \frac{2x + p}{\sqrt{4q - p^2}} + C$$

Placing horizontal arrows between individual parts of equations is possible with practically all environments. This is not the case for vertical arrows, however; doing this requires the \ArrowBetweenLines command from mathtools.

| \ArrowBetweenLines * [arrow type] |
|---|

The starred version lets you place the arrow at the right hand side instead of the default left hand side.

```
\usepackage{mathtools,esint} \newcommand*\diff{\mathop{}\!\mathrm{d}}
\begin{align*}
\iiint\limits_Rf(x,y,z)\diff R
 &= \int\limits_{z_1}^{z_2}
   \int\limits_{\varphi_1(z)}^{\varphi_2(z)}
   \int\limits_{r_1(\varphi,z)}^{r_2(\varphi,z)}
       F(r,\varphi,z)r\diff r\diff\varphi\diff z\\
%\ArrowBetweenLines[\Downarrow]% left hand side
\ArrowBetweenLines*            % right hand side (\Updownarrow is the default)
 &= \int\limits_{\varphi_1(z)}^{\varphi_2(z)}
   \int\limits_{\vartheta_1(\varphi)}^{\vartheta_2(\varphi)}
   \int\limits_{r_1(\vartheta,\varphi)}^{r_2(\vartheta,\varphi)}
       F(r,\vartheta,\varphi)r^2\sin\vartheta\diff r\,r\diff\vartheta\diff\varphi
\end{align*}
```

09-17-13

$$\iiint_R f(x, y, z) \, dR = \int_{z_1}^{z_2} \int_{\varphi_1(z)}^{\varphi_2(z)} \int_{r_1(\varphi,z)}^{r_2(\varphi,z)} F(r, \varphi, z) r \, dr \, d\varphi \, dz$$

$$\updownarrow$$

$$= \int_{\varphi_1(z)}^{\varphi_2(z)} \int_{\vartheta_1(\varphi)}^{\vartheta_2(\varphi)} \int_{r_1(\vartheta,\varphi)}^{r_2(\vartheta,\varphi)} F(r, \vartheta, \varphi) r^2 \sin \vartheta \, dr \, r \, d\vartheta \, d\varphi$$

## 9.17.5  Delimiters

The various ways of creating left-right pairs of parentheses of the same size were described in depth in Section 4.5 on page 38. Especially in complex equation terms, it can be difficult to match

\left...\right pairs. To make this easier, mathtools lets you combine such a pair into one command.

\DeclarePairedDelimiter{[*command*]}{*left*}{*right*}

If for example \DeclarePairedDelimiter\abs\lvert\rvert is defined, a starred version and an optional argument for \abs are defined internally. The starred version automatically uses the \left-\right commands. Alternatively one of \big, \Big, \bigg, or \Bigg can be chosen through the optional argument.

$$|a| \ \left|\frac{a}{b}\right| \ \left|\frac{a}{b}\right|$$
$$\left|\frac{a}{b}\right| \ \left|\frac{a}{b}\right|$$

```
\usepackage{mathtools}
\DeclarePairedDelimiter\abs{\lvert}{\rvert}

\[ \abs{a} ~ \abs{\frac{a}{b}} ~ \abs*{\frac{a}{b}} \]
\[ \abs[\big]{\frac{a}{b}} ~ \abs[\bigg]{\frac{a}{b}} \]
```

09-17-14

## 9.18 relsize — sequences of symbols

Double integrals or double sums frequently occur in math essays.

```
\usepackage{amsmath}
\[ \sum_{i=n}^ma_i\cdot\sum_{j=l}^kb_j = \sum_{i=n}^m\sum_{j=l}^ka_ib_j \]
```

$$\sum_{i=n}^{m} a_i \cdot \sum_{j=l}^{k} b_j = \sum_{i=n}^{m}\sum_{j=l}^{k} a_i b_j$$

09-18-1

The two consecutive sum symbols have the same size by default. To slightly enlarge the first symbol to express the math dependency better, it could be scaled with the \scalebox command from the graphicx package. To do this, it is best to define a \Sum command with an optional parameter for the scaling:

```
\usepackage{amsmath,graphicx}
\newcommand*\Sum[1][1]{\ensuremath\mathop{\scalebox{#1}{$\displaystyle\sum$}}}
\begin{align}
 \Sum_{i=n}^ma_i\cdot\sum_{j=l}^kb_j &= \Sum[1.2]_{i=n}^m\Sum_{j=l}^ka_ib_j
\end{align}
```

$$\sum_{i=n}^{m} a_i \cdot \sum_{j=l}^{k} b_j = \sum_{i=n}^{m}\sum_{j=l}^{k} a_i b_j \tag{9.31}$$

09-18-2

Another way is to use the relsize package, which defines a \mathlarger command that allows larger characters depending on the used math character set.

```
\usepackage{mathptmx,relsize}
\[ \sum_{i=n}^ma_i\cdot\sum_{j=l}^kb_j=\mathlarger{\sum}_{i=n}^m\sum_{j=l}^ka_ib_j \]
```

$$\sum_{in}^{m} a_i \cdot \sum_{jl}^{k} b_j \ \sum_{in}^{m}\sum_{jl}^{k} a_i b_j$$

09-18-3

## 9.19 **xypic** — simple graphics

The \xymatrix command is defined by the xypic package by Kristoffer H. Rose, which is part of the larger xy package. Similar to the amscd package, it can be used to create commutative diagrams. Once they reach a certain complexity, it becomes very difficult to understand the source code (see second example).

09-19-1

$$
\begin{array}{ccc}
A & B & C \\
 & & \\
D & E & F \\
 & & \\
G & H & I
\end{array}
\qquad (9.32)
$$

```
\usepackage{xypic}

\begin{equation}
\xymatrix{%
A\POS[];[d]**\dir{~},[];[dr]**\dir{-} & B & C\\
D & E\POS[];[l]**\dir{.},[];[r]**\dir{~}
  & F\POS[];[dl]**\dir{~}\\
G & H & I}
\end{equation}
```

```
\usepackage[all,line]{xy} \usepackage{graphicx}
\CompileMatrices
\xymatrix{
 & 0 \ar[rr] & & A^p \ar@{^{(}->}[rr]^{f^p} \ar[ld]_{d_A^p}
  \ar'[d][dd]^{\alpha^p} & & B^p \ar@{>>}[rr]^{g^p} \ar[ld]_{d_B^p}
  \ar@/^/@{.>}[ld] \ar'[d][dd]^{\beta^p} & & C^p \ar[rr] \ar[ld]_{d_C^p}
  \ar@/^/@{.>}[ll] \ar'[d][dd]^{\gamma^p} \ar@/_/@{.>}[dd] & & 0 \\
0 \ar[rr] & & A^{p+1} \ar@{^{(}->}[rr]^(.7){f^{p+1}}
  \ar[dd]^(.3){\alpha^{p+1}} \ar@/_/@{.>}[dd] & & B^{p+1}
  \ar@{>>}[rr]^(.7){g^{p+1}} \ar@/^/@{.>}[ll]
  \ar[dd]^(.3){\beta^{p+1}} & & C^{p+1} \ar[rr]
  \ar[dd]^(.3){\gamma^{p+1}} & & 0 & \\
 & 0 \ar'[r][rr] & & A_1^p \ar[ld]_{d_{A_{1}}^p}
  \ar@{^{(}->}'[r]^{f_1^{p}}[rr] & & B_1^p
  \ar[ld]_{d_{B_1}^p} \ar@{>>}'[r]^{g_1^{p}}[rr]
  \ar@/^/@{.>}[ld] & & C_1^p \ar[rr]
  \ar[ld]_{d_{C_1}^p} \ar@/^/@{.>}[ll] & & 0 \\
0 \ar[rr] & & A_1^{p+1} \ar@{^{(}->}[rr]^{f_1^{p+1}} & & B_1^{p+1}
  \ar@{>>}[rr]^{g_1^{p+1}} \ar@/^/@{.>}[ll] & & C_1^{p+1} \ar[rr] & &
  0 & }
```

09-19-2

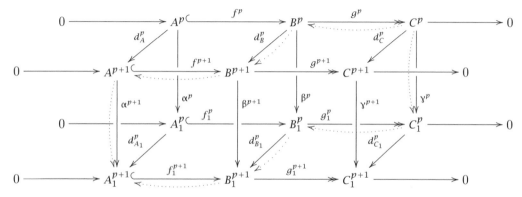

## 9.20 **xlop** — visualisation of the elementary arithmetic operations

The xlop package by Jean-Côme Charpentier lets you typeset simple additions and multiplications as they are taught in primary school. The default output can be changed through national package options.

$2.8 + 1.2 = 4$
$02.8 + 1.2 = 04.0$

$$\begin{array}{r} 1\;_{12.13}\;_{14} \\ - \phantom{0} \\ \;_{10}\;_{15.16}\;7 \\ \hline 0\;\;6.6\;\;7 \end{array}$$

```
\usepackage{xlop}

\opadd[style=text]{02.8}{1.2}\par
\opadd[style=text,deletezero=false]{02.8}{1.2}\par
\opsub[carrysub,lastcarry,columnwidth=2.5ex,
    offsetcarry=-0.4,decimalsepoffset=-3pt,
    deletezero=false]{12.34}{5.67}
```

09-20-1

```
\usepackage{xlop}
\opmul[displayshiftintermediary=shift]{435}{1001205}\hfill
\opmul[displayshiftintermediary=all]{435}{1001205}\hfill
\opmul[displayshiftintermediary=none]{435}{1001205}
```

09-20-2

```
\usepackage{xlop}
\opdiv{25}{7}
```

09-20-3

```
\usepackage{xlop}
\opcopy{0.8}{a}\opcopy{-17}{n}%
\oppower{a}{n}{r}$\opprint{a}^{\opprint{n}} = \opprint{r}$
```

$$0.8^{-17} = 44.408920985006261616945266723632815$$

09-20-4

# Examples

The examples given in this chapter are mostly based on personal experience in areas of math education and questions on the newsgroups news:comp.text,tex and news:de.comp.text.tex. They are not necessarily original or complete.

## 10.1  Matrix

### 10.1.1  Identity matrix

There are several ways of creating an identity matrix. In the following example this is done with the default array environment.

10-01-1

$$
\left(\begin{array}{ccccc}
1 & & & & \\
& 1 & & 0 & \\
& & 1 & & \\
0 & & & 1 & \\
& & & & 1
\end{array}\right)
$$

```
\[ \left(\begin{array}{ccccc}
1                                      \\
   & 1 &   & \parbox[b][0pt][b]{0pt}{\huge0}\\
   &   & 1                             \\
\parbox[b][0pt][c]{0pt}{\huge0} & & & 1    \\
   & & & & 1\end{array}\right) \]
```

### 10.1.2 Linear systems of equations

This example makes good use of the customisation of columns and the spacing between them.

```
\[ \begin{array}{l@{\,=\,}*{5}{l@{\,+\,}}l}
y_1 & x_{11} & x_{12} & x_{13} & \dots & x_{1(n-1)} & x_{1n} \\
y_2 & x_{21} & x_{22} & x_{23} & \dots & x_{2(n-1)} & x_{2n} \\
\ \vdots &\ \vdots &\ \vdots &\ \vdots &\ \ddots &\ \vdots &\ \vdots\\
y_{n-1} & x_{(n-1)1} & x_{(n-1)2} & x_{(n-1)3} & \dots & x_{(n-1)3} & x_{(n-1)n}\\
y_n & x_{n1} & x_{n2} & x_{n3} & \dots & x_{(n-1)(n-1)} & x_{nn}
\end{array} \]
```

$$
\begin{array}{l@{\,=\,}*{5}{l@{\,+\,}}l}
y_1 & x_{11} & x_{12} & x_{13} & \dots & x_{1(n-1)} & x_{1n} \\
y_2 & x_{21} & x_{22} & x_{23} & \dots & x_{2(n-1)} & x_{2n} \\
\vdots & \vdots & \vdots & \vdots & \ddots & \vdots & \vdots \\
y_{n-1} & x_{(n-1)1} & x_{(n-1)2} & x_{(n-1)3} & \dots & x_{(n-1)3} & x_{(n-1)n} \\
y_n & x_{n1} & x_{n2} & x_{n3} & \dots & x_{(n-1)(n-1)} & x_{nn}
\end{array}
$$

10-01-2

The individual parts of the definition `{l@{\,=\,}*{5}{l@{\,+\,}}l}` are explained below:

| | |
|---|---|
| `l` | left-aligned column |
| `@{\,=\,}` | the column separator is an equals sign; before and after it a `\thinmuskip` |
| `*{5}` | repeat the following column definition five times |
| `{l@{\,+\,}}` | the column separator is a plus sign; before and after it a `\thinmuskip` |
| `l` | left-aligned column |

### 10.1.3 Matrix with additional comments

This example uses the `\rotatebox` command from the `rotating` package (or `graphicx`, which loads it automatically). For LaTeX and TeX as a typesetting engine, there is no difference between a table and a matrix. Therefore they can be chained without any problem in order to create complex arrangements. The `\xleftarrow` command requires the `amsmath` package.

$$
\begin{bmatrix}
X_x & Y_x & Z_x & T_x \\
X_y & Y_y & Z_y & T_y \\
X_z & Y_z & Z_z & T_z \\
0 & 0 & 0 & 1
\end{bmatrix}
$$

```
\usepackage{rotating,amsmath}
\newcommand*\rb[1]{\rotatebox{90}{$\xleftarrow{#1}$}}

\begin{tabular}{c}
$\begin{matrix}
  \rb{text1}&\rb{text1}&\rb{text1}&\rb{text1}\\
  \end{matrix}$\\
$\begin{bmatrix}
  X_x & Y_x & Z_x & T_x\\ X_y & Y_y & Z_y & T_y \\
  X_z & Y_z & Z_z & T_z\\ 0  & 0  & 0  & 1
\end{bmatrix}$
\end{tabular}
```

10-01-3

### 10.1.4 Symbolic diagrams with psmatrix

Symbolic or commutative diagrams can be created with various packages. The two examples given here use the `pst-node` and the `pstricks-add` packages from the `pstricks` set of packages.

```
\usepackage{amsmath,amssymb,amsopn,pst-node}
\DeclareMathOperator{\Hom}{Hom}
\DeclareMathOperator{\Mod}{Mod}
\DeclareMathOperator{\obj}{obj}
Let $\mathcal{C}:=\mathcal{D}:=\Mod_A$, $M\in
    \obj(\Mod_A)$ be fixed,
\begin{align*}
    F:=\square\otimes_A M : \Mod_A &\rightarrow \Mod_A, \\
    G:=\Hom_A(M,\square)  : \Mod_A &\rightarrow \Mod_A.
\end{align*}
These two functors are adjoint, by
$h(X,Y)(\phi)(x)(m)\rightarrow\phi(x\otimes m)$:\\[10pt]
$\begin{psmatrix}[colsep=-0.3cm, rowsep=0pt]
  \quad\phi{\;} & & {\;} h(X,Y)(\phi)\quad\\
  \Hom_A(M \otimes_A X, Y) & & \Hom_A(X, \Hom_A(M, Y)) \\[1.5cm]
      & L_A^2(M, X; Y)
\end{psmatrix}$
\ncline{|->}{1,1}{1,3}
\ncline[doubleline=true]{2,1}{2,3}\nbput{h(X,Y)}
\naput[nrot=:U,labelsep=0pt]{$\sim$}
\ncline[doubleline=true]{2,1}{3,2}\naput[nrot=:U,labelsep=0pt]{$\sim$}
\ncline[doubleline=true]{3,2}{2,3}\naput[nrot=:U,labelsep=0pt]{$\sim$}
```

Let $C := \mathcal{D} := \mathrm{Mod}_A$, $M \in \mathrm{obj}(\mathrm{Mod}_A)$ be fixed,

$$F := \square \otimes_A M : \mathrm{Mod}_A \rightarrow \mathrm{Mod}_A,$$
$$G := \mathrm{Hom}_A(M, \square) : \mathrm{Mod}_A \rightarrow \mathrm{Mod}_A.$$

These two functors are adjoint, by $h(X,Y)(\phi)(x)(m) \rightarrow \phi(x \otimes m)$:

```
\usepackage{pstricks-add}
\psset{arrowsize=8pt,arrowlength=1,linewidth=1pt,nodesep=2pt,shortput=tablr}
\large\begin{psmatrix}[colsep=12mm,rowsep=10mm]
        &   & $R_2$              \\
        &   & 0    &   & $R_3$\\
$e_b:S$ & 1 &      & 1 & 0     \\
        &   & 0                  \\
        &   & $R_1$
\end{psmatrix}
\ncline{h-}{1,3}{2,3}<{$e_{r2}$}>{$f_{r2}$}
\ncline{-h}{2,3}{3,2}<{$e_1$}
```

10-01-4

```
\ncline{-h}{3,1}{3,2}^{$e_s$}_{$f_{s}$}
\ncline{-h}{3,2}{4,3}>{$e_3$}<{$f_3$}
\ncline{-h}{4,3}{3,4}>{$e_4$}<{$f_4$}
\ncline{-h}{3,4}{2,3}>{$e_2$}<{$f_2$}
\ncline{-h}{3,4}{3,5}^{$e_5$}
\ncline{-h}{3,5}{2,5}<{$e_{r3}$}>{$f_{r3}$}
\ncline{-h}{4,3}{5,3}<{$e_{r1}$}>{$f_{r1}$}
```

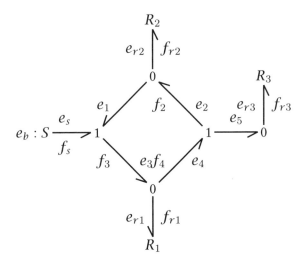

10-01-5

## 10.2 Case distinctions

In cases where the amsmath package cannot be used, the question is how to typeset case distinctions. Usually it's best to use an array environment, with a bit of optimising to improve the horizontal spacing. In the following example two array environments, a $3 \times 1$ followed by a $3 \times 3$, are chained and the column spacing is reduced by 3pt. The vertical alignment of the final column is improved with a \phantom command; it takes up the space its argument would take up without actually outputting it.

```
\begin{equation}\addtolength{\arraycolsep}{-3pt}
I(z)=\delta_{0}\left\{\begin{array}{lcrcl}
  D+z & \quad & -D & \le z\le & -p\\
  D-\frac{1}{2}\left(p-\frac{z^{2}}{p}\right)
      & \quad & -p & \le z\le & \phantom{-}p\\
  D-z & \quad & p & \le z\le & \phantom{-}D
\end{array}\right.
\end{equation}
```

$$I(z) = \delta_0 \begin{cases} D + z & -D \le z \le -p \\ D - \frac{1}{2}\left(p - \frac{z^2}{p}\right) & -p \le z \le p \\ D - z & p \le z \le D \end{cases} \tag{10.1}$$

10-02-1

To number the individual rows of the array environment is more difficult — LaTeX was not designed to be able to do this. But because TeX is a complete programming language, this restriction can be rectified with some programming.

Another way of typesetting cases is to use a `flalign` environment:

```
\usepackage{amsmath,multirow,bigdelim,calc}
\begin{tabular}{rc}
\ldelim\{{2}{1.5cm}[text --- text] & \parbox{\linewidth-2cm-4\tabcolsep}{\vspace*{1ex}
  \begin{flalign*}
    x & = 2\quad\text{if }y >2 & \\ x & = 3\quad\text{if }y \le 2 &
  \end{flalign*}}
\end{tabular}
```

10-02-2

$$\text{text} - \text{text}\begin{cases} x = 2 & \text{if } y > 2 \\ x = 3 & \text{if } y \le 2 \end{cases}$$

A simpler solution can be found in Section 9.9 on page 174, using the `cases` package, which also allows for numbering subequations in cases. Alternatively, you can use the `mathtools` package (cf. Section 9.17 on page 185); it also provides an extended `cases` environment.

## 10.3  Arrays

Large math units should always be split up into smaller ones, but then horizontal and vertical alignment can become a problem. The `array` environment makes it possible to align almost arbitrarily complex units such that the result corresponds to the expectations. The following examples show this for various uses. The source code quite often becomes hard to follow; we have tried to structure it formally here.

```
\usepackage{amsmath}
\[
\begin{array}{rcll}
y & = & x^{2}+bx+c\\
  & = & x^{2}+2\times\dfrac{b}{2}x+c\\
  & = & \underbrace{x^{2}+2\times\dfrac{b}{2}x+\left(\frac{b}{2}\right)^{2}}-
      {\left(\dfrac{b}{2}\right)^{2}+c}\\
  &   & \qquad\left(x+{\dfrac{b}{2}}\right)^{2}\\
  & = & \left(x+\dfrac{b}{2}\right)^{2}-\left(\dfrac{b}{2}\right)^{2}+c
  & \left|+\left({\dfrac{b}{2}}\right)^{2}-c\right.\\
    y+\left(\dfrac{b}{2}\right)^{2}-c & = & \left(x+
    \dfrac{b}{2}\right)^{2} & \left|\strut(\textrm{vertex form})\right.\\
y-y_{S} & = & (x-x_{S})^{2}\\
S(x_{S};y_{S}) & \,\textrm{or}\,,
    & S\left(-\dfrac{b}{2};\,\left(\dfrac{b}{2}\right)^{2}-c\right)
\end{array}
\]
```

$$
\begin{array}{rcl}
y & = & x^2 + bx + c \\[2mm]
& = & x^2 + 2 \times \dfrac{b}{2}x + c \\[2mm]
& = & x^2 + 2 \times \dfrac{b}{2}x + \left(\dfrac{b}{2}\right)^2 - \left(\dfrac{b}{2}\right)^2 + c \\[2mm]
& & \underbrace{\phantom{x^2 + 2 \times \dfrac{b}{2}x + \left(\dfrac{b}{2}\right)^2}}_{\left(x + \frac{b}{2}\right)^2} \\[4mm]
& = & \left(x + \dfrac{b}{2}\right)^2 - \left(\dfrac{b}{2}\right)^2 + c \qquad\qquad \left|\; + \left(\dfrac{b}{2}\right)^2 - c \right. \\[3mm]
y + \left(\dfrac{b}{2}\right)^2 - c & = & \left(x + \dfrac{b}{2}\right)^2 \qquad\qquad\qquad\quad \left|\; \text{(vertex form)} \right. \\[3mm]
y - y_S & = & (x - x_S)^2 \\[2mm]
S(x_S; y_S) & \text{or} & S\left(-\dfrac{b}{2}; \left(\dfrac{b}{2}\right)^2 - c\right)
\end{array}
$$

```
\usepackage{amsmath}
\begin{equation}
\begin{array}{rcl}
\underline{RS} & = & \left(\begin{array}{*8c}
01 & a4 & 55 & 87 & 5a & 58 & db & 9e\\ a4 & 56 & 82 & f3 & 1e & c6 & 68 & e5\\
02 & a1 & fc & c1 & 47 & ae & 3d & 19\\ a4 & 55 & 87 & 5a & 58 & db & 9e & 03
\end{array}\right)\\~\\
\left(\begin{array}{c}  s_{i,0}\\  s_{i,1}\\  s_{i,2}\\  s_{i,3}
\end{array}\right) & = & \underline{RS} \times
\left(\begin{array}{c}  m_{8i+0}\\  m_{8i+1}\\  \cdots\\  m_{8i+6}\\  m_{8i+7}
\end{array}\right)\\~\\
  S_{i} & = & \sum_{j=0}^{3}s_{i,j}\times2^{8j}\qquad i=0,1,\ldots,k-1\\
  S & = & \left(S_{k-1},S_{k-2},\ldots,S_{1},S_{0}\right)
\end{array}
\end{equation}
```

$$
\underline{RS} \;=\; \begin{pmatrix}
01 & a4 & 55 & 87 & 5a & 58 & db & 9e \\
a4 & 56 & 82 & f3 & 1e & c6 & 68 & e5 \\
02 & a1 & fc & c1 & 47 & ae & 3d & 19 \\
a4 & 55 & 87 & 5a & 58 & db & 9e & 03
\end{pmatrix}
$$

$$
\begin{pmatrix} s_{i,0} \\ s_{i,1} \\ s_{i,2} \\ s_{i,3} \end{pmatrix}
\;=\; \underline{RS} \times
\begin{pmatrix} m_{8i+0} \\ m_{8i+1} \\ \cdots \\ m_{8i+6} \\ m_{8i+7} \end{pmatrix}
\tag{10.2}
$$

$$
\begin{aligned}
S_i &= \textstyle\sum_{j=0}^{3} s_{i,j} \times 2^{8j} \qquad i = 0,1,...,k-1 \\
S &= (S_{k-1}, S_{k-2}, ..., S_1, S_0)
\end{aligned}
$$

10-03-3

$$\lim_{n->\infty} q^n = \begin{cases} \text{divergent} & q \le -1 \\ 0 & |q| < 1 \\ 1 & q = 1 \\ \infty & q > 1 \end{cases}$$

```
$\lim\limits_{n->\infty}q^{n}=\left\{%
  \begin{array}{lc@{\kern2pt}c@{\kern2pt}r}
    \textrm{divergent}\  & q & \le & -1\\
        0 & |q| & < & 1\\
        1 & q  & = & 1\\
        \infty  & q & > & 1
  \end{array}\right.$
```

```
\[ \left(
\begin{array}{c@{}c@{}c}
  \begin{array}{|cc|}\hline
    a_{11} & a_{12} \\a_{21} & a_{22} \\\hline
  \end{array} & 0 & 0 \\
  0 & \begin{array}{|ccc|}\hline
        b_{11} & b_{12} & b_{13}\\
        b_{21} & b_{22} & b_{23}\\
        b_{31} & b_{32} & b_{33}\\\hline
      \end{array} & 0 \\
0 & 0 & \begin{array}{|cc|}\hline
        c_{11} & c_{12} \\  c_{21} & c_{22} \\\hline
      \end{array}
\end{array}
\right) \]
```

10-03-4

$$\left( \begin{array}{ccc} \boxed{\begin{array}{cc} a_{11} & a_{12} \\ a_{21} & a_{22} \end{array}} & 0 & 0 \\ 0 & \boxed{\begin{array}{ccc} b_{11} & b_{12} & b_{13} \\ b_{21} & b_{22} & b_{23} \\ b_{31} & b_{32} & b_{33} \end{array}} & 0 \\ 0 & 0 & \boxed{\begin{array}{cc} c_{11} & c_{12} \\ c_{21} & c_{22} \end{array}} \end{array} \right)$$

10-03-5

$$Y = \begin{bmatrix} 0 & 0 & 1 & 0 \\ 1 & 0 & 1 & 0 \\ 1 & 1 & 1 & 1 \\ \hline 2 & 1 & 3 & 1 \end{bmatrix}$$

```
\[ Y= \begin{array}{c}
  \left[\begin{array}{rrrr}
   0 & 0 & 1 & 0\\
   1 & 0 & 1 & 0\\
   1 & 1 & 1 & 1
  \end{array}\right]\\[3ex]\hline
  \begin{array}{rrrr}
   2 & 1 & 3 & 1
  \end{array}
  \end{array} \]
```

## 10.4 Horizontal braces

A horizontal brace under the radicand of a root can be created as follows:

$$z = \sqrt{\underbrace{x^2 + y^2}_{=z^2}}$$

```
\[ z=\sqrt{\underbrace{x^2+y^2}_{\scriptstyle=z^2}}\]
\[ z=\underbrace{\sqrt{x^2+y^2}}_{\scriptstyle=z^2}\]
```
10-04-1

$$z = \underbrace{\sqrt{x^2 + y^2}}_{=z^2}$$

However, neither of these outputs are what is wanted — a brace under the radicand (but not the root symbol) that does not influence the size of the root symbol. So instead use a box of size zero; here \makebox and \widthof (which requires the calc package) are used:

$$z = \sqrt{\underbrace{x^2 + y^2}_{=z^2}}$$

```
\usepackage{calc}

\[
  z =\;\;\underbrace{%
   \makebox[\widthof{~$x^2+y^2$}][r]{%
    $\sqrt{x^2+y^2}$}}_{\scriptstyle=z^2}
\]
```
10-04-2

Overlapping braces such as 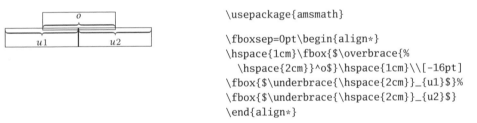 require some tricks. Each one of the three braces is a separate box in the TEX sense and cannot easily be pushed into another box. However, the boxes in the above arrangement look like this:

```
\usepackage{amsmath}

\fboxsep=0pt\begin{align*}
\hspace{1cm}\fbox{$\overbrace{%
  \hspace{2cm}}^o$}\hspace{1cm}\\[-16pt]
\fbox{$\underbrace{\hspace{2cm}}_{u1}$}%
\fbox{$\underbrace{\hspace{2cm}}_{u2}$}
\end{align*}
```
10-04-3

The trick is to insert a negative line feed after the first (top) brace. This does not change even when a math expression is inserted between the braces. The following math reformulation (quadratic explement) uses this trick. Thus it is possible to let \underbrace and \overbrace overlap. Additionally, the individual braces have been typeset in different colours.

```
\usepackage{amsmath,color}
\begin{align}
y &= 2x^2 -3x +5\nonumber\\
  & \hphantom{= \ 2\left(x^2-\frac{3}{2}\,x\right. }%
    \textcolor{blue}{%
      \overbrace{\hphantom{+\left(\frac{3}{4}\right)^2- %
        \left(\frac{3}{4}\right)^2}}^{=0}}\nonumber\\[-11pt]
  &= 2\left(\textcolor{red}{%
    \underbrace{x^2-\frac{3}{2}\,x + \left(\frac{3}{4}\right)^2}}
    \underbrace{-\left(\frac{3}{4}\right)^2 + \frac{5}{2}}\right)\\
```

```
    &= 2\left(\qquad\textcolor{red}{\left(x-\frac{3}{4}\right)^2}
       \qquad + \ \frac{31}{16}\qquad\right)\nonumber\\
  y\textcolor{blue}{{}-\frac{31}{8}}
    &= 2\left(x\textcolor{cyan}{{}-\frac{3}{4}}\right)^2\nonumber\\
    &\Rightarrow S\left(\frac{3}{4}\right|\left.\frac{31}{8}\right)\nonumber
\end{align}
```

`10-04-4`

$$y = 2x^2 - 3x + 5$$

$$= 2\left(x^2 - \frac{3}{2}x + \overbrace{\left(\frac{3}{4}\right)^2 - \left(\frac{3}{4}\right)^2}^{=0} + \frac{5}{2}\right) \tag{10.3}$$

$$= 2\left(\underbrace{\left(x - \frac{3}{4}\right)^2} + \frac{31}{16}\right)$$

$$y - \frac{31}{8} = 2\left(x - \frac{3}{4}\right)^2$$

$$\Rightarrow S\left(\frac{3}{4}\middle|\frac{31}{8}\right)$$

If individual parts of a formula have different heights, the vertical alignment of several \underbrace can be a problem.

```
\usepackage{amsmath}
\begin{equation}
  \binom{x_R}{y_R} = \underbrace{r}_{\text{scaling}}\times%
    \underbrace{\begin{pmatrix}
              \sin\gamma & -\cos\gamma \\ \cos\gamma & \sin\gamma \\
              \end{pmatrix}}_{\text{rotation}}
  \binom{x_K}{y_K} + \underbrace{\binom{t_x}{t_y}}_{\text{translation}}
\end{equation}
```

`10-04-5`

$$\begin{pmatrix}x_R\\y_R\end{pmatrix} = \underbrace{r}_{\text{scaling}} \times \underbrace{\begin{pmatrix}\sin\gamma & -\cos\gamma\\\cos\gamma & \sin\gamma\end{pmatrix}}_{\text{rotation}}\begin{pmatrix}x_K\\y_K\end{pmatrix} + \underbrace{\begin{pmatrix}t_x\\t_y\end{pmatrix}}_{\text{translation}} \tag{10.4}$$

The lack of vertical alignment of the first brace with the other two does not look good. This can be improved relatively easy though, by using \vphantom to achieve equal heights:

```
\usepackage{amsmath}
\begin{equation}
  \binom{x_R}{y_R} = \underbrace{r\vphantom{\binom{A}{B}}}_{\text{scaling}}\times%
    \underbrace{\begin{pmatrix}
              \sin\gamma & -\cos\gamma \\ \cos\gamma & \sin\gamma \\
              \end{pmatrix}}_{\text{rotation}}
  \binom{x_K}{y_K} + \underbrace{\binom{t_x}{t_y}}_{\text{translation}}
\end{equation}
```

`10-04-6`

$$\begin{pmatrix}x_R\\y_R\end{pmatrix} = \underbrace{r}_{\text{scaling}} \times \underbrace{\begin{pmatrix}\sin\gamma & -\cos\gamma\\\cos\gamma & \sin\gamma\end{pmatrix}}_{\text{rotation}}\begin{pmatrix}x_K\\y_K\end{pmatrix} + \underbrace{\begin{pmatrix}t_x\\t_y\end{pmatrix}}_{\text{translation}} \tag{10.5}$$

The remaining point to improve is the horizontal spacing between parts of the equation, which are pulled apart by the long first and last braces. The spacing can be brought back to normal by inserting a negative space through \hspace or the TEX version \kern:

```
\usepackage{amsmath}
\begin{equation}
  \binom{x_R}{y_R} =
    \kern-10pt\underbrace{r\vphantom{\binom{A}{B}}}_{\text{scaling}}\kern-7pt
    \times\underbrace{\begin{pmatrix}
                      \sin\gamma & -\cos\gamma \\ \cos\gamma & \sin\gamma \\
                      \end{pmatrix}}_{\text{rotation}}
  \binom{x_K}{y_K} + \kern-5pt\underbrace{\binom{t_x}{t_y}}_{\text{translation}}
\end{equation}
```

$$\begin{pmatrix} x_R \\ y_R \end{pmatrix} = r \times \underbrace{\begin{pmatrix} \sin\gamma & -\cos\gamma \\ \cos\gamma & \sin\gamma \end{pmatrix}}_{\text{rotation}} \begin{pmatrix} x_K \\ y_K \end{pmatrix} + \underbrace{\begin{pmatrix} t_x \\ t_y \end{pmatrix}}_{\text{translation}} \qquad (10.6)$$

with scaling under $\begin{pmatrix} x_R \\ y_R \end{pmatrix}$

10-04-7

The following example is another relatively complex arrangement of horizontal braces. Without manipulating the source code, we get the following form:

```
\usepackage{amsmath}
\newcommand*\num[1]{\hphantom{#1}} \newcommand*\vsp{\vphantom{\rangle_1}}
\begin{equation*}
  \frac{300}{5069} \underbrace{\longmapsto}_{%
    \substack{\Delta a=271\num9\vsp \\[2pt]
              \Delta b=4579\vsp\\[2pt]
              \text{$1$ iteration}}} \frac{29}{490}%
  \underbrace{\longmapsto \frac{19}{321}\longmapsto}_{%
    \substack{\Delta a=10\num{9}=\langle271\rangle_{29}\num{20}\\[2pt]
              \Delta b=169=\langle4579\rangle_{490}\\[2pt]
              \text{$2$ iterations}}}\frac{9}{152}
  \underbrace{\longmapsto \frac{8}{135}\longmapsto\dots\longmapsto}_{%
    \substack{\Delta a=1\num{7}=\langle10\rangle_{9}\num{119}\\[2pt]
              \Delta b=17=\langle169\rangle_{152}\\[2pt]
              \text{$8$ iterations}}}\frac{1}{16}
  \underbrace{\longmapsto\dots\longmapsto}_{%
    \substack{\Delta a=0=\langle1\rangle_{1}\num{76} \\[2pt]
              \Delta b=1=\langle17\rangle_{16} \\[2pt]
              \text{$8$ iterations}}}\frac{1}{1}
\end{equation*}
```

$$\frac{300}{5069} \underset{\substack{\Delta a=271 \\ \Delta b=4579 \\ 1\text{ iteration}}}{\longmapsto} \frac{29}{490} \underset{\substack{\Delta a=10=\langle271\rangle_{29} \\ \Delta b=169=\langle4579\rangle_{490} \\ 2\text{ iterations}}}{\longmapsto} \frac{19}{321} \longmapsto \frac{9}{152} \underset{\substack{\Delta a=1=\langle10\rangle_9 \\ \Delta b=17=\langle169\rangle_{152} \\ 8\text{ iterations}}}{\longmapsto} \frac{8}{135} \longmapsto \cdots \longmapsto \frac{1}{16} \underset{\substack{\Delta a=0=\langle1\rangle_1 \\ \Delta b=1=\langle17\rangle_{16} \\ 8\text{ iterations}}}{\longmapsto} \cdots \longmapsto \frac{1}{1}$$

10-04-8

Again the layout is not optimal. However, we can improve the horizontal spacing of the elements with the \mathclap command (defined in Section 6.12 on page 108, part of the mathtools package) and the vertical alignment of the braces with \vphantom once again. The advantage of \mathclap over \makebox is that you don't need to leave math mode.

```
\usepackage{mathtools}
\newcommand*\num[1]{\hphantom{#1}} \newcommand*\vsp{\vphantom{\rangle_1}}
\begin{equation*}
  \frac{300}{5069} \underbrace{\longmapsto\vphantom{\frac{1}{1}}}_{%
    \mathclap{\substack{\Delta a=271\num9\vsp \\[2pt]
                        \Delta b=4579\vsp\\[2pt]
                        \text{$1$ iteration}}}}\frac{29}{490}
  \underbrace{\longmapsto \frac{19}{321}\longmapsto}_{%
    \mathclap{\substack{\Delta a=10\num{9}=\langle271\rangle_{29}\num{20}\\[2pt]
                        \Delta b=169=\langle4579\rangle_{490}\\[2pt]
                        \text{$2$ iterations}}}}\frac{9}{152}
  \underbrace{\longmapsto \frac{8}{135}\longmapsto\dots\longmapsto}_{%
    \substack{\Delta a=1\num{7}=\langle10\rangle_{9}\num{119}\\[2pt]
              \Delta b=17=\langle169\rangle_{152}\\[2pt]
              \text{$8$ iterations}}}\frac{1}{16}
  \underbrace{\longmapsto\dots\longmapsto\vphantom{\frac{8}{135}}}_{%
    \substack{\Delta a=0=\langle1\rangle_{1}\num{76} \\[2pt]
              \Delta b=1=\langle17\rangle_{16} \\[2pt]
              \text{$8$ iterations}}}\frac{1}{1}
\end{equation*}
```

10-04-9

## 10.5 Integrals

Correct positioning of limits can be difficult for multiple integrals. These examples have good layouts:

```
\usepackage{eucal,amsmath,esint} \newcommand*\Q[2]{\frac{\partial#1}{\partial#2}}
The \emph{first theorem of Green} is:
\[
  \underset{\mathcal{G}\quad}\iiint\!\left[u\nabla^{2}v+\left(\nabla u,
  \nabla v\right)\right]d^{3}V=\underset{\mathcal{S}\quad}\oiint u\Q{v}{n}d^{2}A
\]
The \emph{second theorem of Green} is:
\[
  \underset{{\mathcal{G}\quad}}\iiint\!\left[u\nabla^{2}v-v\nabla^{2}u\right]d^{3}V
    =\underset{\mathcal{S}\quad}\oiint\left(u\Q{v}{n}-v\Q{u}{n}\right)d^{2}A
\]
```

The *first theorem of Green* is:

$$\iiint\limits_{\mathcal{G}} \left[ u\nabla^2 v + (\nabla u, \nabla v) \right] d^3V = \oiint\limits_{\mathcal{S}} u\frac{\partial v}{\partial n} d^2A$$

The *second theorem of Green* is:

$$\iiint\limits_{\mathcal{G}} \left[ u\nabla^2 v - v\nabla^2 u \right] d^3V = \oiint\limits_{\mathcal{S}} \left( u\frac{\partial v}{\partial n} - v\frac{\partial u}{\partial n} \right) d^2A$$

```
\usepackage{amsmath,esint} \newcommand*\diff{\mathop{}\!\mathrm{d}}
\begin{align*}
\biggl(\int_{-\infty}^\infty e^{-x^2}\,dx\biggr)^2
  &= \int_{-\infty}^\infty\int_{-\infty}^\infty e^{-(x^2+y^2)}\diff x\diff y \\
  &= \int_0^{2\pi}\int_0^\infty e^{-r^2}r\diff r\diff\theta         \\
  &= \int_0^{2\pi}\biggl(-{e^{-r^2}\over2}\bigg\vert_{r=0}^{r=\infty}\,\biggr)\diff\theta\
  &= \pi                                    \tag*{q.e.d.}
\end{align*}
```

$$\left( \int_{-\infty}^\infty e^{-x^2}\, dx \right)^2 = \int_{-\infty}^\infty \int_{-\infty}^\infty e^{-(x^2+y^2)}\, dx\, dy$$

$$= \int_0^{2\pi} \int_0^\infty e^{-r^2} r\, dr\, d\theta$$

$$= \int_0^{2\pi} \left( -\frac{e^{-r^2}}{2} \bigg|_{r=0}^{r=\infty} \right) d\theta$$

$$= \pi \qquad\qquad\qquad \text{q.e.d.}$$

## 10.6 Sum and product

The commands \sum and \prod need some special handling to get the proper size and the correct setting of limits.

$$\prod_{k\geq 0} \frac{1}{(1-q^kz)} = \frac{\sum\limits_{n\geq 0} z^n}{\prod\limits_{1\leq k\leq n}(1-q^k)}$$

```
\usepackage{amsmath}
\let\dst\displaystyle

\begin{align*}
\prod_{k\ge0}\dfrac{1}{(1-q^kz)} &=
    \dfrac{\dst\sum_{n\ge0}z^n}
          {\dst\prod_{1\le k\le n}(1-q^k)}
\end{align*}
```

```
\usepackage{amsmath}
\usepackage[sans]{dsfont}
```

```
\begin{align*}
  \sum_{i=n}^m a_i &= \sum_{j=n}^m a_j \tag{index renaming}\\
  \sum_{i=n}^m a_i &= \sum_{i=n-l}^{m-l} a_{i+l} &\forall\, l\in\mathds{Z}{}
          \tag{index shift},\\
  \sum_{i=n}^m a_i+\sum_{i=m+1}^k a_i &= \sum_{i=n}^k a_i,
                      &\text{if } n\le m<k,\\
  \sum_{i=n}^m a_i+\sum_{i=n}^m b_i &= \sum_{i=n}^m (a_i+b_i),\\
  \sum_{i=n}^m c\times a_i &= c\times\sum_{i=n}^m a_i & \forall\, c\in\mathds{R}{,}\\
  \sum_{i=n}^m a_i \times\sum_{j=l}^k b_j &=
                      \sum_{i=n}^m\sum_{j=l}^k a_i b_j
\end{align*}
```

10-06-2

$$\sum_{i=n}^m a_i = \sum_{j=n}^m a_j \qquad\qquad \text{(index renaming)}$$

$$\sum_{i=n}^m a_i = \sum_{i=n-l}^{m-l} a_{i+l} \qquad \forall\, l \in \mathbb{Z}, \qquad \text{(index shift)}$$

$$\sum_{i=n}^m a_i + \sum_{i=m+1}^k a_i = \sum_{i=n}^k a_i, \qquad \text{if } n \le m < k,$$

$$\sum_{i=n}^m a_i + \sum_{i=n}^m b_i = \sum_{i=n}^m (a_i + b_i),$$

$$\sum_{i=n}^m c \times a_i = c \times \sum_{i=n}^m a_i \qquad \forall\, c \in \mathbb{R},$$

$$\sum_{i=n}^m a_i \times \sum_{j=l}^k b_j = \sum_{i=n}^m \sum_{j=l}^k a_i b_j$$

```
\usepackage{amsmath}
\let\dst\displaystyle
\[ \binom{n}{k} = \dfrac{\dst\prod_{i=1}^n i}
  {\dst\prod_{i=1}^k i\dst\prod_{i=1}^{n-k} i} \]
The Euclidean algorithm: if $a>b$ are integers then $\gcd(a,b)=\gcd(a\bmod b,b)$.
If $\prod_{i=1}^n p^{e_i}_i$ is the prime factorization of $x$ then
\[ S(x)=\sum_{d\vert x}d=\prod_{i=1}^n{p^{e_i+1}_i-1\over p_i-1}. \]
```

10-06-3

$$\binom{n}{k} = \frac{\prod\limits_{i=1}^n i}{\prod\limits_{i=1}^k i \prod\limits_{i=1}^{n-k} i}$$

The Euclidean algorithm: if $a > b$ are integers then $\gcd(a, b) = \gcd(a \bmod b, b)$. If $\prod_{i=1}^n p_i^{e_i}$ is the prime factorization of $x$ then

$$S(x) = \sum_{d|x} d = \prod_{i=1}^n \frac{p_i^{e_i+1} - 1}{p_i - 1}.$$

## 10.7 Vertical alignment

### 10.7.1 Across several paragraphs

Several lines of a displayed equation can be aligned vertically as long as the environment is not left and page breaks have been activated through \allowdisplaybreaks. However, alignment is a bit more difficult for alternating formula — text — formula etc. because TeX does not insert page breaks within an \intertext. The following example shows this; the text is placed on the next page despite there being enough space left on the left page.

```
\usepackage{amsmath,color}\newcommand\diff{\mathop{}\!\mathrm{d}}\allowdisplaybreaks
\newcommand\demoText{Now there is some completely pointless text with completely
pointless content that is only here to show that there is no problem with the
{\color{blue}colour}.}
\begin{align}
y &= \int_a^bf(x)\diff x = \left.F(x)\right|_a^b\\
\intertext{\demoText}
y &= \int_a^bf(x)\diff x = \left.F(x)\right|_a^b\\
\intertext{\demoText}
y &= \int_a^bf(x)\diff x = \left.F(x)\right|_a^b\\
\intertext{\demoText}
y &= \int_a^bf(x)\diff x = \left.F(x)\right|_a^b\\
\intertext{\demoText}
y &= \int_a^bf(x)\diff x = \left.F(x)\right|_a^b
\end{align}
```

10-07-1

$$y = \int_a^b f(x)\,dx = F(x)|_a^b \quad (7)$$

Now there is some completely pointless text with completely pointless content that is only here to show that there is no problem with the colour.

$$y = \int_a^b f(x)\,dx = F(x)|_a^b \quad (8)$$

Now there is some completely pointless text with completely pointless content that is only here to show that there is no problem with the colour.

$$y = \int_a^b f(x)\,dx = F(x)|_a^b \quad (9)$$

Now there is some completely pointless text with completely pointless content that is only here to show that there is no problem with the colour.

$$y = \int_a^b f(x)\,dx = F(x)|_a^b \quad (10)$$

In fact, in the example above, the alignment of the equals sign would not have been a problem even if the whole equation had been split into its individual parts and the text and the math kept separate. However, there is a genuine problem if we want to intersperse text between equations (like the fololwing set) that have left and right hand sides of unequal length:

10-07-2

$$f(x) = a$$
$$g(x) = x^2 - 4x$$
$$f(x) - g(x) = x^2 + x^3 + x$$
$$g = x^2 + x^3 + x^4 + x^5 + b$$

```
\usepackage{amsmath}

\begin{align*}
   f(x) &= a\\
   g(x) &= x^2-4x\\
 f(x)-g(x) &= x^2+x^3+x\\
     g &= x^2+x^3+x^4+x^5+b
\end{align*}
```

In this case, we need to take the longest left and right hand side expressions as reference lengths, and use \hphantom. In the coding below, the first line of each equation is only for alignment and stays invisible otherwise because it has no height. Nevertheless the corresponding line feed has to be corrected; this is done by modifying the preceding vertical space \abovedisplayshortskip or \abovedisplayskip. It is reduced by 7mm, which was determined through trial and error. After the equations, those values need to be reset so that following equations appear as normal.

```
\hphantom{\mbox{$f(x)-g(x)$}} & \hphantom{\mbox{$= x^2+x^3+x^4+x^5+b$}}
```

```
\usepackage{amsmath,color} \AtBeginDocument{% da sonst nicht wirksam!
\addtolength\abovedisplayshortskip{-7mm}\addtolength\abovedisplayskip{-7mm}}
Now there is some completely pointless text with completely pointless
    content that is only here to show that there is no problem with the
    {\color{red}colour}.
\begin{align}
\hphantom{\mbox{$f(x)-g(x)$}} & \hphantom{\mbox{$= x^2+x^3+x^4+x^5+b$}}\nonumber\\
f(x) &= a\\ g(x) &= x^2-4x
\end{align}
Now there is some completely pointless text with completely pointless
    content that is only here to show that there is no problem with the
    {\color{red}colour}.
\begin{align}
\hphantom{\mbox{$f(x)-g(x)$}} & \hphantom{\mbox{$= x^2+x^3+x^4+x^5+b$}}\nonumber\\
f(x)-g(x) &= x^2+x^3+x
\end{align}
Now there is some completely pointless text with completely pointless
    content that is only here to show that there is no problem with the
    {\color{red}colour}.
\begin{align}
\hphantom{\mbox{$f(x)-g(x)$}} & \hphantom{\mbox{$= x^2+x^3+x^4+x^5+b$}}\nonumber\\
g(x) &= x^2+x^3+x^4+x^5+b
\end{align}
Now there is some completely pointless text with completely pointless
    content that is only here to show that there is no problem with the
    {\color{red}colour}.
```

Now there is some completely pointless text with completely pointless content that is only here to show that there is no problem with the colour.

$$f(x) = a \tag{10.11}$$
$$g(x) = x^2 - 4x \tag{10.12}$$

10-07-3

Now there is some completely pointless text with completely pointless content that is only here to show that there is no problem with the colour.

$$f(x) - g(x) = x^2 + x^3 + x \tag{10.13}$$

Now there is some completely pointless text with completely pointless content that is only here to show that there is no problem with the colour.

$$g(x) = x^2 + x^3 + x^4 + x^5 + b \tag{10.14}$$

Now there is some completely pointless text with completely pointless content that is only here to show that there is no problem with the colour.

Now a page break can occur in the middle of the text as the example doesn't use \intertext any more.

### 10.7.2 List environments

A similar alignment problem arises when putting equations in one of the three list environments. Without additional consideration itemize yields

- first function:
  $P_1 = \sum_a \in A$

- another function:
  $\sin(P_1) = \text{const.}$

- and a bit different:
  $P_3 + P_2 - P_1 = 0$

```
\begin{itemize}
\item first function: \\
   $P_1 = \displaystyle\sum_a \in A$
\item another function: \\
   $\sin\left(P_1\right) = \mathrm{const.}$
\item and a bit different: \\
   $P_{3}+P_{2}-P_{1} = 0$
\end{itemize}%
```

10-07-4

The \makebox command makes it possible to make the left hand sides of the equations the same width and therefore align the equals signs vertically. The reference size is the equation with the widest left hand side expression; i.e. the last one.

- first function:
  $P_1 = \sum_a \in A$

- another function:
  $\sin(P_1) = \text{const.}$

- and a bit different:
  $P_3 + P_2 - P_1 = 0$

```
\newsavebox\lW \sbox\lW{$P_{3}+P_{2}-P_{1}$}

\begin{itemize}
\item first function: \\
   $\makebox[\wd\lW][r]{$P_1$} =
      \displaystyle\sum_a \in A$
\item another function: \\
   $\makebox[\wd\lW][r]{%
      $\sin\left(P_1\right)$} = \mathrm{const.}$
\item and a bit different: \\
   $P_{3}+P_{2}-P_{1} = 0$
\end{itemize}
```

10-07-5

### 10.7.3 Additional column for comments

The next example was created by Hartmut Henkel and makes a good case for the power of LaTeX when it comes to aligning text and equations in unusual ways. This example requires the document class to be loaded with the `fleqn` option (cf. Section 3.4.3 on page 24). It was created with 0.5\textwidth for the additional text here, but this can be changed arbitrarily.

```
\usepackage{amsmath}
\makeatletter
\@fleqntrue\let\old@mathmargin=\@mathmargin\@mathmargin=-1sp
\let\oldmathindent=\mathindent\let\mathindent=\@mathmargin
\newsavebox{\myendhook}        % for the tables
\def\tagform@#1{{(\maketag@@@{\ignorespaces#1\unskip\@@italiccorr)}
  \makebox[0pt][r]{%            % aligned after the line number
    \makebox[0.5\textwidth][l]{\usebox{\myendhook}}}%
  \global\sbox{\myendhook}{}% empty box
}}
\makeatother
\sbox{\myendhook}{\footnotesize% save table
\begin{tabular}{@{}ll}
  $a_0$   & Bohr radius ($\mathrm{= 0{,}53\,\mbox{\AA}}$)\\
  $e$     & elementary charge\\
$N_{si}$ & number of silicium atoms\\
          & per unit volume\\
     $m$ & atom mass\\
     $Z$ & atomic number\\
\end{tabular}}
%
\begin{equation}
\varepsilon = \frac{E\times 4\times\pi\times \varepsilon_{0}
    \times a_0 \times \left( Z_i^{\frac{2}{3}} + Z_{Si}^{\frac{2}{3}}
    \right)^{-\frac{1}{2}}} {Z_i\times Z_{Si}\times e^2\times\left(1+
    \frac{m_i}{m_{Si}}\right)}\,;
\end{equation}
%
\sbox{\myendhook}{abc}
\begin{equation} a^2+b^2 = c^2 \end{equation}
\begin{equation} z = 9 \end{equation}
```

10-07-6

$$\varepsilon = \frac{E \times 4 \times \pi \times \varepsilon_0 \times a_0 \times \left( Z_i^{\frac{2}{3}} + Z_{Si}^{\frac{2}{3}} \right)^{-\frac{1}{2}}}{Z_i \times Z_{Si} \times e^2 \times \left( 1 + \frac{m_i}{m_{Si}} \right)} \;;$$

| | |
|---|---|
| $a_0$ | Bohr radius ($= 0{,}53$ Å) |
| $e$ | elementary charge |
| $N_{si}$ | number of silicium atoms |
| | per unit volume |
| $m$ | atom mass |
| $Z$ | atomic number |

(10.15)

$$a^2 + b^2 = c^2 \qquad\qquad \text{abc} \qquad (10.16)$$

$$z = 9 \qquad (10.17)$$

The example requires the `amsmath` package; otherwise the LaTeX command that positions the equation number has to be redefined.

### 10.7.4 **Phantom instructions**

When a set of equations are aligned at the equals signs, we need to consider what happens if one of the equations is too long to fit on a single line. The part of the equation that is typeset on the continued line does not have an equals sign, so does not automatically align itself. Therefore we need to use a corresponding \phantom instruction and an additional \mathrel definition to reserve the appropriate space by treating the "phantom sign" like a normal equals sign. In the following example, both minus signs at the beginning of the equations are aligned correctly by implementing this.

```
\usepackage{amsmath}
\begin{align*}
\hat H\left(r,R\right)
  &= -\frac{1}{2}\sum\limits_{i=1}^N\nabla_i^2 -\frac{1}{2}\sum\limits_{I=1}^M
     \frac{1}{A_I}\nabla_I^2 +\sum\limits_{i=1}^N\sum\limits_{j>i}^N
     \frac{1}{\left|r_i-r_j\right|}\\
  &\mathrel{\phantom{=}}-\sum\limits_{i=1}^N \sum\limits_{I=1}^M
     \frac{Z_I}{\left|r_i-R_I \right|}+\sum\limits_{I=1}^M \sum\limits_{J>I}^M
     \frac{Z_I ZJ}{\left|R_I-R_J\right|}
\end{align*}
```

$$\hat{H}\left(r,R\right) = -\frac{1}{2}\sum_{i=1}^{N}\nabla_i^2 - \frac{1}{2}\sum_{I=1}^{M}\frac{1}{A_I}\nabla_I^2 + \sum_{i=1}^{N}\sum_{j>i}^{N}\frac{1}{\left|r_i - r_j\right|}$$

$$-\sum_{i=1}^{N}\sum_{I=1}^{M}\frac{Z_I}{\left|r_i - R_I\right|} + \sum_{I=1}^{M}\sum_{J>I}^{M}\frac{Z_I Z_J}{\left|R_I - R_J\right|}$$

10-07-7

## 10.8 **Node connections**

The following example is a typical use of the comprehensive command package pstricks, or rather pst-node in this case. [27]

```
\usepackage{pst-node,amsmath}
\definecolor{lila}{rgb}{0.6,0.2,0.5}\definecolor{darkyellow}{rgb}{1,0.9,0}
\psset{nodesep=3pt}\def\xstrut{\vphantom{\dfrac{(A)^1}{(B)^1}}}
The binding energy in the liquid drop model consists of the following terms:
\begin{itemize}
\item the \rnode{b}{surface term}
\item the \rnode{a}{volume term}\\[0.75cm]
\begin{equation}
E = \rnode[t]{ae}{\psframebox*[fillcolor=darkyellow,
  linestyle=none]{\xstrut a_vA}} + \rnode[t]{be}{\psframebox*[fillcolor=lightgray,
  linestyle=none]{\xstrut -a_fA^{2/3}}} +
\rnode[t]{ce}{\psframebox*[fillcolor=green,linestyle=none]{\xstrut -
  a_c\dfrac{Z(Z-1)}{A^{1/3}}}} +
\rnode[t]{de}{\psframebox*[fillcolor=cyan,
  linestyle=none]{\xstrut -a_s\frac{(A-2Z)^2}{A}}} +
\rnode[t]{ee}{\psframebox*[fillcolor=yellow,
  linestyle=none]{\xstrut E_p}}
```

```
\end{equation}\\
\item the \rnode{c}{Coulomb term}
\item the \rnode{d}{asymmetry term}
\item and a \rnode{e}{pairing term}
\end{itemize}
\nccurve[angleA=-90,angleB=90]{->}{a}{ae}\nccurve[angleB=45]{->}{b}{be}
\nccurve[angleB=-90]{->}{c}{ce}          \nccurve[angleB=-90]{->}{d}{de}
\nccurve[angleB=-90]{->}{e}{ee}
```

10-08-1

The binding energy in the liquid drop model consists of the following terms:

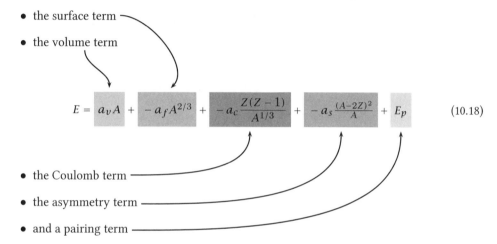

- the surface term
- the volume term

$$E = a_v A + -a_f A^{2/3} + -a_c \frac{Z(Z-1)}{A^{1/3}} + -a_s \frac{(A-2Z)^2}{A} + E_p \tag{10.18}$$

- the Coulomb term
- the asymmetry term
- and a pairing term

## 10.9  Special alignment of displayed formulae

### Parallel formulae

Usually displayed formulae can't be placed next to each other; each one takes up all of the horizontal width. However, it can be done by using a table with column type m and no inter-column spacing:

```
\usepackage{amsmath,array}
\AtBeginDocument{\setlength\tabcolsep{0pt}\renewcommand\arraystretch{0}
\setlength\abovedisplayskip{0pt}
\setlength\abovedisplayshortskip{0pt}
\setlength\belowdisplayskip{0pt}
\setlength\belowdisplayshortskip{0pt}}
\newcommand*{\diff}{\mathop{}\!\mathrm{d}}
\begin{tabular}{@{}*{2}{m{0.5\linewidth}}@{}}
\begin{align}\oint E\diff s=0\end{align} &
\begin{align}\nabla\times B=0\end{align}   \\[-5pt]
\begin{align} a =\frac{c}{d} \end{align} &
\begin{align} b = 1 \end{align}            \\[-5pt]
\begin{align} c =1 \end{align}         &
\begin{align} \int 2x\diff x = x^2+C \end{align}
\end{tabular}
```

$$\oint E\,\mathrm{d}s = 0 \qquad (10.19) \qquad\qquad \nabla \times B = 0 \qquad (10.20)$$

$$a = \frac{c}{d} \qquad (10.21) \qquad\qquad b = 1 \qquad (10.22)$$

$$c = 1 \qquad (10.23) \qquad\qquad \int 2x\,\mathrm{d}x = x^2 + C \qquad (10.24)$$

10-09-1

However, the above layout is only possible by making significant changes to the regular vertical spacing. To keep these changes local to this table, everything is enclosed in a \bgroup...\egroup sequence. All relevant spacings are set to 0pt and at the end of the first two rows of the table an additional negative spacing of 5pt is inserted. The m column type takes care of the centred vertical alignment of the cells of the table, though this is not important in this example, however, as there is not text in the adjacent columns. Depending on the document class, it may be necessary to not set the spacings to absolute zero.

Often the tabled equations should be handled as subequations. As shown in Section 6.11.2 on page 107, you can use the subequations environment of the amsmath package for this. This environment only affects how equations are numbered, it does not care what the content is. Therefore a table can be part of the environment as well:

```
\usepackage{amsmath,array}
\AtBeginDocument{\setlength\tabcolsep{0pt}
\renewcommand\arraystretch{0}
\setlength\abovedisplayskip{0pt}
\setlength\abovedisplayshortskip{0pt}
\setlength\belowdisplayskip{0pt}
\setlength\belowdisplayshortskip{0pt}}
\newcommand*{\diff}{\mathop{}\!\mathrm{d}}

\begin{subequations}
\begin{tabular}{@{}*{2}{m{0.5\linewidth}}@{}}
\begin{align} \oint E \mathrm{d}s=0 \end{align}
  & \begin{align} \nabla\times B=0 \label{sub:foo}\end{align}\\
\begin{align} a =\frac{c}{d} \end{align}
  & \begin{align} b = 1 \end{align}                \\
\begin{align} c =1 \label{sub:baz}\end{align}
  & \begin{align} \int 2x\diff x = x^2+C \end{align}
\end{tabular}\par
All subequations can be referenced in the usual way. For example,
equations~\eqref{sub:foo} and~\eqref{sub:baz} are given correctly with their
letters. The label has to be set correctly of course.
\end{subequations}
```

$$\oint E\,\mathrm{d}s = 0 \qquad (10.25a) \qquad\qquad \nabla \times B = 0 \qquad (10.25b)$$

$$a = \frac{c}{d} \qquad (10.25c) \qquad\qquad b = 1 \qquad (10.25d)$$

$$c = 1 \qquad (10.25e) \qquad\qquad \int 2x\,\mathrm{d}x = x^2 + C \qquad (10.25f)$$

10-09-2

All subequations can be referenced in the usual way. For example, equations (10.25b) and (10.25e) are given correctly with their letters. The label has to be set correctly of course.

### Formulae in tables

A displayed formula can't be put into a table column of type l, r or c. An explicit \parbox, minipage, or a column of type p, b or m is required. The latter two column types are only available when the array package has been loaded. In the following example, a displayed equation has been put into a p, b and m column respectively in conjunction with a normal l column. Usually the m (middle) column is the best choice when combined with columns which contain normal text without line breaks.

```
\usepackage{amsmath,array}
\newcommand*{\diff}{\mathop{}\!\mathrm{d}}
\begin{tabular}{|l|p{0.7\linewidth}|}\hline
p column\,\, & \begin{align} \oint E \diff s=0 \end{align}\\\hline
\end{tabular}\\
\begin{tabular}{|l|b{0.7\linewidth}|}\hline
b column\,\, & \begin{align} \oint E \diff s=0 \end{align}\\\hline
\end{tabular}\\
\begin{tabular}{|l|m{0.7\linewidth}|}\hline
m column & \begin{align} \oint E \diff s=0 \end{align}\\\hline
\end{tabular}
```

10-09-3

| p column | |
|---|---|
| | $$\oint E \, \mathrm{d}s = 0 \qquad (10.26)$$ |

| | |
|---|---|
| | $$\oint E \, \mathrm{d}s = 0 \qquad (10.27)$$ |
| b column | |

| m column | $$\oint E \, \mathrm{d}s = 0 \qquad (10.28)$$ |
|---|---|

The spacing between rows in tables or array environments can be influenced by redefining the \arraystretch command; the default value is 1. In this case all of the rows of a table or matrix are affected. Individual rows can easily be stretched with an invisible line; in math mode this line should be symmetric to the base line or math axis. The following example shows how to change a line with height only into a line with height and depth. The second one can be used inside math to get more space above and below the current line (cf. Example 10-09-7 on the next page).

10-09-4

```
A\rule{1cm}{0.5pt}\rule{5pt}{1cm}\quad
   \rule[-0.5cm]{5pt}{1cm}\rule{1cm}{0.5pt}B
```

The use of the optional argument for \rule drags the line down to give it both height and depth. This is important because formulae are always aligned centred vertically to the math axis. Take the following example:

$$\tan\alpha = \frac{\sin\alpha}{\cos\alpha}$$

$$\tan 2\alpha = \frac{2\tan\alpha}{1 - \tan^2\alpha}$$

$$\tan 3\alpha = \frac{3\tan\alpha - \tan^3\alpha}{1 - 3\tan^2\alpha}$$

```
\begin{tabular}{|p{5cm}|}\hline
\[ \tan \alpha=
   \frac{\sin\alpha}{\cos\alpha} \]\\
\[ \tan2\alpha=
   \frac{2\tan\alpha}{1-\tan^2\alpha}\]\\
\[ \tan3\alpha=\frac{3\tan\alpha
  -\tan^3\alpha}{1-3\tan^2\alpha} \]\\\hline
\end{tabular}
```

`10-09-5`

The spacing is more or less OK, at least not too small. This is because displayed formulae are used. This is often not possible however because the `array` environment has to be used. In this case, the result would look like the following:

$$\tan\alpha = \frac{\sin\alpha}{\cos\alpha}$$
$$\tan 2\alpha = \frac{2\tan\alpha}{1 - \tan^2\alpha}$$
$$\tan 3\alpha = \frac{3\tan\alpha - \tan^3\alpha}{1 - 3\tan^2\alpha}$$

```
\usepackage{amsmath}

\begin{tabular}{|l|}\hline
$\tan\alpha=\dfrac{\sin\alpha}
   {\cos\alpha} $\\
$\tan2\alpha=\dfrac{2\tan\alpha}
   {1-\tan^2\alpha} $\\
$\tan3\alpha=\dfrac{3\tan\alpha-\tan^3\alpha}
   {1-3\tan^2\alpha} $\\\hline
\end{tabular}
```

`10-09-6`

The lack of additional vertical space makes the formulae appear very crowded. For a better result we can redefine \arraystretch to increase the line spacing. More ways to modify tables can be found in detail in [29].

$$\tan\alpha = \frac{\sin\alpha}{\cos\alpha}$$

$$\tan 2\alpha = \frac{2\tan\alpha}{1 - \tan^2\alpha}$$

$$\tan 3\alpha = \frac{3\tan\alpha - \tan^3\alpha}{1 - 3\tan^2\alpha}$$

```
\usepackage{amsmath}
\renewcommand\arraystretch{2.2}

\begin{tabular}{|l|}\hline
$\tan\alpha=\dfrac{\sin\alpha}
   {\cos\alpha} $\\
$\tan2\alpha=\dfrac{2\tan\alpha}
   {1-\tan^2\alpha} $\\
\rule[-3ex]{0pt}{6ex}%
$\tan3\alpha=\dfrac{3\tan\alpha-\tan^3\alpha}
   {1-3\tan^2\alpha} $\\\hline
\end{tabular}
```

`10-09-7`

### Formulae in a list environment

Displayed formulae in list environments like itemize, enumerate, or description are not output at the same height as the label of the environment because of the additional vertical spacing:

10-09-8

- $y = f(x)$

- 

$$g(x) = \int f(x)\mathrm{d}x \qquad (10.29)$$

```
\begin{itemize}
\item $ y = f(x) $
\item \begin{equation} g(x) = \int f(x)
   \mathrm{d}x \end{equation}
\end{itemize}
```

While formulae in inline mode are no problem at all, displayed formulae have to be shifted upwards manually by the inserted vertical distance \abovedisplayshortskip. If you are needing to do this frequently, you should define a new environment instead.

10-09-9

- 

$$y = f(x) \qquad (10.30)$$

- $y = f(x)$

- 

$$g(x) = \int f(x)\mathrm{d}x \qquad (10.31)$$

```
\def\itemMath#1{%
   \raisebox{-\abovedisplayshortskip}{%
      \parbox{\linewidth}{%
         \begin{equation}#1\end{equation}}}}

\begin{itemize}
\item \itemMath{y = f(x)}
\item $ y = f(x) $
\item \itemMath{g(x)=\int f(x)\mathrm{d}x}
\end{itemize}
```

### Handling of long to very long formulae

The following formula is extremely long horizontally as well as vertically. This requires both patience and care when typing the source code, and also consideration which abbreviations to use to help you keep track of it. Page breaks will need to be activated through \allowdisplaybreaks, and you will need to take care of breaking lines that are too long manually. The breqn package, which also inserts line breaks into displayed formulae, does not work satisfactorily yet.

```
\allowdisplaybreaks
```

$$A = \lim_{n \to \infty} U = \lim_{n \to \infty} \sum_{i=0}^{n-1} (\Delta x \times f(a + i \times \Delta x))$$

$$= \lim_{n \to \infty} (\Delta x \times f(a) + \Delta x \times f(a + \Delta x) + \Delta x \times f(a + 2 \times \Delta x) + \Delta x \times f(a + 3 \times \Delta x) + \ldots +$$

$$\Delta x \times f(a + (n - 1) \times \Delta x))$$

$$= \lim_{n \to \infty} \Delta x \times (f(a) + f(a + \Delta x) + f(a + 2 \times \Delta x) + f(a + 3 \times \Delta x) + \ldots +$$

$$+ f(a + (n - 1) \times \Delta x))$$

$$= \lim_{n \to \infty} \Delta x \left( a^2 + (a + \Delta x)^2 + (a + 2 \times \Delta x)^2 + (a + 3 \times \Delta x)^2 + \ldots +$$

$$+ (a + (n - 1) \times \Delta x)^2 \right)$$

$$= \lim_{n \to \infty} \Delta x \left( a^2 + \left( a^2 + 2a\Delta x + (\Delta x)^2 \right) + \left( a^2 + 2 \times 2a\Delta x + 2^2 (\Delta x)^2 \right) \right.$$
$$\left. + \left( a^2 + 2 \times 3a\Delta x + 3^2 (\Delta x)^2 \right) + \ldots + \left( a^2 + 2 \times (n-1)a\Delta x + (n-1)^2 (\Delta x)^2 \right) \right)$$

$$= \lim_{n \to \infty} \Delta x \left( na^2 + 2a\Delta x (1 + 2 + 3 + \ldots + (n-1)) + (\Delta x)^2 \left( 1^2 + 2^2 + 3^2 + \ldots + +(n-1)^2 \right) \right)$$

$$= \lim_{n \to \infty} \Delta x \left( na^2 + 2a\Delta x \frac{n(n-1)}{2} + (\Delta x)^2 \frac{n(2n-1)(n-1)}{6} \right)$$

$$= \lim_{n \to \infty} \frac{b-a}{n} \left( na^2 + 2a\frac{b-a}{n}\frac{n(n-1)}{2} + \left(\frac{b-a}{n}\right)^2 \frac{n(2n-1)(n-1)}{6} \right)$$

$$= \lim_{n \to \infty} \frac{b-a}{n} \left( na^2 + a(b-a)(n-1) + \frac{(b-a)^2}{n}\frac{(2n-1)(n-1)}{6} \right)$$

$$= \lim_{n \to \infty} (b-a) \left( a^2 + a(b-a)\frac{n-1}{n} + \frac{(b-a)^2}{n^2}\frac{(2n-1)(n-1)}{6} \right)$$

$$= \lim_{n \to \infty} (b-a) \left( a^2 + a(b-a)\left( 1 - \underbrace{\frac{1}{n}} \right) + (b-a)^2\frac{1}{6}\left( 2 - \underbrace{\frac{3}{n}} + \underbrace{\frac{1}{n^2}} \right) \right)$$

series converging to zero $n \to \infty$

$$= (b-a) \left( a^2 + a(b-a) + (b-a)^2\frac{2}{6} \right) = (b-a) \left( a^2 + ab - a^2 + \frac{b^2 - 2ab + a^2}{3} \right)$$

$$= \frac{1}{3}(b-a) \left( 3ab + b^2 - 2ab + a^2 \right) = \frac{1}{3}(b-a) \left( b^2 + ab + a^2 \right) = \frac{1}{3}\left( b^3 - a^3 \right)$$

```
\usepackage{amsmath}
\begin{align}
 \sum_{i=1}^\infty \frac{1}{i} &= 1+\frac{1}{2}+\frac{1}{3}+\frac{1}{4}
  +\frac{1}{5}+\frac{1}{6}+\frac{1}{7}+\frac{1}{8}+\frac{1}{9}
  +\cdots+\frac{1}{16}+\cdots\\
 & \ge 1+\frac{1}{2}+\underbrace{\frac{1}{4}+\frac{1}{4}}_{\frac{1}{2}}
    +\underbrace{\frac{1}{8}+\frac{1}{8}+\frac{1}{8}+\frac{1}{8}}_{\frac{1}{2}}
    +\underbrace{\frac{1}{16}+\cdots+\frac{1}{16}}_{\frac{1}{2}}+\cdots\\
 &= 1+\frac{1}{2}+\frac{1}{2}+\frac{1}{2}+\frac{1}{2}+\cdots\to\infty
    \quad\rule{1ex}{1ex}
\end{align}
```

$$\sum_{i=1}^{\infty} \frac{1}{i} = 1 + \frac{1}{2} + \frac{1}{3} + \frac{1}{4} + \frac{1}{5} + \frac{1}{6} + \frac{1}{7} + \frac{1}{8} + \frac{1}{9} + \cdots + \frac{1}{16} + \cdots \qquad (10.32)$$

10-09-10

$$\geq 1 + \frac{1}{2} + \underbrace{\frac{1}{4} + \frac{1}{4}}_{\frac{1}{2}} + \underbrace{\frac{1}{8} + \frac{1}{8} + \frac{1}{8} + \frac{1}{8}}_{\frac{1}{2}} + \underbrace{\frac{1}{16} + \cdots + \frac{1}{16}}_{\frac{1}{2}} + \cdots \qquad (10.33)$$

$$= 1 + \frac{1}{2} + \frac{1}{2} + \frac{1}{2} + \frac{1}{2} + \cdots \to \infty \quad \blacksquare \qquad (10.34)$$

### Polynomial division

The following solution requires the \hspace command, which moves the individual parts of the division horizontally relatively to the equals sign.

10-09-11

$$
\begin{aligned}
(x^5 + x^2) : (x^3 + x^2) &= x^2 - x + 1\\
\underline{-(x^5 + x^4)}&\\
-x^4 + x^2&\\
\underline{-(-x^4 - x^3)}&\\
x^3 + x^2&\\
\underline{-(x^3 + x^2)}&\\
0&
\end{aligned}
$$

```
\usepackage{amsmath}

\begin{align*}
 (x^{5}+x^{2}):(x^{3}+x^{2}) &= x^{2}-x+1\\
 \underline{{-(x^{5}+x^{4})}}\hspace{9ex}\\
 -x^{4}+x^{2}\hspace{5ex}\\
 \underline{{-(-x^{4}-x^{3})}}\hspace{4.5ex}\\
 x^{3}+x^{2}\hspace{1.5ex}\\
 \underline{{-(x^{3}+x^{2})}}\hspace{1ex}}\\
 0\hspace{5ex}
\end{align*}
```

### Apostils for equations

The normal apostils (marginpar) are similar to floating environments; LATEX will place them where it makes most sense typographically. This is not necessarily where you would expect them to be placed, however. In the following example, the apostil is created without using the \marginpar command. It therefore will definitely appear to the left of the equation.

```
\usepackage{amsmath,ragged2e}
\newcommand\eqnmarg[2]{\noindent\hspace*{-\marginparsep}\llap{\parbox[c][.5\height]{%
  \marginparwidth}{\RaggedRight#2}}\hspace{\marginparsep}%
  \begin{minipage}[c]{.99\linewidth}#1\end{minipage}}
\eqnmarg{\begin{align}
    & \frac {\sigma_{m,y,d}} {k_m \times f_{m,y,d}}
        + \frac {\sigma_{c,0,d}}{k_{c,y} \times f_{c,0,d}} \leq 1,0 \\[1ex]
    & \qquad\text{with } k_{c,y} = \min
    \left(\frac{1}{k+\sqrt{k^2-\lambda_{rel,c}^2}}; 1 \right) &&\nonumber
\end{align}}{\tiny This is a remark for the adjacent equation.}\par
\eqnmarg{\begin{align}
    & \frac {\sigma_{m,y,d}} {k_m \times f_{m,y,d}} \\[1ex]
    & \qquad\text{ with }k_{c,y} = \min &&\nonumber
\end{align}}{\tiny This is a remark for the adjacent equation as well.}
```

10-09-12

This is a remark for the adjacent equation.

$$
\frac{\sigma_{m,y,d}}{k_m \times f_{m,y,d}} + \frac{\sigma_{c,0,d}}{k_{c,y} \times f_{c,0,d}} \leq 1,0 \tag{10.35}
$$

$$
\text{with } k_{c,y} = \min\left(\frac{1}{k + \sqrt{k^2 - \lambda_{rel,c}^2}}; 1\right)
$$

This is a remark for the adjacent equation as well.

$$
\frac{\sigma_{m,y,d}}{k_m \times f_{m,y,d}} \tag{10.36}
$$

$$
\text{with } k_{c,y} = \min
$$

### Frames with column separators

The ways mentioned earlier for framing expressions don't work if the object to frame contains the column separator &. However, you may want to frame a whole line, especially with a displayed equation. The trick here is to extend the definition of the \boxed command to move the normal box left by the width of the left hand side of the equation. This example defines both new and existing versions of the \boxed command:

```
\usepackage{amsmath} \newcommand*\CMD[1]{\texttt{\textbackslash#1}}
\newcommand*\aboxed[2]{\rlap{\boxed{#1#2}}\phantom{\hskip\fboxrule\hskip\fboxsep#1}}
\begin{align*}
  F &= \vec\nabla\Phi \\
    &= \frac{\partial\Phi}{\partial x}\hat\imath + [\ldots] \\
    &= (3x^3+y)\hat\imath + [\ldots] \\
  \aboxed{F}{= 20\hat\imath + [\ldots]}    \tag*{\CMD{aboxed}}\\
  \boxed{F=20\hat\imath + [\ldots]}        \tag*{\CMD{boxed}}\\
  \rlap{\boxed{F=20\hat\imath + [\ldots]}} \tag*{\CMD{rlap}\{\CMD{boxed}\}}\\
  \aboxed{F}{= 20\hat\imath + [\ldots]}    \tag*{\CMD{aboxed}}\\
  \ldots &= \ldots
\end{align*}
```

10-09-13

$$F = \vec\nabla\Phi$$
$$= \frac{\partial\Phi}{\partial x}\hat{\imath} + [\dots]$$
$$= (3x^3 + y)\hat{\imath} + [\dots]$$
$$\boxed{F = 20\hat{\imath} + [\dots]} \qquad \text{\aboxed}$$
$$\boxed{F = 20\hat{\imath} + [\dots]} \qquad \text{\boxed}$$
$$\boxed{F = 20\hat{\imath} + [\dots]} \qquad \text{\rlap\{\boxed\}}$$
$$\boxed{F = 20\hat{\imath} + [\dots]} \qquad \text{\aboxed}$$
$$\dots = \dots$$

## 10.10 Tricks

This tip for creating a double horizontal line within one of the amsmath environments is by Nina Mazumdar and Philip Taylor.

$$4x + 6y - 2z = -4$$
$$x - y + 2z = 9$$
$$\overline{\phantom{xxxxxxxxxxxx}}$$
$$5x + 5y = 5$$
$$x + y = 1$$
$$\overline{\overline{\phantom{xxxxxxxxxxxx}}}$$
$$y = 1 - x$$

10-10-1

```
\usepackage{amsmath}

\begin {align*}
    4x+   6y -2z &= -4\\
    x    -y +2z &= 9 \\[-1.4 ex]%
                    \cline{1-2}
    5x + 5y      &= 5 \\
    x + y        &= 1 \\[-1.4 ex]%
\multispan{2}{%
    \leaders\hbox{$\mkern-2mu\Relbar\mkern-2mu$}\hfil}\\
              y &= 1-x
\end {align*}
```

In longer formulae, it can become confusing when placing \left...\right combinations. To simply this, the brace symbols can be made active so that they use the appropriate prefix internally themselves.

```
\usepackage{amsmath,amssymb}
\makeatletter
\def\resetMathstrut@{\setbox\z@\hbox{%
  \mathchardef\@tempa\mathcode'\[\relax
  \def\@tempb##1"##2##3{\the\textfont"##3\char"}%
  \expandafter\@tempb\meaning\@tempa \relax}%
  \ht\Mathstrutbox@\ht\z@ \dp\Mathstrutbox@\dp\z@}
\makeatother
\def\Active#1#2{\catcode'#2\active\bgroup\uccode'\~'#2\uppercase{\egroup #1~}}
\def\Math #1#2{\mathcode'#2 "8000\bgroup\afterassignment\egroup \Active {#1}#2}
\Math\def({\left(}        \Math\def){\right)}
    \def\{{\left\lbrace}      \def\}{\right\rbrace}
\[ (\frac{3}{4}) \{\frac{3}{4}\} \]


\begin{align*}
  & \left.
  \begin{aligned}
  r(x)=&\frac{a_{11}}{x-\lambda_1}+\frac{a_{12}}{(x-\lambda_1)^2}+\ldots+
  \frac{a_{1n_1}}{(x-\lambda_1)^{n_1}}+\ldots\\
  & +\frac{a_{r1}}{x-\lambda_r}+\frac{a_{r2}}{(x-\lambda_r)^2}+\ldots+
  \frac{a_{rn_r}}{(x-\lambda_r)^{n_r}}+
  \end{aligned} \}~\mathbb{R}\\
  & \left.
  \begin{aligned}
  \phantom{r(x)=}
  & +\frac{\alpha_{11}x+\beta_{11}}{x^2+A_1x+B_1}+\frac{\alpha_{12}x+
  \beta_{12}}{(x^2+A_1x+B_1)^2}+\ldots+\frac{\alpha_{1m_1}x+
  \beta_{1m_1}}{(x^2+A_1x+B_1)^{m_1}}+\ldots\\
  & +\frac{\alpha_{s1}x+\beta_{s1}}{x^2+A_sx+B_s}+\frac{\alpha_{s2}x+
  \beta_{s2}}{(x^2+A_sx+B_s)^2}+\ldots+
  \frac{\alpha_{sm_s}x+\beta_{sm_s}}{(x^2+A_sx+B_s)^{m_s}}+\ldots
  \end{aligned} \}~\mathbb{C}
\end{align*}
```

10-10-2

$$\left(\frac{3}{4}\right) \left\{\frac{3}{4}\right\}$$

$$
\left.
\begin{aligned}
r\,(x) = {} & \frac{a_{11}}{x-\lambda_1} + \frac{a_{12}}{(x-\lambda_1)^2} + \ldots + \frac{a_{1n_1}}{(x-\lambda_1)^{n_1}} + \ldots \\
& + \frac{a_{r1}}{x-\lambda_r} + \frac{a_{r2}}{(x-\lambda_r)^2} + \ldots + \frac{a_{rn_r}}{(x-\lambda_r)^{n_r}} +
\end{aligned}
\right\} \; \mathbb{R}
$$

$$
\left.
\begin{aligned}
& + \frac{\alpha_{11}x+\beta_{11}}{x^2+A_1x+B_1} + \frac{\alpha_{12}x+\beta_{12}}{(x^2+A_1x+B_1)^2} + \ldots + \frac{\alpha_{1m_1}x+\beta_{1m_1}}{(x^2+A_1x+B_1)^{m_1}} + \ldots \\
& + \frac{\alpha_{s1}x+\beta_{s1}}{x^2+A_sx+B_s} + \frac{\alpha_{s2}x+\beta_{s2}}{(x^2+A_sx+B_s)^2} + \ldots + \frac{\alpha_{sm_s}x+\beta_{sm_s}}{(x^2+A_sx+B_s)^{m_s}} + \ldots
\end{aligned}
\right\} \; \mathbb{C}
$$

Here is a definition for a special \pm sign (plus/minus):

$$+$$
$$(-)$$
$$y = \overset{+}{\underset{(-)}{}} \cdots$$

```
\usepackage{amsmath}
\def\pM{\ensuremath{\genfrac{}{}{0pt}{}{+}%
     {\scriptstyle(\kern-1pt-\kern-1pt)}}}

{\Large$\pM$}\\    $y=\pM\cdots$
```

10-10-3

Unexpected spacing may occur in a math expression when other commands form groups that correspond to a subequation for TEX. In the following example, \textcolor gives wrong spacings because it puts its argument into a group. This makes it formally equal to the preceding \alpha of type \mathord and therefore *no* space is inserted after the preceding \alpha. This is rectified by inserting an empty group before \sin such that the type \mathop of \sin is taken into account again and leads to the corresponding space after \alpha.

$$\cos \alpha \sin \beta$$
$$\cos \alpha\sin \beta$$
$$\cos \alpha \sin \beta$$

```
\usepackage{xcolor,amsmath}

\begin{align*}
  \cos\alpha\sin\beta\\% correct
  \cos\alpha\textcolor{magenta}{\sin\beta}\\% wrong
  \cos\alpha\textcolor{magenta}{{}\sin\beta}% correct
\end{align*}
```

10-10-4

Braces under a matrix usually appear to be too large. However, an extension (idea by Andreas Matthias) lets you specify a spacing as an optional argument by which to shorten the length of the brace.

```
\usepackage{amsmath,amssymb}
\makeatletter
\def\underbraceMatrix{\@ifnextchar[{\underbraceM@i}{\underbraceM@i[4pt]}}
\def\underbraceM@i[#1]#2{\underbraceM@ii{#2}{#1}}%
\def\underbraceM@ii#1#2{\mathop{\vtop{\m@th\ialign{##\crcr
   $\hfil\displaystyle{#1}\hfil$\crcr\noalign{\kern3\p@\nointerlineskip}%
   \kern#2\upbracefill\kern#2\crcr\noalign{\kern3\p@}}}}\limits}
\makeatother
\[ \underbrace{%
  \left(\begin{array}{cc}
   A_{\cal E E} & A_{\cal E I}\\ A_{\cal I E} & A_{\cal I I}
  \end{array}\right)}_{\textstyle=A}
  \underbraceMatrix{%
  \left(\begin{array}{cc}
   A_{\cal E E} & A_{\cal E I}\\ A_{\cal I E} & A_{\cal I I}
\end{array}\right)}_{\textstyle=A}
\underbraceMatrix[1em]{%
\left(\begin{array}{cc}
  A_{\cal E E} & A_{\cal E I}\\ A_{\cal I E} & A_{\cal I I}
\end{array}\right)}_{\textstyle=A} \]
```

$$\underbrace{\begin{pmatrix} A_{\cal EE} & A_{\cal EI} \\ A_{\cal IE} & A_{\cal II} \end{pmatrix}}_{=A} \underbrace{\begin{pmatrix} A_{\cal EE} & A_{\cal EI} \\ A_{\cal IE} & A_{\cal II} \end{pmatrix}}_{=A} \underbrace{\begin{pmatrix} A_{\cal EE} & A_{\cal EI} \\ A_{\cal IE} & A_{\cal II} \end{pmatrix}}_{=A}$$

10-10-5

10-10-6

$$4x + 6y - 2z = -4$$
$$x - y + 2z = 9$$
$$\overline{\phantom{xxxxxxxxxxxx}}$$
$$5x + 5y = 5$$
$$x + y = 1$$

$$4x + 6y - 2z = -4$$
$$x - y + 2z = 9$$
$$\overline{\phantom{xxxxxxxxxxxx}}$$
$$5x + 5y = 5$$
$$x + y = 1$$

```
\usepackage{amsmath}
\makeatletter
\newcommand*{\spl@}[1]{
  \[ \begin{split} #1 \end{split} \] }
\newenvironment{spl}
  {\collect@body\spl@}
  {}
\makeatother

\[
  \begin{split}
    4x+  6y -2z &= -4\\
     x   -y +2z &= 9 \\[-1.4 ex]
        \cline{1-2}
    5x + 5y      &= 5 \\
     x + y       &= 1 \\[-1.4 ex]
  \end{split}
\]
\begin{spl}% the new defined environment
  4x+  6y -2z &= -4\\
   x   -y +2z &= 9 \\[-1.4 ex]
      \cline{1-2}
  5x + 5y      &= 5 \\
   x + y       &= 1 \\[-1.4 ex]
\end{spl}
```

```
\usepackage{amsmath}
\newlength{\myVSpace}\setlength{\myVSpace}{1ex}
\newcommand\xstrut{\rule[-0.5\myVSpace]{0pt}{\myVSpace}}
\begin{align*}
\left.\begin{aligned}
\xstrut\varphi\left(z_1\right) &= P\frac{-i}{2\left(s_1-s_2\right)\left(1+is_1\right)}
  \left[1-\frac{z_1}{\sqrt{z_1^2-R^2\left(1+s_1^2\right)}}\right]\\
\xstrut\psi\left(z_2\right) &= P\frac{+i}{2\left(s_1-s_2\right)\left(1+is_2\right)}
  \left[1-\frac{z_2}{\sqrt{z_2^2-R^2\left(1+s_2^2\right)}}\right]
\end{aligned}\right\} \text{some text}
\end{align*}

\addtolength{\myVSpace}{3cm}% see preamble for definition
\begin{align*}
\left.\begin{aligned}
\xstrut\varphi\left(z_1\right) &= P\frac{-i}{2\left(s_1-s_2\right)\left(1+is_1\right)}
  \left[1-\frac{z_1}{\sqrt{z_1^2-R^2\left(1+s_1^2\right)}}\right]\\
\xstrut\psi\left(z_2\right) &= P\frac{+i}{2\left(s_1-s_2\right)\left(1+is_2\right)}
  \left[1-\frac{z_2}{\sqrt{z_2^2-R^2\left(1+s_2^2\right)}}\right]
\end{aligned}\right\} \text{some text}
\end{align*}
```

$$\varphi\left(z_{1}\right)=P\frac{-i}{2\left(s_{1}-s_{2}\right)\left(1+is_{1}\right)}\left[1-\frac{z_{1}}{\sqrt{z_{1}^{2}-R^{2}\left(1+s_{1}^{2}\right)}}\right]$$

$$\psi\left(z_{2}\right)=P\frac{+i}{2\left(s_{1}-s_{2}\right)\left(1+is_{2}\right)}\left[1-\frac{z_{2}}{\sqrt{z_{2}^{2}-R^{2}\left(1+s_{2}^{2}\right)}}\right]$$

some text

$$\varphi\left(z_{1}\right)=P\frac{-i}{2\left(s_{1}-s_{2}\right)\left(1+is_{1}\right)}\left[1-\frac{z_{1}}{\sqrt{z_{1}^{2}-R^{2}\left(1+s_{1}^{2}\right)}}\right]$$

some text

$$\psi\left(z_{2}\right)=P\frac{+i}{2\left(s_{1}-s_{2}\right)\left(1+is_{2}\right)}\left[1-\frac{z_{2}}{\sqrt{z_{2}^{2}-R^{2}\left(1+s_{2}^{2}\right)}}\right]$$

# Fonts and math

## 11.1  Math fonts

The math fonts consist in general of five families. For the default font Computer Modern these are

**family 0 — Computer Modern Roman (cmr)**  This is the default font type "roman", which is used for upright letters (used in math mode for operator names) and for the Greek uppercase letters and other symbols. Table 11.1 on page 245 contains all characters of this font.

<table>
<tr><td>11-01-1</td><td>

mod log sin

Φ Ψ Γ Υ

; = ()[] : +...

</td><td>

`$\bmod\ \log\ \sin$`

`$\Phi\ \Psi\ \Gamma\ \Upsilon$`

`$ ; = ( ) [ ] : + \ldots$`

</td></tr>
</table>

**family 1 — Computer Modern Math Italic (cmmi)**  This is one of the special math fonts as well and is frequently confused with the text font cmi; the difference is very small for some characters. The letter "a" is completely different though:

<table>
<tr><td>11-01-2</td><td>

(a) $a - a - $a$ - a - a$ $(a)(a)$

</td><td>

`\fontencoding{OT1}\fontfamily{cmr}\selectfont`
`(a)\ $a-\mathrm{a}-\mathsf{a}-\mathit{a}-a$`
`(\textit{a})(\textsl{a})`

</td></tr>
</table>

The text font cmit has the common ligations that the math font cmmi does not have.

*Duffy Duffy* Duffy Duffy          `\textit{Duffy} $Duffy$ Duffy $\mathrm{Duffy}$`

11-01-3

An advantage of `cmmi` is the presence of the old-style digits (cf. Table 11.3 on page 246 and Table 11.13 on page 252). This is rather odd though as those digits are actually only used in text mode.

0123456789 − 0123456789          `$0123456789 - \fam1 0123456789$`

11-01-4

Other useful characters of the `cmmi` font can be taken from Table 11.3 on page 246.

**family 2 − Computer Modern Symbols (cmsy)** This symbol font contains the calligraphic alphabet of capital letters (cf. Table 11.4 on page 246 and Table 11.14 on page 253).

$\mathcal{ABCDEFGHIJKLMNOPQRSTUVXYZ}$          `$\fam2 ABCDEFGHIJKLMNOPQRSTUVXYZ$`

11-01-5

It also contains many more symbols, for example:

$\cap \cup \ominus \otimes \triangle \exists \forall \subset \le \succ \leftarrow \dots$          `$\cap\cup\ominus\otimes\bigtriangleup\exists\;` `\forall\subset\le\succ\leftarrow\dots$`

11-01-6

**family 3 − Computer Modern Extensibles (cmex)** This font contains practically all of the arbitrarily stretchable characters, the delimiters, which have been used frequently in the previous chapters. These characters are not contained in the font as such, they are assembled by TeX when they are created from the individual characters. In particular, the height of parentheses is not limited. This is also the reason why this character set is very TeX-specific and cannot be used with other systems.

**family 4 − $\mathcal{A}_{\mathcal{M}}\mathcal{S}$ symbol fonts (msam and msbm)** Many of the math characters are not part of standard LaTeX and are therefore are not loaded automatically. All of these fonts have been developed by the $\mathcal{A}_{\mathcal{M}}\mathcal{S}$ and made publicly available. A roundup can be found in tables 11.6 and 11.7 on page 248.

The math fonts have the same five parameters (attributes) as the text fonts, but cannot be adjusted as easily individually. Outside math mode, the fonts can be accessed as usual; for example:

`{\fontencoding{OML}\fontfamily{cmm}\selectfont\char000}` to display $\Gamma$.

The `\mathversion` command is used to choose a whole set of math fonts; they usually match each other. For `\mathversion{normal}`, Table 11.2 on the next page shows a summary of the choice of fonts for this version.

A new version can be defined with the `\DeclareMathVersion` command. The assignment is then done through commands:

Table 11.1: Examples for \fontencoding and \fontfamily of the math fonts

| encoding | family | example |
|---|---|---|
| OML | cmm | $\Gamma\Delta\Theta\Lambda\Xi\Pi\Sigma\Upsilon\Phi\Psi\Omega\alpha\beta\gamma\delta\epsilon\zeta\eta\theta\iota\kappa\lambda\mu\nu\xi\pi\rho\sigma\tau\upsilon\phi\chi\psi\omega$ |
| OT1 | cmr | $\Gamma\Delta\Theta\Lambda\Xi\Pi\Sigma\Upsilon\Phi\Psi\Omega$ ffi fi fl ffi ffl ıȷ ` ´ ˇ ˘ ¯ ˚ ¸ ßæœøÆŒØ-!" |
| OMS | cmsy | $-\cdot\times *\div\diamond\pm\mp\oplus\ominus\otimes\oslash\odot\bigcirc\circ\bullet\asymp\equiv\subseteq\supseteq\leq\geq\preceq\succeq\sim\approx\subset\supset\ll\gg\prec\succ\leftarrow$ |
| OMX | cmex | $()\lbrack\rbrack\lfloor\rfloor\lceil\rceil\lbrace\rbrace\langle\rangle\Vert\backslash\wedge$ |
| U | msa | ⊡⊞⊠◻■.◇◆◡◠⇌⇍⊟⊢⊣⊦⊸↠↞⊏⊐⇈⇊↿↾↦↤⇆⇄↻ |
| U | msb | ≨≩≨≩≴≵≴≵≶≷≸≹≼≽≼≽≾≿⋨⋩⋨⋩≰≱≮≯≁≆ |

Table 11.2: Summary of the fonts for \mathversion{*normal*}

| | internal fonts | | external font attributes | | | |
|---|---|---|---|---|---|---|
| alphabet | symbol font | | | | | |
| \mathnormal | letters | | OML | cmm | m | it |
| \mathrm | operators | | OT1 | cmr | m | n |
| \mathcal | symbols | | OMS | cmsy | m | n |
| | largesymbols | | | | | |
| \mathbf | | | OT1 | cmr | bx | n |
| \mathsf | | | OT1 | cmss | m | n |
| \mathit | | | OT1 | cmr | m | it |
| \mathtt | | | OT1 | cmtt | m | n |

```
\DeclareMathAlphabet{alphabet}{encoding}{family}{series}{shape}
\SetMathAlphabet{alphabet}{encoding}{family}{series}{shape}
\DeclareMathDelimiter{macro name}{type}{font1}{no.}{font2}{no.}
\DeclareMathAccent{macro name}{type}{font}{no.}
\DeclareSymbolFontAlphabet{alphabet}{symbol font}
\DeclareMathSymbol{macro name}{type}{font}{no.}
\DeclareMathRadical{macro name}{font1}{no.}{font2}{no.}
\DeclareMathSizes{D}{T}{S}{SS}
```

The file fontmath.ltx[1] contains all the definitions for standard LaTeX. Only a simple example is given here; in practice it makes no sense, but it does show how to do it and the result very clearly:

---

[1] To be found at $TEXMF/tex/latex/base/.

```
\DeclareMathVersion{foo}
\SetMathAlphabet{\mathbf}{foo}{OMS}{cmsy}{m}{n}
\DeclareMathAlphabet{\mathxy}{OT1}{cmr}{bx}{n}
\SetMathAlphabet{\mathxy}{foo}{OT1}{}{}{}
[$ \mathbf{y=f(x)}     $] [(bold)
\mathversion{bold}     $ \mathbf{y=f(x)} $] [(foo)
\mathversion{foo}      $ \mathbf{y=f(x)} $
                       $ \mathxy{y=f(x)} $] [(normal)
\mathversion{normal} $ \mathbf{y=f(x)} $]
```

$$[\mathbf{y = f(x)}] \, [(\text{bold}) \ \mathbf{y = f(x)}] \, [(\text{foo}) \ \dagger = \{ (\S) \, y = f(x) \}$$
$$[(\text{normal}) \ \mathbf{y = f(x)}]$$

11-01-7

The definitions always have to be in the preamble. First a new version is defined through `\DeclareMathVersion{foo}`; the settings of the version "normal" are taken as default. The next line redefines the alphabet `\mathbf`; the bold characters are substituted for simple symbols. Additionally the new alphabet `\mathxy` is defined. It has the properties of `\mathbf` except for the font (cmr instead of hvrm) and is assigned to the new version foo.

## 11.2 Combination of text and math fonts

Many combinations of fonts are possible because of the separation of text and math fonts; however, they do not always look good in the document. The number of freely available text and math fonts that match each other is limited. For this book, the Lucida was used for the text and mathematical part. A license is available at `http://tug.org/store/lucida` with a discount for members of a TeX users group. Table 11.3 shows a summary of matching text and math fonts (after `http://home.vr-web.de/was/mathfonts.html`).

**Table 11.3**: Matching text and math fonts (after [23])

| text font | matching math font(s) |
|---|---|
| Aldus | Euler |
| Berthold Baskerville | BA-Math |
| New Baskerville | MathTime Professional |
| Bera Serif | Lucida New Math |
| Bitstream Charter | CH-Math, Lucida New Math |
| ITC Charter | Lucida New Math |
| CM-Bright | CM-Bright |
| Concrete | Concrete, Euler |
| Helvetica | HV-Math |
| Lucida Bright | Lucida New Math |
| Minion | Euler, MathTime |
| MinionPro | MnSymbol |
| New Century Schoolbook | Fourier |
| Palatino | mathpazo (PSNFSS), PA-Math, Euler |

continued…

... continued

| text font | matching math font(s) |
|---|---|
| Sabon | Euler |
| Syntax | Euler |
| Tekton | Informal Math |
| Times | `mathptmx` (PSNFSS), TM-Math, MathTime, MathTime Pro |
| Monotype TimesNR | MathTime Professional |
| Times Ten | MathTime Professional |
| Utopia | Fourier |

## 11.2.1  Standard CM Fonts with `cm-super`

Computer Modern is *the* standard font which was designed by Donald E. Knuth for TeX. The font set contains serif, sans-serif, monotype and matching math fonts. The math symbols are very extensive and almost complete. There are Type 1 versions of Computer Modern by Blue Sky Research, which were made publicly available by the $\mathcal{A}_{\mathcal{M}}\!\mathcal{S}$ and a number of publishers and other companies. The best Type 1 variant nowadays is Latin Modern by Bogusław Jackowski and Janusz M. Nowacki; it is very close to and in some cases better as CM and still under development.

```
\usepackage[T1]{fontenc}
\input{fontDemo-E}
```

11-02-1

> **Theorem 1 (Residue Theorem).** Let $f$ be analytic in the region $G$ except for the isolated singularities $a_1, a_2, \ldots, a_m$. If $\gamma$ is a closed rectifiable curve in $G$ which does not pass through any of the points $a_k$ and if $\gamma \approx 0$ in $G$ then
>
> $$\operatorname*{Res}_{z=a} f(z) = \operatorname*{Res}_{a} f = \frac{1}{2\pi\mathrm{i}} \int_C f(z)\,\mathrm{d}z,$$
>
> where $C \subset D\backslash\{a\}$ is a closed line $n(C, a) = 1$ (e. g. a counterclockwise circle loop).
>
> $A\Lambda\Delta\nabla BCD\Sigma EF\Gamma GHIJ K L M N O\Theta\Omega P\Phi\Pi\Xi QRST U V W X Y \Upsilon\Psi Z$ ABCDabcd1234
> $a\alpha b\beta c\partial d\delta e\epsilon\varepsilon f\zeta\xi g\gamma h\hbar\iota i j k\kappa l\ell\lambda mn\eta\theta\vartheta o\sigma\varsigma\phi\varphi\wp p\rho\varrho qr st\tau\pi u\mu\nu\upsilon v w\omega\varpi$
>
> $\boldsymbol{xyz\infty \propto \emptyset y = f(x)}$   $\qquad \sum \int \prod \prod \int \sum \sum_a^b \int_a^b \prod_a^b \sum_a^b \int_a^b \prod_a^b$

### Latin Modern as Type-1-Font

The development of the lm fonts in the current version is said to be complete, though small corrections of ligations are still being made.

```
\usepackage[T1]{fontenc} \usepackage{lmodern}
\input{fontDemo-E}
```

---

**Theorem 1 (Residue Theorem).** Let $f$ be analytic in the region $G$ except for the isolated singularities $a_1, a_2, \ldots, a_m$. If $\gamma$ is a closed rectifiable curve in $G$ which does not pass through any of the points $a_k$ and if $\gamma \approx 0$ in $G$ then

$$\operatorname*{Res}_{z=a} f(z) = \operatorname*{Res}_{a} f = \frac{1}{2\pi i} \int_C f(z) \, dz,$$

where $C \subset D\backslash\{a\}$ is a closed line $n(C, a) = 1$ (e.g. a counterclockwise circle loop).

ΑΛΔ∇BCDΣEFΓGHIJ$KLMN$O$\Theta$Ω$P\Phi\Pi\Xi$Q$RST$U$VW$X$Y$ϒΨZ ABCDabcd1234

$a\alpha b\beta c\partial d\delta e\epsilon\varepsilon f\zeta\xi g\gamma h\hbar\imath\jmath k\kappa l\ell\lambda mn\eta\theta\vartheta o\sigma\varsigma\phi\varphi\wp\rho\varrho q r s t\tau\pi u\mu\nu\upsilon\nu w\omega\varpi$

$\boldsymbol{xyz\infty \propto \emptyset y = f(x)}$

$\sum \int \Pi \quad \prod \int \sum \quad \sum_a^b \int_a^b \Pi_a^b \quad \sum_a^b \int_a^b \prod_a^b$

---

### Computer Modern Bright

cm-super by Vladimir Volovich contains Type 1 versions of the Commputer Modern and hfbright by Harald Harders contains versions of the math font in T1 encoding.

```
\usepackage[T1]{fontenc} \usepackage{cmbright}
\input{fontDemo-E}
```

---

**Theorem 1 (Residue Theorem).** Let $f$ be analytic in the region $G$ except for the isolated singularities $a_1, a_2, \ldots, a_m$. If $\gamma$ is a closed rectifiable curve in $G$ which does not pass through any of the points $a_k$ and if $\gamma \approx 0$ in $G$ then

$$\operatorname*{Res}_{z=a} f(z) = \operatorname*{Res}_{a} f = \frac{1}{2\pi i} \int_C f(z) \, dz,$$

where $C \subset D\backslash\{a\}$ is a closed line $n(C, a) = 1$ (e.g. a counterclockwise circle loop).

ΑΛΔ∇BCDΣEFΓGHIJKLMNOΘΩPΦΠΞQRSTUVWXYϒΨZ ABCDabcd1234

$a\alpha b\beta c\partial d\delta e\epsilon\varepsilon f\zeta\xi g\gamma h\hbar\imath\jmath k\kappa l\ell\lambda mn\eta\theta\vartheta o\sigma\varsigma\phi\varphi\wp\rho\varrho q r s t\tau\pi u\mu\nu\upsilon\nu w\omega\varpi$

$xyz\infty \propto \emptyset y = f(x)$

$\sum \int \Pi \quad \prod \int \sum \quad \sum_a^b \int_a^b \Pi_a^b \quad \sum_a^b \int_a^b \prod_a^b$

## 11.2.2  Concrete and Euler or Concrete Math

The eulervm package by Walter Schmidt implements virtual fonts for Euler.Ulrik Vieth developed the Concrete math fonts to match the Concrete text fonts.

```
\usepackage[T1]{fontenc} \usepackage{ccfonts,eulervm}
\input{fontDemo-E}
```

11-02-4

**Theorem 1 (Residue Theorem).** Let f be analytic in the region G except for the isolated singularities $a_1, a_2, \ldots, a_m$. If $\gamma$ is a closed rectifiable curve in G which does not pass through any of the points $a_k$ and if $\gamma \approx 0$ in G then

$$\operatorname*{Res}_{z=a} f(z) = \operatorname*{Res}_a f = \frac{1}{2\pi i} \int_C f(z)\, dz,$$

where $C \subset D \backslash \{a\}$ is a closed line $n(C, a) = 1$ (e. g. a counterclockwise circle loop).

AΛΔ∇BCDΣEFΓGHIJKLMNOΘΩPΦΠΞQRSTUVWXYϒΨZ ABCDabcd1234
aαβc∂dδeεεfζξgγhℏιιijkκlℓλmnηθϑoσσφφ℘pρqrstτπuμνυvwωϖ

xyz∞ ∝ Øy = f(x)      $\sum \int \prod \prod \int \sum \sum_a^b \int_a^b \prod_a^b \sum_a^{b} \int_a^{b} \prod^{b}$

The ccfonts package sets the text font to Concrete and the math font to the Concrete math fonts if eulervm was not loaded. The Concrete font has no bold version; instead, Computer Modern is used. Non-bold text is done with the cm–super Concrete fonts.

```
\usepackage[T1]{fontenc} \usepackage{ccfonts}
\input{fontDemo-E}
```

11-02-5

**Theorem 1 (Residue Theorem).** Let $f$ be analytic in the region $G$ except for the isolated singularities $a_1, a_2, \ldots, a_m$. If $\gamma$ is a closed rectifiable curve in $G$ which does not pass through any of the points $a_k$ and if $\gamma \approx 0$ in $G$ then

$$\operatorname*{Res}_{z=a} f(z) = \operatorname*{Res}_a f = \frac{1}{2\pi i} \int_C f(z)\, dz,$$

where $C \subset D \backslash \{a\}$ is a closed line $n(C, a) = 1$ (e. g. a counterclockwise circle loop).

AΛΔ∇BCDΣEFΓGHIJ$KLMNO$ΘΩPΦΠΞQRST$UVWXY$ϒΨZ ABCDabcd1234
aαβc∂dδeεεfζξgγhℏιιijkκlℓλmnηθϑoσσφφ℘pρqrstτπuμνυvwωϖ

$xyz\infty \propto \emptyset y = f(x)$      $\sum \int \prod \prod \int \sum \sum_a^b \int_a^b \prod_a^b \sum_a^{b} \int_a^{b} \prod^{b}$

### 11.2.3 Bookman and Kerkis

The Kerkis font is an extension of Bookman especially for Greek characters. Math symbols are contained in the font, but are not used by LaTeX. The kmath package really only loads the txfonts package.

```
\usepackage[OT1]{fontenc} \usepackage{kmath,kerkis}
\input{fontDemo-E}
```

---

**Theorem 1 (Residue Theorem).** Let $f$ be analytic in the region $G$ except for the isolated singularities $a_1, a_2, \ldots, a_m$. If $\gamma$ is a closed rectifiable curve in $G$ which does not pass through any of the points $a_k$ and if $\gamma \approx 0$ in $G$ then

$$\operatorname*{Res}_{z=a} f(z) = \operatorname*{Res}_{a} f = \frac{1}{2\pi i} \int_C f(z)\,dz,$$

where $C \subset D\backslash\{a\}$ is a closed line $n(C, a) = 1$ (e. g. a counterclockwise circle loop).

ΑΛΔ∇BCDΣEFΓGHIJ*KLMN*OΘΩPΦΠΞQRST*UVWXY*ΥΨZ ABCDabcd1234
*aabβc∂dδeεεfζξgyhħiijkκlℓ∧mnηϑϑoοςφφ℘pρρqrsttπuμνυυwωϖ*

$xyz\infty \propto \emptyset y = f(x)$   $\Sigma \int \Pi \prod \int \sum \Sigma_a^b \int_a^b \Pi_a^b \sum_a^b \int_a^b \prod_a^b$

11-02-6

---

### 11.2.4 Palatino and **pxfonts** or **mathpazo**

Young Ryu created the pxfonts package, which provides math symbols for Palatino. For the sans-serif version, a different scaling than the package default is chosen here.

```
\usepackage[T1]{fontenc} \usepackage{pxfonts}
\DeclareFontFamily{T1}{pxss}{} \DeclareFontShape{T1}{pxss}{m}{n}{<->s * [0.90]t1xss}{}
\input{fontDemo-E}
```

---

**Theorem 1 (Residue Theorem).** Let $f$ be analytic in the region $G$ except for the isolated singularities $a_1, a_2, \ldots, a_m$. If $\gamma$ is a closed rectifiable curve in $G$ which does not pass through any of the points $a_k$ and if $\gamma \approx 0$ in $G$ then

$$\operatorname*{Res}_{z=a} f(z) = \operatorname*{Res}_{a} f = \frac{1}{2\pi i} \int_C f(z)\,dz,$$

where $C \subset D\backslash\{a\}$ is a closed line $n(C, a) = 1$ (e. g. a counterclockwise circle loop).

ΑΛΔ∇BCDΣEFΓGHIJ*KLMN*OΘΩPΦΠΞQRST*UVWXY*ΥΨZ ABCDabcd1234
*aabβc∂dδeεεfζξgyhħiijkκlℓ∧mnηϑϑoοςφφ℘pρρqrsttπuμνυυwωϖ*

$xyz\infty \propto \emptyset y = f(x)$   $\Sigma \int \Pi \prod \int \sum \Sigma_a^b \int_a^b \Pi_a^b \sum_a^b \int_a^b \prod_a^b$

11-02-7

Diego Puga developed the Pazo math fonts, which provide Greek letters and letter-like symbols matching Palatino. The `mathpazo` package by Diego Puga and Walter Schmidt uses Palatino for Latin characters, Pazo for Greek and other letter-like characters, and Computer Modern for geometric symbols.

```
\usepackage[T1]{fontenc} \usepackage[osf,sc]{mathpazo}
\input{fontDemo-E}
```

11-02-8

**Theorem 1 (Residue Theorem).** Let $f$ be analytic in the region $G$ except for the isolated singularities $a_1, a_2, \ldots, a_m$. If $\gamma$ is a closed rectifiable curve in $G$ which does not pass through any of the points $a_k$ and if $\gamma \approx 0$ in $G$ then

$$\operatorname*{Res}_{z=a} f(z) = \operatorname*{Res}_a f = \frac{1}{2\pi i} \int_C f(z)\, dz,$$

where $C \subset D \setminus \{a\}$ is a closed line $n(C, a) = 1$ (e. g. a counterclockwise circle loop).

$A\Lambda\Delta\nabla BCD\Sigma EF\Gamma GHIJKLMNO\Theta\Omega P\Phi\Pi\Xi QRSTUVWXYY\Psi Z$ ABCDabcd1234
$a\alpha b\beta c\partial d\delta e\epsilon\varepsilon f\zeta\xi g\gamma h\hbar\iota ijk\kappa\ell\lambda mn\eta\theta\vartheta o\sigma\varsigma\phi\varphi\wp p\rho\varrho qrst\tau\pi u\mu v\nu v\upsilon w\omega\varpi$

$xyz\infty \propto \varnothing y = f(x)$

$\Sigma\int\Pi \prod\int\sum \Sigma_a^b \int_a^b \Pi_a^b \sum_a^b \int_a^b \prod_a^b$

The matching math Type 1 fonts for Palatino are not freely available, but can be bought for example at Micropress. A corresponding adaption of LaTeX is available through the `pamathx` package by Walter Schmidt.

```
\usepackage[T1]{fontenc} \usepackage{pamathx}
\input{fontDemo-E}
```

11-02-9

**Theorem 1 (Residue Theorem).** Let $f$ be analytic in the region $G$ except for the isolated singularities $a_1, a_2, \ldots, a_m$. If $\gamma$ is a closed rectifiable curve in $G$ which does not pass through any of the points $a_k$ and if $\gamma \approx 0$ in $G$ then

$$\operatorname*{Res}_{z=a} f(z) = \operatorname*{Res}_a f = \frac{1}{2\pi i} \int_C f(z)\, dz,$$

where $C \subset D \setminus \{a\}$ is a closed line $n(C, a) = 1$ (e. g. a counterclockwise circle loop).

$A\Lambda\Delta\nabla BCD\Sigma EF\Gamma GHIJKLMNO\Theta\Omega P\Phi\Pi\Xi QRSTUVWXYY\Psi Z$ ABCDabcd1234
$a\alpha b\beta c\partial d\delta e\epsilon\varepsilon f\zeta\xi g\gamma h\hbar\iota ijk\kappa\ell\lambda mn\eta\theta\vartheta o\sigma\varsigma\phi\varphi\wp p\rho\varrho qrst\tau\pi u\mu v\nu v\upsilon w\omega\varpi$

$xyz\infty \propto \varnothing y = f(x)$

$\Sigma\int\Pi \prod\int\sum \Sigma_a^b \int_a^b \Pi_a^b \sum_a^b \int_a^b \prod_a^b$

The TeXGyre font project also provides a free Palatino font family. The symbols provided by the `mathpazo` package are used for the math characters.

```
\usepackage[T1]{fontenc} \usepackage{mathpazo,tgpagella}
\input{fontDemo-E}
```

11-02-10

**Theorem 1 (Residue Theorem).** Let $f$ be analytic in the region $G$ except for the isolated singularities $a_1, a_2, \ldots, a_m$. If $\gamma$ is a closed rectifiable curve in $G$ which does not pass through any of the points $a_k$ and if $\gamma \approx 0$ in $G$ then

$$\operatorname*{Res}_{z=a} f(z) = \operatorname*{Res}_{a} f = \frac{1}{2\pi i} \int_C f(z)\,dz,$$

where $C \subset D \setminus \{a\}$ is a closed line $n(C, a) = 1$ (e.g. a counterclockwise circle loop).

ΑΛΔ∇BCDΣEFΓGHIJ*KLMN*OΘΩPΦΠΞQRST*UVWXY*ΥΨZ ABCDabcd1234
*aαbβc∂dδeεεfζξgγhℏiιjkκlℓλmnηθϑoσςφϕ℘pρϱqrstτπuμνυvwωϖ*

*xyz*∞ ∝ ∅*y* = *f*(*x*)     $\sum \int \prod \prod \int \sum \Sigma_a^b \int_a^b \Pi_a^b \sum_a \int_a \prod_a$

## 11.2.5  Times and `txfonts` or `mathptmx`

Young Ryu created the txfonts collection, which apart from Greek and other symbols also contains a complete set of geometric symbols including the `amsmath` symbols.

```
\usepackage[T1]{fontenc} \usepackage{txfonts}
\input{fontDemo-E}
```

11-02-11

**Theorem 1 (Residue Theorem).** Let $f$ be analytic in the region $G$ except for the isolated singularities $a_1, a_2, \ldots, a_m$. If $\gamma$ is a closed rectifiable curve in $G$ which does not pass through any of the points $a_k$ and if $\gamma \approx 0$ in $G$ then

$$\operatorname*{Res}_{z=a} f(z) = \operatorname*{Res}_{a} f = \frac{1}{2\pi i} \int_C f(z)\,dz,$$

where $C \subset D \setminus \{a\}$ is a closed line $n(C, a) = 1$ (e.g. a counterclockwise circle loop).

ΑΛΔ∇BCDΣEFΓGHIJ*KLMN*OΘΩPΦΠΞQRST*UVWXY*ΥΨZ ABCDabcd1234
*aαbβc∂dδeεεfζξgγhℏiιjkκlℓλmnηθϑoσςφϕ℘pρϱqrstτπuμνυvwωϖ*

*xyz*∞ ∝ ∅*y* = *f*(*x*)     $\sum \int \prod \prod \int \sum \Sigma_a^b \int_a^b \Pi_a^b \sum_a \int_a \prod_a$

The `mathptmx` package by Sebastian Rahtz and Walter Schmidt uses Times for Latin letters and Symbol for Greek and other symbols.

```
\usepackage[T1]{fontenc} \usepackage{mathptmx}
\input{fontDemo-E}
```

11-02-12

**Theorem 1 (Residue Theorem).** Let $f$ be analytic in the region $G$ except for the isolated singularities $a_1, a_2, \ldots, a_m$. If $\gamma$ is a closed rectifiable curve in $G$ which does not pass through any of the points $a_k$ and if $\gamma \approx 0$ in $G$ then

$$\operatorname*{Res}_{z=a} f(z) = \operatorname*{Res}_a f = \frac{1}{2\pi i} \int_C f(z)\, dz,$$

where $C \subset D \backslash \{a\}$ is a closed line $n(C,a) = 1$ (e. g. a counterclockwise circle loop).

ΑΛΔ∇BCDΣEFΓGHIJ*KLMNO*ΘΩΡΦΠΞ℧RST*UVWXY*ΥΨZ ABCDabcd1234
$a\alpha b\beta c\partial d\delta e\varepsilon \varepsilon f\zeta \xi g\gamma h\hbar\iota i jk\kappa l\ell \lambda mn\eta\,\theta\,\vartheta o\sigma\varsigma\phi\varphi\wp p\rho\varrho qrst\,\tau\pi u\mu\nu\upsilon w\omega\varpi$

$xyz\infty \propto \emptyset y = f(x)$ 　　　　　 $\Sigma\int\Pi \;\prod\int\sum \Sigma_a^b \int_a^b \Pi_a^b \sum_a^b \int_a^b \prod_a^b$

The TeXGyre font project also provides a free Times font family. The symbols provided by the `mathptmx` package are used for the math characters.

```
\usepackage[T1]{fontenc} \usepackage{mathptmx,tgtermes}
\input{fontDemo-E}
```

11-02-13

**Theorem 1 (Residue Theorem).** Let $f$ be analytic in the region $G$ except for the isolated singularities $a_1, a_2, \ldots, a_m$. If $\gamma$ is a closed rectifiable curve in $G$ which does not pass through any of the points $a_k$ and if $\gamma \approx 0$ in $G$ then

$$\operatorname*{Res}_{z=a} f(z) = \operatorname*{Res}_a f = \frac{1}{2\pi i} \int_C f(z)\, dz,$$

where $C \subset D \backslash \{a\}$ is a closed line $n(C,a) = 1$ (e. g. a counterclockwise circle loop).

ΑΛΔ∇BCDΣEFΓGHIJ*KLMNO*ΘΩΡΦΠΞ℧RST*UVWXY*ΥΨZ ABCDabcd1234
$a\alpha b\beta c\partial d\delta e\varepsilon \varepsilon f\zeta \xi g\gamma h\hbar\iota i jk\kappa l\ell \lambda mn\eta\,\theta\,\vartheta o\sigma\varsigma\phi\varphi\wp p\rho\varrho qrst\,\tau\pi u\mu\nu\upsilon w\omega\varpi$

$xyz\infty \propto \emptyset y = f(x)$ 　　　　　 $\Sigma\int\Pi \;\prod\int\sum \Sigma_a^b \int_a^b \Pi_a^b \sum_a^b \int_a^b \prod_a^b$

### 11.2.6 Helvetica, hvmath, and Lucida New Math

There are no matching free math fonts for the text font Helvetica, so Times is often used.

```
\usepackage[T1]{fontenc} \usepackage{mathptmx} \usepackage[scaled]{helvet}
\input{fontDemo-E}
```

---

**Theorem 1 (Residue Theorem).** Let $f$ be analytic in the region $G$ except for the isolated singularities $a_1, a_2, \ldots, a_m$. If $\gamma$ is a closed rectifiable curve in $G$ which does not pass through any of the points $a_k$ and if $\gamma \approx 0$ in $G$ then

$$\operatorname*{Res}_{z=a} f(z) = \operatorname{Res}_a f = \frac{1}{2\pi i} \int_C f(z) \, dz,$$

where $C \subset D \backslash \{a\}$ is a closed line $n(C, a) = 1$ (e. g. a counterclockwise circle loop).

ΑΛΔ∇BCDΣEFΓGHIJ*KLMN*OΘΩPΦΠΞQRST*UVWX*YΥΨZ ABCDabcd1234
$a\alpha b\beta c\partial d\delta e\epsilon\varepsilon f\zeta\xi g\gamma h\hbar\iota\imath j k\kappa l\ell\lambda m n\eta\theta\vartheta o\sigma\varsigma\phi\varphi\wp p\rho\varrho q r s t\tau\pi u\mu v\nu\upsilon w\omega\varpi$

$xyz\infty \propto \emptyset y = f(x)$

$\Sigma \int \Pi \prod \int \sum \Sigma_a^b \int_a^b \Pi_a^b \sum_a^b \int_a^b \prod_a^b$

---

The TeXGyre font project also provides a free Helvetica font family. The symbols provided by the lucbmath (Lucida New Math) package are used for the math characters.

```
\usepackage[T1]{fontenc} \usepackage[expert,lucidasmallscale]{lucbmath}
\usepackage{tgheros}
\input{fontDemo-E}
```

---

**Theorem 1 (Residue Theorem).** Let $f$ be analytic in the region $G$ except for the isolated singularities $a_1, a_2, \ldots, a_m$. If $\gamma$ is a closed rectifiable curve in $G$ which does not pass through any of the points $a_k$ and if $\gamma \approx 0$ in $G$ then

$$\operatorname*{Res}_{z=a} f(z) = \operatorname{Res}_a f = \frac{1}{2\pi i} \int_C f(z) \, dz,$$

where $C \subset D \backslash \{a\}$ is a closed line $n(C, a) = 1$ (e. g. a counterclockwise circle loop).

ΑΛΔ∇BCDΣEFΓGHIJ*KLMN*OΘΩPΦΠΞQRST*UVWX*YΥΨZ ABCDabcd1234
$a\alpha b\beta c\partial d\delta e\epsilon\varepsilon f\zeta\xi g\gamma h\hbar\iota\imath j k\kappa l\ell\lambda m n\eta\theta\vartheta o\sigma\varsigma\phi\varphi\wp p\rho\varrho q r s t\tau\pi u\mu v\nu\upsilon w\omega\varpi$

$xyz\infty \propto \emptyset y = f(x)$

$\Sigma \int \Pi \prod \int \sum \Sigma_a^b \int_a^b \Pi_a^b \sum_a^b \int_a^b \prod_a^b$

---

A matching commercial version is available for example from Micropress. The package hvmath by Walter Schmidt adapts LATEX for it.

```
\usepackage[T1]{fontenc} \usepackage{hvmath}
\input{fontDemo-E}
```

11-02-16

**Theorem 1 (Residue Theorem).** Let $f$ be analytic in the region $G$ except for the isolated singularities $a_1, a_2, \ldots, a_m$. If $\gamma$ is a closed rectifiable curve in $G$ which does not pass through any of the points $a_k$ and if $\gamma \approx 0$ in $G$ then

$$\operatorname*{Res}_{z=a} f(z) = \operatorname*{Res}_a f = \frac{1}{2\pi i} \int_C f(z) \, dz,$$

where $C \subset D\backslash\{a\}$ is a closed line $n(C, a) = 1$ (e. g. a counterclockwise circle loop).

ΑΛΔ∇BCDΣEFΓGHIJKLMNOΘΩPΦΠΞQRSTUVWXYYΨZ ABCDabcd1234
aαbβc∂dδeεεfζξgγhħιιιjkκlℓλmnηθϑoοςφφ℘pρϱqrstτπμννυwωϖ

$xyz\infty \propto \emptyset y = f(x)$

$\Sigma \int \Pi \prod \int \sum \Sigma_a^b \int_a^b \Pi_a^b \sum_a \int_a^b \prod_a^b$

## 11.2.7 MinionPro and MnSymbol

The MnSymbol package contains almost all symbols that match the commercial font Adobe MinionPro.

```
\usepackage[T1]{fontenc} \usepackage[lf,swash,minionint]{MinionPro}
\input{fontDemo-E}
```

11-02-17

**Theorem 1 (Residue Theorem).** Let $f$ be analytic in the region $G$ except for the isolated singularities $a_1, a_2, \ldots, a_m$. If $\gamma$ is a closed rectifiable curve in $G$ which does not pass through any of the points $a_k$ and if $\gamma \approx 0$ in $G$ then

$$\operatorname*{Res}_{z=a} f(z) = \operatorname*{Res}_a f = \frac{1}{2\pi i} \int_C f(z) \, dz,$$

where $C \subset D\backslash\{a\}$ is a closed line $n(C, a) = 1$ (e. g. a counterclockwise circle loop).

ΑΛΔ∇BCDΣEFΓGHIJKLMNOΘΩPΦΠΞQRSTUVWXYYΨZ ABCDabcd1234
aαbβc∂dδeεεfζξgγhħιιιjkκlℓλmnηθϑoοςφφ℘pρρqrstτπμμνννυwω@

$xyz\infty \propto \emptyset y = f(x)$

$\Sigma \int \Pi \prod \int \sum \Sigma_a^b \int_a^b \Pi_a^b \sum_a \int_a^b \prod_a^b$

### 11.2.8 URW Garamond and Math Design

The Math Design fonts for Garamond are very extensive with Greek letters, symbols from Computer Modern and the $\mathcal{AMS}$ symbols.

```
\usepackage[T1]{fontenc} \usepackage[garamond]{mathdesign}
\input{fontDemo-E}
```

11-02-18

> **Theorem 1 (Residue Theorem).** Let $f$ be analytic in the region $G$ except for the isolated singularities $a_1, a_2, \ldots, a_m$. If $\gamma$ is a closed rectifiable curve in $G$ which does not pass through any of the points $a_k$ and if $\gamma \approx 0$ in $G$ then
>
> $$\operatorname*{Res}_{z=a} f(z) = \operatorname*{Res}_{a} f = \frac{1}{2\pi i} \int_C f(z)\,dz,$$
>
> where $C \subset D\backslash\{a\}$ is a closed line $n(C,a) = 1$ (e. g. a counterclockwise circle loop).
>
> $\Lambda\Lambda\triangle\nabla BCD\Sigma EF\Gamma GHIJ KLMNO\Theta\Omega P\Phi\Pi\Xi QRSTU V W X Y\Upsilon\Psi Z$ ABCDabcd1234
> $a\alpha b\beta c\partial d\delta e\epsilon\varepsilon f\zeta\xi g\gamma h\hbar\iota\imath jkx ll\lambda mn\eta\theta\vartheta o\sigma\varsigma\phi\varphi\wp pp\varrho qrst\tau\pi u\mu v\nu\upsilon w\omega\varpi$
>
> $xyz\infty\propto\emptyset y = f(x)$       $\Sigma\int\Pi\prod\int\sum\Sigma_a^b\int_a^b\Pi_a^b\sum_a^b\int_a^b\prod_a^b$

### 11.2.9 Utopia and Math Design

The Math Design fonts for Utopia are very extensive and contain Greek letters and standard and $\mathcal{AMS}$ symbols.

```
\usepackage[T1]{fontenc} \usepackage[utopia]{mathdesign}
\input{fontDemo-E}
```

11-02-19

> **Theorem 1 (Residue Theorem).** Let $f$ be analytic in the region $G$ except for the isolated singularities $a_1, a_2, \ldots, a_m$. If $\gamma$ is a closed rectifiable curve in $G$ which does not pass through any of the points $a_k$ and if $\gamma \approx 0$ in $G$ then
>
> $$\operatorname*{Res}_{z=a} f(z) = \operatorname*{Res}_{a} f = \frac{1}{2\pi i} \int_C f(z)\,dz,$$
>
> where $C \subset D\backslash\{a\}$ is a closed line $n(C,a) = 1$ (e. g. a counterclockwise circle loop).
>
> $\Lambda\Lambda\triangle\nabla BCD\Sigma EF\Gamma GHIJ KLMNO\Theta\Omega P\Phi\Pi\Xi QRSTU V W X Y\Upsilon\Psi Z$ ABCDabcd1234
> $a\alpha b\beta c\partial d\delta e\epsilon\varepsilon f\zeta\xi g\gamma h\hbar\iota\imath jk\kappa ll\lambda mn\eta\theta\vartheta o\sigma\varsigma\phi\varphi\wp pp\varrho qrst\tau\pi u\mu v\nu\upsilon w\omega\varpi$
>
> $xyz\infty\propto\emptyset y = f(x)$       $\Sigma\int\Pi\prod\int\sum\Sigma_a^b\int_a^b\Pi_a^b\sum_a^b\int_a^b\prod_a^b$

## 11.2.10 Utopia and Fourier

The Fourier font is freely available and is an extension of Adobe's Fourier.

`\usepackage[T1]{fontenc} \usepackage[expert]{fourier}`
`\input{fontDemo-E}`

11-02-20

**Theorem 1 (Residue Theorem).** Let $f$ be analytic in the region $G$ except for the isolated singularities $a_1, a_2, \ldots, a_m$. If $\gamma$ is a closed rectifiable curve in $G$ which does not pass through any of the points $a_k$ and if $\gamma \approx 0$ in $G$ then

$$\operatorname*{Res}_{z=a} f(z) = \operatorname*{Res}_{a} f = \frac{1}{2\pi i} \int_C f(z)\,\mathrm{d}z,$$

where $C \subset D \setminus \{a\}$ is a closed line $n(C, a) = 1$ (e. g. a counterclockwise circle loop).

ΑΛΔ∇BCDΣEFΓGHIJ$KLMN$OΘΩPΦΠΞQRST$UVWXY$ΥΨZ ABCDabcd1234
$a\alpha b\beta c\partial d\delta e\epsilon\varepsilon f\zeta\xi g\gamma h\hbar\iota\imath j k\kappa l\ell\lambda m n\eta\theta\vartheta o\sigma\varsigma\phi\varphi\wp p\rho\varrho q r s t\tau\pi u\mu\nu\upsilon v\omega\varpi$

$xyz\infty\propto\emptyset y = f(x)$

$\Sigma\int\Pi\ \prod\int\ \sum\ \Sigma_a^b\int_a^b\Pi_a^b\ \sum_a^b\int_a^b\prod_a^b$

## 11.2.11 New Century Schoolbook and Fourier

The corresponding package `fouriernc` automatically loads `fourier` for the math font and extends it by the PostScript text font New Century Schoolbook.

`\usepackage[T1]{fontenc} \usepackage{fouriernc}`
`\input{fontDemo-E}`

11-02-21

**Theorem 1 (Residue Theorem).** Let $f$ be analytic in the region $G$ except for the isolated singularities $a_1, a_2, \ldots, a_m$. If $\gamma$ is a closed rectifiable curve in $G$ which does not pass through any of the points $a_k$ and if $\gamma \approx 0$ in $G$ then

$$\operatorname*{Res}_{z=a} f(z) = \operatorname*{Res}_{a} f = \frac{1}{2\pi i} \int_C f(z)\,\mathrm{d}z,$$

where $C \subset D \setminus \{a\}$ is a closed line $n(C, a) = 1$ (e. g. a counterclockwise circle loop).

ΑΛΔ∇BCDΣEFΓGHIJ$KLMN$OΘΩPΦΠΞQRST$UVWXY$ΥΨZ ABCDabcd1234
$a\alpha b\beta c\partial d\delta e\epsilon\varepsilon f\zeta\xi g\gamma h\hbar\iota\imath j k\kappa l\ell\lambda m n\eta\theta\vartheta o\sigma\varsigma\phi\varphi\wp p\rho\varrho q r s t\tau\pi u\mu\nu\upsilon v\omega\varpi$

$\boldsymbol{xyz\infty\propto\emptyset y = f(x)}$

$\Sigma\int\Pi\ \prod\int\ \sum\ \Sigma_a^b\int_a^b\Pi_a^b\ \sum_a^b\int_a^b\prod_a^b$

### 11.2.12 Bera Serif and Lucida New Math

The package bera by Walter Schmidt loads the entire font family for serif, sans-serif and monotypes.

```
\usepackage[T1]{fontenc} \usepackage{bera}
\usepackage[expert,lucidasmallscale]{lucbmath}
\input{fontDemo-E}
```

---

**Theorem 1 (Residue Theorem).** Let $f$ be analytic in the region $G$ except for the isolated singularities $a_1, a_2, \ldots, a_m$. If $\gamma$ is a closed rectifiable curve in $G$ which does not pass through any of the points $a_k$ and if $\gamma \approx 0$ in $G$ then

$$\operatorname*{Res}_{z=a} f(z) = \operatorname*{Res}_a f = \frac{1}{2\pi i} \int_C f(z)\, dz,$$

where $C \subset D\backslash\{a\}$ is a closed line $n(C, a) = 1$ (e.g. a counterclockwise circle loop).

ΑΛΔ∇BCDΣEFΓGHIJ*KLMNO*ΘΩPΦΠΞQRST*UVW XYY*ΨZ ABCDabcd1234
$a\alpha b\beta c\partial d\delta e\epsilon\varepsilon f\zeta\xi g y h\hbar\iota\iota j k\kappa l\ell\lambda m n\eta\theta\vartheta o\sigma\varsigma\phi\varphi\wp p\rho\varrho q r s t\tau\pi u\mu\nu\upsilon v w\omega\varpi$

$\boldsymbol{xyz}\infty \propto \emptyset\boldsymbol{y} = \boldsymbol{f(x)}$  $\qquad \Sigma\int\Pi \ \prod\int \ \sum \ \Sigma_a^b \int_a^b \Pi_a^b \ \sum_a^b \int_a^b \prod_a^b$

11-02-22

---

### 11.2.13 Lucida Bright and Lucida New Math

The commercial Lucida can be bought for a low price at http://www.tug.org or http://www.pctex.com. Walter Schmidt created a new package lucimatx, which is only available at http://www.pctex.com.

```
\usepackage[T1]{fontenc} \usepackage[expert,lucidasmallscale]{lucidabr}
\input{fontDemo-E}
```

---

**Theorem 1 (Residue Theorem).** Let $f$ be analytic in the region $G$ except for the isolated singularities $a_1, a_2, \ldots, a_m$. If $\gamma$ is a closed rectifiable curve in $G$ which does not pass through any of the points $a_k$ and if $\gamma \approx 0$ in $G$ then

$$\operatorname*{Res}_{z=a} f(z) = \operatorname*{Res}_a f = \frac{1}{2\pi i} \int_C f(z)\, dz,$$

where $C \subset D\backslash\{a\}$ is a closed line $n(C, a) = 1$ (e.g. a counterclockwise circle loop).

ΑΛΔ∇BCDΣEFΓGHIJ*KLMNO*ΘΩPΦΠΞQRST*UVW XYY*ΨZ ABCDabcd1234
$a\alpha b\beta c\partial d\delta e\epsilon\varepsilon f\zeta\xi g y h\hbar\iota\iota j k\kappa l\ell\lambda m n\eta\theta\vartheta o\sigma\varsigma\phi\varphi\wp p\rho\varrho q r s t\tau\pi u\mu\nu\upsilon v w\omega\varpi$

$\boldsymbol{xyz}\infty \propto \emptyset\boldsymbol{y} = \boldsymbol{f(x)}$  $\qquad \Sigma\int\Pi \ \prod\int \ \sum \ \Sigma_a^b \int_a^b \Pi_a^b \ \sum_a^b \int_a^b \prod_a^b$

11-02-23

### 11.2.14  Iwona

The `iwona` package by Marcin Woliński supports the sans-serif fonts developed by Janusz M. Nowacki, which have text as well as math characters.

```
\usepackage[T1]{fontenc} \usepackage[math]{iwona}
\input{fontDemo-E}
```

11-02-24

**Theorem 1 (Residue Theorem).** Let $f$ be analytic in the region $G$ except for the isolated singularities $a_1, a_2, \ldots, a_m$. If $\gamma$ is a closed rectifiable curve in $G$ which does not pass through any of the points $a_k$ and if $\gamma \approx 0$ in $G$ then

$$\operatorname*{Res}_{z=a} f(z) = \operatorname*{Res}_{a} f = \frac{1}{2\pi i} \int_C f(z)\,dz,$$

where $C \subset D\backslash\{a\}$ is a closed line $n(C, a) = 1$ (e.g. a counterclockwise circle loop).

AΛΔ∇BCDΣEFΓGHIJ$K$$L$$M$$N$O$\Theta$$\Omega$PΦ$\Pi$Ξ$Q$RST$U$$V$$W$$X$$Y$ΥΨZ ABCDabcd1234

$a\alpha b\beta c\partial d\delta e\epsilon\varepsilon f\zeta\xi g\gamma h\hbar i\imath j k\kappa l\ell\lambda m n\eta\theta\vartheta o\sigma\varsigma\phi\varphi\wp p\rho\varrho q r s t\tau\pi u\mu\nu\upsilon\nu w\omega\varpi$

$xyz\infty \propto \emptyset y = f(x)$

$$\Sigma \smallint \Pi \prod \int \sum \sum_a^b \int_a^b \Pi_a^b \sum_a^b \int_a^b \prod_a^b$$

### 11.2.15  Antykwa Półtawskiego and CM

This font does not have any own math fonts as yet and requires the font encoding QX or OT4.

```
\usepackage[QX]{fontenc} \usepackage{antpolt}
\input{fontDemo-E}
```

11-02-25

**Theorem 1 (Residue Theorem).** Let $f$ be analytic in the region $G$ except for the isolated singularities $a_1, a_2, \ldots, a_m$. If $\gamma$ is a closed rectifiable curve in $G$ which does not pass through any of the points $a_k$ and if $\gamma \approx 0$ in $G$ then

$$\operatorname*{Res}_{z=a} f(z) = \operatorname*{Res}_{a} f = \frac{1}{2\pi i} \int_C f(z)\,dz,$$

where $C \subset D\backslash\{a\}$ is a closed line $n(C, a) = 1$ (e.g. a counterclockwise circle loop).

AΛΔ∇BCDΣEFΓGHIJ$KLMN$O$\Theta\Omega$PΦ$\Pi$Ξ$QR$ST$UVWXY$ΥΨZ ABCDabcd1234

$a\alpha b\beta c\partial d\delta e\epsilon\varepsilon f\zeta\xi g\gamma h\hbar i\imath i j k\kappa l\ell\lambda m n\eta\theta\vartheta o\sigma\varsigma\phi\varphi\wp p\rho\varrho q r s t\tau\pi u\mu\nu\upsilon\nu w\omega\varpi$

$xyz\infty \propto \emptyset y = f(x)$

$$\Sigma \smallint \Pi \prod \int \sum \sum_a^b \int_a^b \Pi_a^b \sum_a^b \int_a^b \prod_a^b$$

### 11.2.16 Antykwa Toruńska

The `anttor` package by Marcin Woliński supports the fonts developed by Janusz M. Nowacki, which contain text as well as math characters.

```
\usepackage[T1]{fontenc} \usepackage[math]{anttor}
\input{fontDemo-E}
```

---

**Theorem 1 (Residue Theorem).** Let $f$ be analytic in the region $G$ except for the isolated singularities $a_1, a_2, \ldots, a_m$. If $\gamma$ is a closed rectifiable curve in $G$ which does not pass through any of the points $a_k$ and if $\gamma \approx 0$ in $G$ then

$$\operatorname*{Res}_{z=a} f(z) = \operatorname*{Res}_{a} f = \frac{1}{2\pi i} \int_C f(z)\, dz,$$

where $C \subset D \backslash \{a\}$ is a closed line $n(C, a) = 1$ (e.g. a counterclockwise circle loop).

ΑΛΔ∇BCDΣEFΓGHIJKLMNOΘΩPΦΠΞQRSTUVWXYΥΨZ ABCDabcd1234
aαbβc∂d∂eεεfζξgγhħιιjkκlℓλmnηθϑoσςφφ℘pρϱqrstττπuμνννυwωῶ

$xyz\infty \propto \emptyset y = f(x)$     $\sum \int \prod \prod \int \sum \Sigma_a^b \int_a^b \Pi_a^b \sum_a^b \int_a^b \prod_a^b$

---

## 11.3 Character tables

Only the Computer Modern fonts of standard LaTeX and the Lucida fonts used for this book are listed.

| short name | denotation | page | short name | denotation | page |
|---|---|---|---|---|---|
| cmr | CM Roman | 245 | hlcrm | LucidaNewMath-Roman | 251 |
| cmti | CM Italic | 246 | hlcrim | LucidaNewMath-Italic | 252 |
| cmmi | CM Math Italic | 246 | hlcry | LucidaNewMath-Symbol | 253 |
| cmsy | CM Symbol | 246 | hlcrv | LucidaNewMath-Extension | 255 |
| cmex | CM Ex-Symbols | 247 | hlcrie8r | LucidaCalligraphy-Italic | 255 |
| msam | AMS Symbol a | 248 | hlhr8r | LucidaBright | 256 |
| msam | AMS Symbol b | 248 | | | |
| line | Line Symbols | 249 | | | |
| linew | Line Symbols bold | 249 | | | |
| lcircle | Circle Symbols | 250 | | | |
| lcirclew | Circle Symbols bold | 250 | | | |

| | ′0 | ′1 | ′2 | ′3 | ′4 | ′5 | ′6 | ′7 | |
|---|---|---|---|---|---|---|---|---|---|
| ′00x | Γ | Δ | Θ | Λ | Ξ | Π | Σ | Υ | "0x |
| ′01x | Φ | Ψ | Ω | ff | fi | fl | ffi | ffl | |
| ′02x | ı | ȷ | ` | ´ | ˇ | ˘ | ¯ | ˚ | "1x |
| ′03x | ¸ | ß | æ | œ | ø | Æ | Œ | Ø | |
| ′04x | ´ | ! | ” | # | $ | % | & | ’ | "2x |
| ′05x | ( | ) | * | + | , | - | . | / | |
| ′06x | 0 | 1 | 2 | 3 | 4 | 5 | 6 | 7 | "3x |
| ′07x | 8 | 9 | : | ; | < | = | > | ? | |
| ′10x | @ | A | B | C | D | E | F | G | "4x |
| ′11x | H | I | J | K | L | M | N | O | |
| ′12x | P | Q | R | S | T | U | V | W | "5x |
| ′13x | X | Y | Z | [ | “ | ] | ˆ | ˙ | |
| ′14x | ‘ | a | b | c | d | e | f | g | "6x |
| ′15x | h | i | j | k | l | m | n | o | |
| ′16x | p | q | r | s | t | u | v | w | "7x |
| ′17x | x | y | z | – | — | ” | ˜ | ¨ | |
| | "8 | "9 | "A | "B | "C | "D | "E | "F | |

Figure 11.1: The cmr font (Computer Modern Roman).

| | ′0 | ′1 | ′2 | ′3 | ′4 | ′5 | ′6 | ′7 | |
|---|---|---|---|---|---|---|---|---|---|
| ′00x | $\Gamma$ | $\Delta$ | $\Theta$ | $\Lambda$ | $\Xi$ | $\Pi$ | $\Sigma$ | $\Upsilon$ | "0x |
| ′01x | $\Phi$ | $\Psi$ | $\Omega$ | ff | fi | fl | ffi | ffl | |
| ′02x | $\imath$ | $\jmath$ | ` | ´ | ˇ | ˘ | ¯ | ˚ | "1x |
| ′03x | ¸ | ß | æ | œ | ø | Æ | Œ | Ø | |
| ′04x | ´ | ! | ” | # | £ | % | & | ’ | "2x |
| ′05x | ( | ) | * | + | , | - | . | / | |
| ′06x | 0 | 1 | 2 | 3 | 4 | 5 | 6 | 7 | "3x |
| ′07x | 8 | 9 | : | ; | < | = | > | ? | |
| ′10x | @ | A | B | C | D | E | F | G | "4x |
| ′11x | H | I | J | K | L | M | N | O | |
| ′12x | P | Q | R | S | T | U | V | W | "5x |
| ′13x | X | Y | Z | [ | “ | ] | ˆ | ˙ | |
| ′14x | ‘ | a | b | c | d | e | f | g | "6x |
| ′15x | h | i | j | k | l | m | n | o | |
| ′16x | p | q | r | s | t | u | v | w | "7x |
| ′17x | x | y | z | – | — | ” | ˜ | ¨ | |
| | "8 | "9 | "A | "B | "C | "D | "E | "F | |

Figure 11.2: The cmti font (Computer Modern Italic).

|      | '0 | '1 | '2 | '3 | '4 | '5 | '6 | '7 |      |
|------|----|----|----|----|----|----|----|----|------|
| '00x | $\Gamma$ | $\Delta$ | $\Theta$ | $\Lambda$ | $\Xi$ | $\Pi$ | $\Sigma$ | $\Upsilon$ | "0x |
| '01x | $\Phi$ | $\Psi$ | $\Omega$ | $\alpha$ | $\beta$ | $\gamma$ | $\delta$ | $\epsilon$ |  |
| '02x | $\zeta$ | $\eta$ | $\theta$ | $\iota$ | $\kappa$ | $\lambda$ | $\mu$ | $\nu$ | "1x |
| '03x | $\xi$ | $\pi$ | $\rho$ | $\sigma$ | $\tau$ | $\upsilon$ | $\phi$ | $\chi$ |  |
| '04x | $\psi$ | $\omega$ | $\varepsilon$ | $\vartheta$ | $\varpi$ | $\varrho$ | $\varsigma$ | $\varphi$ | "2x |
| '05x | $\leftharpoonup$ | $\leftharpoondown$ | $\rightharpoonup$ | $\rightharpoondown$ | $\smile$ | $\frown$ | $\triangleright$ | $\triangleleft$ |  |
| '06x | 0 | 1 | 2 | 3 | 4 | 5 | 6 | 7 | "3x |
| '07x | 8 | 9 | . | , | < | / | > | $\star$ |  |
| '10x | $\partial$ | $A$ | $B$ | $C$ | $D$ | $E$ | $F$ | $G$ | "4x |
| '11x | $H$ | $I$ | $J$ | $K$ | $L$ | $M$ | $N$ | $O$ |  |
| '12x | $P$ | $Q$ | $R$ | $S$ | $T$ | $U$ | $V$ | $W$ | "5x |
| '13x | $X$ | $Y$ | $Z$ | $\flat$ | $\natural$ | $\sharp$ | $\smile$ | $\frown$ |  |
| '14x | $\ell$ | $a$ | $b$ | $c$ | $d$ | $e$ | $f$ | $g$ | "6x |
| '15x | $h$ | $i$ | $j$ | $k$ | $l$ | $m$ | $n$ | $o$ |  |
| '16x | $p$ | $q$ | $r$ | $s$ | $t$ | $u$ | $v$ | $w$ | "7x |
| '17x | $x$ | $y$ | $z$ | $\imath$ | $\jmath$ | $\wp$ | $\vec{}$ | $\frown$ |  |
|      | "8 | "9 | "A | "B | "C | "D | "E | "F |  |

**Figure 11.3:** The cmmi font (Computer Modern Math Italic).

|      | '0 | '1 | '2 | '3 | '4 | '5 | '6 | '7 |      |
|------|----|----|----|----|----|----|----|----|------|
| '00x | $-$ | $\cdot$ | $\times$ | $*$ | $\div$ | $\diamond$ | $\pm$ | $\mp$ | "0x |
| '01x | $\oplus$ | $\ominus$ | $\otimes$ | $\oslash$ | $\odot$ | $\bigcirc$ | $\circ$ | $\bullet$ |  |
| '02x | $\asymp$ | $\equiv$ | $\subseteq$ | $\supseteq$ | $\leq$ | $\geq$ | $\preceq$ | $\succeq$ | "1x |
| '03x | $\sim$ | $\approx$ | $\subset$ | $\supset$ | $\ll$ | $\gg$ | $\prec$ | $\succ$ |  |
| '04x | $\leftarrow$ | $\rightarrow$ | $\uparrow$ | $\downarrow$ | $\leftrightarrow$ | $\nearrow$ | $\searrow$ | $\simeq$ | "2x |
| '05x | $\Leftarrow$ | $\Rightarrow$ | $\Uparrow$ | $\Downarrow$ | $\Leftrightarrow$ | $\nwarrow$ | $\swarrow$ | $\propto$ |  |
| '06x | $\prime$ | $\infty$ | $\in$ | $\ni$ | $\triangle$ | $\triangledown$ | $/$ | $\vert$ | "3x |
| '07x | $\forall$ | $\exists$ | $\neg$ | $\emptyset$ | $<$ | $\Im$ | $>$ | $\perp$ |  |
| '10x | $\aleph$ | $\mathcal{A}$ | $\mathcal{B}$ | $\mathcal{C}$ | $\mathcal{D}$ | $\mathcal{E}$ | $\mathcal{F}$ | $\mathcal{G}$ | "4x |
| '11x | $\mathcal{H}$ | $\mathcal{I}$ | $\mathcal{J}$ | $\mathcal{K}$ | $\mathcal{L}$ | $\mathcal{M}$ | $\mathcal{N}$ | $\mathcal{O}$ |  |
| '12x | $\mathcal{P}$ | $\mathcal{Q}$ | $\mathcal{R}$ | $\mathcal{S}$ | $\mathcal{T}$ | $\mathcal{U}$ | $\mathcal{V}$ | $\mathcal{W}$ | "5x |
| '13x | $\mathcal{X}$ | $\mathcal{Y}$ | $\mathcal{Z}$ | $\cup$ | $\cap$ | $\uplus$ | $\wedge$ | $\vee$ |  |
| '14x | $\vdash$ | $\dashv$ | $\lfloor$ | $\rfloor$ | $\lceil$ | $\rceil$ | $\{$ | $\}$ | "6x |
| '15x | $\langle$ | $\rangle$ | $\mid$ | $\parallel$ | $\updownarrow$ | $\Updownarrow$ | $\backslash$ | $\wr$ |  |
| '16x | $\sqrt{}$ | $\amalg$ | $\nabla$ | $\int$ | $\sqcup$ | $\sqcap$ | $\sqsubseteq$ | $\sqsupseteq$ | "7x |
| '17x | $\S$ | $\dagger$ | $\ddagger$ | $\P$ | $\clubsuit$ | $\diamondsuit$ | $\heartsuit$ | $\spadesuit$ |  |
|      | "8 | "9 | "A | "B | "C | "D | "E | "F |  |

**Figure 11.4:** The cmsy font (Computer Modern Symbol).

| | ´0 | ´1 | ´2 | ´3 | ´4 | ´5 | ´6 | ´7 | |
|---|---|---|---|---|---|---|---|---|---|
| ´00x | ( | ) | [ | ] | ⌊ | ⌋ | ⌈ | ⌉ | ˝0x |
| ´01x | { | } | ⟨ | ⟩ | \| | ‖ | / | \ | |
| ´02x | ( | ) | ( | ) | [ | ] | ⌊ | ⌋ | ˝1x |
| ´03x | ⌈ | ⌉ | { | } | ⟨ | ⟩ | / | \ | |
| ´04x | ( | ) | [ | ] | ⌊ | ⌋ | ⌈ | ⌉ | ˝2x |
| ´05x | { | } | ⟨ | ⟩ | / | \ | / | \ | |
| ´06x | ( | \ | ⌈ | ⌉ | ⌊ | ⌋ | \| | \| | ˝3x |
| ´07x | ( | ) | ⌊ | ⌋ | < | } | > | \| | |
| ´10x | \ | / | \| | \| | ⟨ | ⟩ | ⊔ | ⊔ | ˝4x |
| ´11x | ∮ | ∮ | ⊙ | ⊙ | ⊕ | ⊕ | ⊗ | ⊗ | |
| ´12x | Σ | Π | ∫ | ∪ | ∩ | ⊎ | ∧ | ∨ | ˝5x |
| ´13x | Σ | Π | ∫ | ∪ | ∩ | ⊎ | ∧ | ∨ | |
| ´14x | ⨆ | ⨆ | ⌢ | ⌢ | ⌢ | ~ | ~ | ~ | ˝6x |
| ´15x | [ | ] | ⌊ | ⌋ | ⌈ | ⌉ | { | } | |
| ´16x | √ | √ | √ | √ | √ | \| | ⌈ | ‖ | ˝7x |
| ´17x | ↑ | ↓ | ⌢ | ⌢ | ⌢ | ⌣ | ⇑ | ⇓ | |
| | ˝8 | ˝9 | ˝A | ˝B | ˝C | ˝D | ˝E | ˝F | |

**Figure 11.5**: The cmex font (Computer Modern Extended Symbols).

| | ′0 | ′1 | ′2 | ′3 | ′4 | ′5 | ′6 | ′7 | |
|---|---|---|---|---|---|---|---|---|---|
| ′00x | ⊡ | ⊞ | ⊠ | □ | ■ | ▪ | ◇ | ◆ | ″0x |
| ′01x | ↻ | ↺ | ⇌ | ⇋ | ⊟ | ⊩ | ⊪ | ⊫ | |
| ′02x | ↠ | ↞ | ⇐ | ⇛ | ⇑ | ⇓ | ↿ | ↾ | ″1x |
| ′03x | ↾ | ⇂ | ↣ | ↢ | ⇆ | ⇄ | ↱ | ↳ | |
| ′04x | ⇝ | ⬳ | ↫ | ↬ | ≗ | ≈ | ≳ | ⪸ | ″2x |
| ′05x | ⊸ | ∴ | ∵ | ≑ | ≜ | ⋨ | ≲ | ⪅ | |
| ′06x | ⩽ | ⩾ | ⋞ | ⋟ | ⪏ | ≦ | ⋜ | ≶ | ″3x |
| ′07x | ⑊ | ⁃ | ≓ | ≒ | ‹ | ≥ | › | ≳ | |
| ′10x | ⊏ | ⊐ | ▷ | ◁ | ⊵ | ⊴ | ★ | ◊ | ″4x |
| ′11x | ▼ | ▶ | ◀ | → | ← | △ | ▲ | ▽ | |
| ′12x | ⧢ | ⋚ | ⋛ | ⪋ | ⪌ | ⅄ | ⇒ | ⇐ | ″5x |
| ′13x | ✓ | ⋎ | ⋏ | ⩚ | ∠ | ⦜ | ◁ | α | |
| ′14x | ⌣ | ⌢ | ⋐ | ⋑ | ⊌ | ⋒ | ⋋ | ⋌ | ″6x |
| ′15x | ⋌ | ⋋ | ⊆ | ⊇ | ≏ | ≎ | ⋘ | ⋙ | |
| ′16x | ⌐ | ¬ | ® | Ⓢ | ⋔ | † | ⌣ | ≚ | ″7x |
| ′17x | ⌞ | ⌟ | ✠ | ℭ | ⊤ | ⊙ | ⊛ | ⊖ | |
| | ″8 | ″9 | ″A | ″B | ″C | ″D | ″E | ″F | |

Figure 11.6: The msam font (AMS Symbol a).

| | ′0 | ′1 | ′2 | ′3 | ′4 | ′5 | ′6 | ′7 | |
|---|---|---|---|---|---|---|---|---|---|
| ′00x | ≨ | ≩ | ⊊ | ⊋ | ⪇ | ⪈ | ⪉ | ⪊ | ″0x |
| ′01x | ≨ | ≩ | ⊊ | ⊋ | ⪇ | ⪈ | ⪵ | ⪶ | |
| ′02x | ⋨ | ⋩ | ⋦ | ⋧ | ≰ | ≱ | ⪹ | ⪺ | ″1x |
| ′03x | ⋨ | ⋩ | ⋦ | ⋧ | ≉ | ≠ | ╱ | ╲ | |
| ′04x | ⊊ | ⊋ | ⊈ | ⊉ | ⊊ | ⊋ | ⊈ | ⊉ | ″2x |
| ′05x | ⊊ | ⊋ | ⊈ | ⊉ | ∦ | ∤ | ⫮ | ⋌ | |
| ′06x | ↚ | ↛ | ⊬ | ⊭ | ⊯ | ⊄ | ⊅ | ↮ | ″3x |
| ′07x | ↚ | ↛ | ⇍ | ⇏ | ‹ | ↮ | › | ∅ | |
| ′10x | ∄ | 𝔸 | 𝔹 | ℂ | 𝔻 | 𝔼 | 𝔽 | 𝔾 | ″4x |
| ′11x | ℍ | 𝕀 | 𝕁 | 𝕂 | 𝕃 | 𝕄 | ℕ | 𝕆 | |
| ′12x | ℙ | ℚ | ℝ | 𝕊 | 𝕋 | 𝕌 | 𝕍 | 𝕎 | ″5x |
| ′13x | 𝕏 | 𝕐 | ℤ | ⌢ | ⌢ | ⌣ | ⌣ | | |
| ′14x | ⊣ | ⴄ | | | | | ℧ | ð | ″6x |
| ′15x | ≅ | ⊐ | ⌐ | ⌐ | ⋖ | ⋗ | ⋉ | ⋊ | |
| ′16x | ∣ | ∥ | ╲ | ∼ | ≈ | ≊ | ≋ | ⩰ | ″7x |
| ′17x | ⌢ | ⌢ | Ϝ | ϰ | 𝕜 | ℏ | ℏ | ϶ | |
| | ″8 | ″9 | ″A | ″B | ″C | ″D | ″E | ″F | |

Figure 11.7: The msbm font (AMS Symbol b).

| | ´0 | ´1 | ´2 | ´3 | ´4 | ´5 | ´6 | ´7 | |
|---|---|---|---|---|---|---|---|---|---|
| ´00x | | | | | | | | | ˝0x |
| ´01x | | | | | | | | | |
| ´02x | | | | | | | | | ˝1x |
| ´03x | | | | | | | | | |
| ´04x | | | | | | | | | ˝2x |
| ´05x | | | | | | | | | |
| ´06x | | | | | | | | | ˝3x |
| ´07x | | | | < | | > | | | |
| ´10x | | | | | | | | | ˝4x |
| ´11x | | | | | | | | | |
| ´12x | | | | | | | | | ˝5x |
| ´13x | | | | | | | | | |
| ´14x | | | | | | | | | ˝6x |
| ´15x | | | | | | | | | |
| ´16x | | | | | | | | | ˝7x |
| ´17x | | | | | | | | | |
| | ˝8 | ˝9 | ˝A | ˝B | ˝C | ˝D | ˝E | ˝F | |

Figure 11.8: The line font (line symbols).

| | ´0 | ´1 | ´2 | ´3 | ´4 | ´5 | ´6 | ´7 | |
|---|---|---|---|---|---|---|---|---|---|
| ´00x | | | | | | | | | ˝0x |
| ´01x | | | | | | | | | |
| ´02x | | | | | | | | | ˝1x |
| ´03x | | | | | | | | | |
| ´04x | | | | | | | | | ˝2x |
| ´05x | | | | | | | | | |
| ´06x | | | | | | | | | ˝3x |
| ´07x | | | | < | | > | | | |
| ´10x | | | | | | | | | ˝4x |
| ´11x | | | | | | | | | |
| ´12x | | | | | | | | | ˝5x |
| ´13x | | | | | | | | | |
| ´14x | | | | | | | | | ˝6x |
| ´15x | | | | | | | | | |
| ´16x | | | | | | | | | ˝7x |
| ´17x | | | | | | | | | |
| | ˝8 | ˝9 | ˝A | ˝B | ˝C | ˝D | ˝E | ˝F | |

Figure 11.9: The linew font (line symbols).

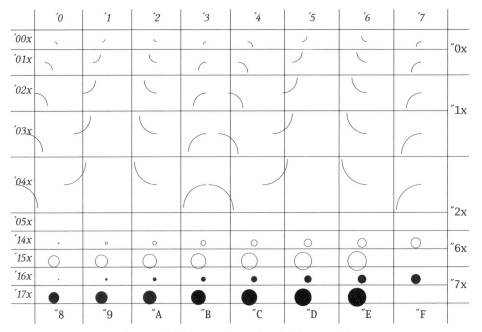

**Figure 11.10:** The `lcircle` font (circle symbols).

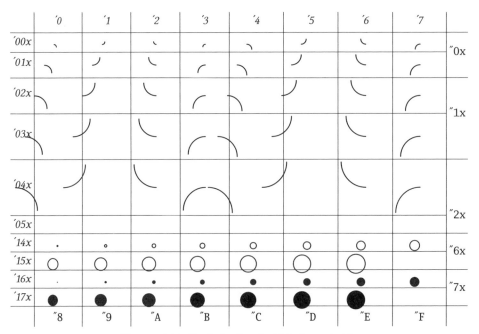

**Figure 11.11:** The `lcirclew` font (circle symbols).

| | ´0 | ´1 | ´2 | ´3 | ´4 | ´5 | ´6 | ´7 | |
|---|---|---|---|---|---|---|---|---|---|
| ´00x | Γ | Δ | Θ | Λ | Ξ | Π | Σ | Υ | "0x |
| ´01x | Φ | Ψ | Ω | α | β | γ | δ | ϵ | |
| ´02x | ζ | η | θ | ι | κ | λ | μ | ν | "1x |
| ´03x | ξ | π | ρ | σ | τ | υ | ϕ | χ | |
| ´04x | ψ | ω | ε | ϑ | ϖ | ϱ | ς | φ | "2x |
| ´05x | ← | ⟵ | → | ⟶ | ' | ' | ▷ | ◁ | |
| ´06x | 0 | 1 | 2 | 3 | 4 | 5 | 6 | 7 | "3x |
| ´07x | 8 | 9 | . | , | < | / | > | ⋆ | |
| ´10x | ∂ | A | B | C | D | E | F | G | "4x |
| ´11x | H | I | J | K | L | M | N | O | |
| ´12x | P | Q | R | S | T | U | V | W | "5x |
| ´13x | X | Y | Z | ♭ | ♮ | ♯ | ⌣ | ⌢ | |
| ´14x | ℓ | a | b | c | d | e | f | g | "6x |
| ´15x | h | i | j | k | l | m | n | o | |
| ´16x | p | q | r | s | t | u | v | w | "7x |
| ´17x | x | y | z | ı | ȷ | ℘ | ⃗ | ⌢ | |
| ´20x | | o | ⟦ | ⟧ | ( | ) | [ | ] | "8x |
| ´21x | ∬ | ∭ | ∮ | ∯ | ∰ | ⨍ | ⨜ | ⨛ | |
| ´22x | ⊣ | ⊃ | ℧ | Ɛ | ∁ | ⊐ | λ | ⊤ | "9x |
| ´23x | ℊ | ℏ | ⤳ | ϰ | ∅ | ℏ | ƛ | ℓ | |
| ´24x | | | | | | | | | "Ax |
| ´25x | | | | ` | ⊫ | | | | |
| ´30x | | | | | | ⊷ | ⊶ | ⊸ | "Cx |
| ´31x | ` | ® | Ⓢ | ✠ | ⊖ | ╱ | ⋈ | ⋉ | |
| ´32x | ⋋ | ⋌ | ⋎ | ⊪ | ⊢ | ⊣ | ⊨ | ⊣ | "Dx |
| ´33x | ╲ | ⊤ | ⪎ | ⪍ | ⋖ | ⋗ | ⋘ | ⋙ | |
| ´34x | ⩽ | ⩾ | ≦ | ≧ | ⪕ | ⪖ | ⪅ | ⪆ | "Ex |
| ´35x | ⪋ | ⪌ | ⪙ | ⪚ | ≋ | ≊ | ⊆ | ⊇ | |
| ´36x | ◀ | ▶ | ◊ | ⋔ | ∣ | ∥ | ⋈ | ⋇ | "Fx |
| ´37x | ⌣ | ≌ | ∈ | ∍ | | | | | |
| | "8 | "9 | "A | "B | "C | "D | "E | "F | |

**Figure 11.12:** The `hlcrm` font (LucidaNewMath-Roman).

| | ´0 | ´1 | ´2 | ´3 | ´4 | ´5 | ´6 | ´7 | |
|---|---|---|---|---|---|---|---|---|---|
| ´00x | $\Gamma$ | $\Delta$ | $\Theta$ | $\Lambda$ | $\Xi$ | $\Pi$ | $\Sigma$ | $\Upsilon$ | "0x |
| ´01x | $\Phi$ | $\Psi$ | $\Omega$ | $\alpha$ | $\beta$ | $\gamma$ | $\delta$ | $\epsilon$ | |
| ´02x | $\zeta$ | $\eta$ | $\theta$ | $\iota$ | $\kappa$ | $\lambda$ | $\mu$ | $\nu$ | "1x |
| ´03x | $\xi$ | $\pi$ | $\rho$ | $\sigma$ | $\tau$ | $\upsilon$ | $\phi$ | $\chi$ | |
| ´04x | $\psi$ | $\omega$ | $\varepsilon$ | $\vartheta$ | $\varpi$ | $\varrho$ | $\varsigma$ | $\varphi$ | "2x |
| ´05x | ↼ | ↽ | ⇀ | ⇁ | ` | ´ | ▷ | ◁ | |
| ´06x | $0$ | $1$ | $2$ | $3$ | $4$ | $5$ | $6$ | $7$ | "3x |
| ´07x | $8$ | $9$ | . | , | < | / | > | $\star$ | |
| ´10x | $\partial$ | $A$ | $B$ | $C$ | $D$ | $E$ | $F$ | $G$ | "4x |
| ´11x | $H$ | $I$ | $J$ | $K$ | $L$ | $M$ | $N$ | $O$ | |
| ´12x | $P$ | $Q$ | $R$ | $S$ | $T$ | $U$ | $V$ | $W$ | "5x |
| ´13x | $X$ | $Y$ | $Z$ | $\flat$ | $\natural$ | $\sharp$ | $\smile$ | $\frown$ | |
| ´14x | $\ell$ | $a$ | $b$ | $c$ | $d$ | $e$ | $f$ | $g$ | "6x |
| ´15x | $h$ | $i$ | $j$ | $k$ | $l$ | $m$ | $n$ | $o$ | |
| ´16x | $p$ | $q$ | $r$ | $s$ | $t$ | $u$ | $v$ | $w$ | "7x |
| ´17x | $x$ | $y$ | $z$ | $\imath$ | $\jmath$ | $\wp$ | $\vec{\ }$ | $\frown$ | |
| ´20x | | $o$ | ⟦ | ⟧ | ( | ) | [ | ] | "8x |
| ´21x | $\iint$ | $\iiint$ | $\oint$ | $\oiint$ | $\oiiint$ | $f$ | $\oint$ | $\oint$ | |
| ´22x | ⊣ | ⊃ | ℧ | $\mathcal{E}$ | ∁ | ⊐ | ⋋ | ⊤ | "9x |
| ´23x | $g$ | $\hbar$ | $\mathfrak{z}$ | $\varkappa$ | $\emptyset$ | $\hbar$ | $\lambda$ | $\ell$ | |
| ´24x | | | | | | | | | "Ax |
| ´25x | | | | ˋ | | ⊫ | | | |
| ´30x | | | | | | ⊶ | ⊷ | ⊸ | "Cx |
| ´31x | ⍀ | ® | Ⓢ | ✠ | ⊖ | ⁄ | ⋉ | ⋊ | |
| ´32x | ⋋ | ⋌ | ⋎ | ⫤ | ⊢ | ⊣ | ⊩ | ⫣ | "Dx |
| ´33x | ⟍ | ⊤ | ≉ | ≈̈ | ⋖ | ⋗ | ⋘ | ⋙ | |
| ´34x | ≦ | ≧ | ≶ | ≷ | ⋚ | ⋛ | ≼ | ≽ | "Ex |
| ´35x | ≦̸ | ≧̸ | ⋚̸ | ⋛̸ | ≈̰ | ≈̱ | ⊆ | ⊇ | |
| ´36x | ◀ | ▶ | ◊ | ⋔ | ∣ | ∥ | ⋈ | ⧚ | "Fx |
| ´37x | ⌣ | ⊴ | ∈ | ∋ | | | | | |
| | "8 | "9 | "A | "B | "C | "D | "E | "F | |

**Figure 11.13:** The `hlcrim` font (LucidaNewMath-Italic).

| | ´0 | ´1 | ´2 | ´3 | ´4 | ´5 | ´6 | ´7 | |
|---|---|---|---|---|---|---|---|---|---|
| ´00x | − | · | × | ∗ | ÷ | ◇ | ± | ∓ | ˝0x |
| ´01x | ⊕ | ⊖ | ⊗ | ⊘ | ⊙ | ◯ | ∘ | • | |
| ´02x | ≍ | ≡ | ⊆ | ⊇ | ≤ | ≥ | ⪯ | ⪰ | ˝1x |
| ´03x | ∼ | ≈ | ⊂ | ⊃ | ≪ | ≫ | ≺ | ≻ | |
| ´04x | ← | → | ↑ | ↓ | ↔ | ↗ | ↘ | ≃ | ˝2x |
| ´05x | ⇐ | ⇒ | ⇑ | ⇓ | ⇔ | ↖ | ↙ | ∝ | |
| ´06x | ′ | ∞ | ∈ | ∋ | △ | ▽ | / | ˌ | ˝3x |
| ´07x | ∀ | ∃ | ¬ | ∅ | < | ℑ | > | ⊥ | |
| ´10x | ℵ | 𝒜 | ℬ | 𝒞 | 𝒟 | ℰ | ℱ | 𝒢 | ˝4x |
| ´11x | ℋ | ℐ | 𝒥 | 𝒦 | ℒ | ℳ | 𝒩 | 𝒪 | |
| ´12x | 𝒫 | 𝒬 | ℛ | 𝒮 | 𝒯 | 𝒰 | 𝒱 | 𝒲 | ˝5x |
| ´13x | 𝒳 | 𝒴 | 𝒵 | ∪ | ∩ | ⊎ | ∧ | ∨ | |
| ´14x | ⊢ | ⊣ | ⌊ | ⌋ | ⌈ | ⌉ | { | } | ˝6x |
| ´15x | ⟨ | ⟩ | \| | ‖ | ↕ | ⇕ | \ | ≀ | |
| ´16x | √ | ∐ | ∇ | ∫ | ⊔ | ⊓ | ⊑ | ⊒ | ˝7x |
| ´17x | § | † | ‡ | ¶ | ♣ | ♦ | ♥ | ♠ | |
| ´20x | | ‾ | + | = | ⋏ | ⋎ | ∆ | ∏ | ˝8x |
| ´21x | ∑ | ∔ | ∟ | ∠ | ∡ | ∢ | ⊾ | ⊿ | |
| ´22x | ∴ | ∵ | ∶ | ∷ | ∸ | ∹ | ∺ | ∻ | ˝9x |
| ´23x | ∿ | ≂ | ⋘ | ≅ | ⋙ | ≊ | ≋ | ≌ | |
| ´24x | | | | | | | | | ˝Ax |
| ´25x | | | | ′ | √ | | | | |
| ´30x | | | | | | ∽ | ∾ | ≎ | ˝Cx |
| ´31x | ≏ | ≐ | ≑ | ≒ | ≓ | ≔ | ≕ | ≖ | |
| ´32x | ≗ | ≘ | ≙ | ≚ | ≛ | ≜ | ≝ | ≞ | ˝Dx |
| ´33x | ≟ | ≣ | ≦ | ≧ | ≲ | ≳ | ≶ | ≷ | |
| ´34x | ≾ | ≿ | ⊌ | ⊍ | ⊏ | ⊐ | ⊚ | ⊛ | ˝Ex |
| ´35x | ⊝ | ⦶ | ⊞ | ⊟ | ⊠ | ⊡ | ⊨ | ⫤ | |
| ´36x | ⊩ | ∺ | ⋞ | ⋟ | ⊴ | ⊵ | ⊼ | ⊽ | ˝Fx |
| ´37x | ⋐ | ⋑ | ⋒ | ⋓ | | | | | |
| | ˝8 | ˝9 | ˝A | ˝B | ˝C | ˝D | ˝E | ˝F | |

**Figure 11.14:** The hlcry font (LucidaNewMath-Symbol).

| | ´0 | ´1 | ´2 | ´3 | ´4 | ´5 | ´6 | ´7 | |
|---|---|---|---|---|---|---|---|---|---|
| ´06x | ( | \ | ⌈ | ⌉ | ⌊ | ⌋ | │ | │ | "3x |
| ´07x | ⌈ | ⌉ | ⌊ | ⌋ | < | { | > | │ | |
| ´10x | \ | / | │ | │ | ⟨ | ⟩ | ⊔ | ⊔ | "4x |
| ´11x | ∮ | ∮ | ⊙ | ⊙ | ⊕ | ⊕ | ⊗ | ⊗ | |
| ´12x | Σ | Π | ∫ | ∪ | ∩ | ⊎ | ∧ | ∨ | "5x |
| ´13x | Σ | Π | ∫ | ∪ | ∩ | ⊎ | ∧ | ∨ | |
| ´14x | ⊔ | ⊔ | ︿ | ⌒ | ⌒ | ~ | ~ | ~ | "6x |
| ´15x | [ | ] | ⌊ | ⌋ | ⌈ | ⌉ | { | } | |
| ´16x | √ | √ | √ | √ | √ | │ | ⌈ | ‖ | "7x |
| ´17x | ↑ | ↓ | ⌒ | ⌒ | ⌣ | ⌣ | ⇑ | ⇓ | |
| ´20x | | | ⟦ | ⟧ | ⟦ | ⟧ | ⟦ | ⟧ | "8x |
| ´21x | ⟦ | ⟧ | ⟦ | ⟧ | ⟦ | ⟧ | ‖ | ‖ | |
| ´22x | ∯ | ∯ | ∫ | ∮ | ∯ | ⌠ | ⌡ | │ | "9x |
| ´23x | | | | | | | | √ | |
| ´24x | | | | | | | | | "Ax |
| ´25x | | | | | | | | | |
| ´30x | | | | | | ⌒ | ⌒ | ~ | "Cx |
| ´31x | ~ | | | | | | | | |
| ´32x | Γ | Δ | Θ | Λ | Ξ | Π | Σ | Υ | "Dx |
| ´33x | Φ | Ψ | Ω | | | | | | |
| ´36x | | | | | | | | | "Fx |
| ´37x | | | | | | | | | |
| | "8 | "9 | "A | "B | "C | "D | "E | "F | |

**Figure 11.15:** The `hlcrv` font (LucidaNewMath-Extension).

| | '0 | '1 | '2 | '3 | '4 | '5 | '6 | '7 | |
|---|---|---|---|---|---|---|---|---|---|
| '06x | O | 1 | 2 | 3 | 4 | 5 | 6 | 7 | "3x |
| '07x | 8 | 9 | : | ; | < | = | > | ? | |
| '10x | @ | A | B | C | D | E | F | G | "4x |
| '11x | H | I | J | K | L | M | N | O | |
| '12x | P | Q | R | S | T | U | V | W | "5x |
| '13x | X | Y | Z | [ | \ | ] | ^ | _ | |
| '14x | ' | a | b | c | d | e | f | g | "6x |
| '15x | h | i | j | k | l | m | n | o | |
| '16x | p | q | r | s | t | u | v | w | "7x |
| '17x | x | y | z | { | \| | } | ~ | | |
| '20x | | ʃ | ' | f | „ | … | † | ‡ | "8x |
| '21x | ˆ | ‰ | Š | ‹ | Œ | Ω | √ | ≈ | |
| '22x | | | | " | " | · | – | — | "9x |
| '23x | ~ | ™ | š | › | œ | Δ | ◊ | ÿ | |
| '24x | | ¡ | ¢ | £ | ¤ | ¥ | ¦ | § | "Ax |
| '25x | ¨ | © | ª | « | ¬ | | ® | ¯ | |
| '26x | ° | ± | ² | ³ | ´ | µ | ¶ | · | "Bx |
| '27x | ¸ | ¹ | º | » | ¼ | ½ | ¾ | ¿ | |
| '30x | À | Á | Â | Ã | Ä | Å | Æ | Ç | "Cx |
| '31x | È | É | Ê | Ë | Ì | Í | Î | Ï | |
| '32x | Ð | Ñ | Ò | Ó | Ô | Õ | Ö | × | "Dx |
| '33x | Ø | Ù | Ú | Û | Ü | Ý | Þ | ß | |
| '34x | à | á | â | ã | ä | å | æ | ç | "Ex |
| '35x | è | é | ê | ë | ì | í | î | ï | |
| '36x | ð | ñ | ò | ó | ô | õ | ö | ÷ | "Fx |
| '37x | ø | ù | ú | û | ü | ý | þ | ÿ | |
| | "8 | "9 | "A | "B | "C | "D | "E | "F | |

Figure 11.16: The hlcrie8r font (LucidaCalligraphy-Italic).

| | ′0 | ′1 | ′2 | ′3 | ′4 | ′5 | ′6 | ′7 | |
|---|---|---|---|---|---|---|---|---|---|
| ′06x | 0 | 1 | 2 | 3 | 4 | 5 | 6 | 7 | ″3x |
| ′07x | 8 | 9 | : | ; | < | = | > | ? | |
| ′10x | @ | A | B | C | D | E | F | G | ″4x |
| ′11x | H | I | J | K | L | M | N | O | |
| ′12x | P | Q | R | S | T | U | V | W | ″5x |
| ′13x | X | Y | Z | [ | \ | ] | ^ | _ | |
| ′14x | ' | a | b | c | d | e | f | g | ″6x |
| ′15x | h | i | j | k | l | m | n | o | |
| ′16x | p | q | r | s | t | u | v | w | ″7x |
| ′17x | x | y | z | { | \| | } | ~ | | |
| ′20x | € | ƒ | , | $f$ | „ | … | † | ‡ | ″8x |
| ′21x | ˆ | ‰ | Š | ‹ | Œ | Ω | √ | ≈ | |
| ′22x | | | | " | " | · | – | — | ″9x |
| ′23x | ~ | ™ | š | › | œ | Δ | ◇ | Ÿ | |
| ′24x | | ¡ | ¢ | £ | ¤ | ¥ | ¦ | § | ″Ax |
| ′25x | ¨ | © | ª | « | ¬ | - | ® | ¯ | |
| ′26x | ° | ± | ² | ³ | ´ | µ | ¶ | · | ″Bx |
| ′27x | ¸ | ¹ | º | » | ¼ | ½ | ¾ | ¿ | |
| ′30x | À | Á | Â | Ã | Ä | Å | Æ | Ç | ″Cx |
| ′31x | È | É | Ê | Ë | Ì | Í | Î | Ï | |
| ′32x | Ð | Ñ | Ò | Ó | Ô | Õ | Ö | × | ″Dx |
| ′33x | Ø | Ù | Ú | Û | Ü | Ý | Þ | ß | |
| ′34x | à | á | â | ã | ä | å | æ | ç | ″Ex |
| ′35x | è | é | ê | ë | ì | í | î | ï | |
| ′36x | ð | ñ | ò | ó | ô | õ | ö | ÷ | ″Fx |
| ′37x | ø | ù | ú | û | ü | ý | þ | ÿ | |
| | ″8 | ″9 | ″A | ″B | ″C | ″D | ″E | ″F | |

**Figure 11.17:** The `hlhr8r` font (LucidaBright).

# Tables

**Table A.1**: File extensions and their meaning

| | |
|---|---|
| `.aux` | (auxiliary) auxiliary file, contains links etc. |
| `.bbl` | (bibliography) auxiliary file, contains bibliographic entries |
| `.bib` | (bibliography) contains the database |
| `.blg` | (bibliography) log file, contains the output of the BibTeX run |
| `.cfg` | (configure) TeX file with configuration information |
| `.clo` | (class options) TeX file with definitions for the document class and the corresponding class options |
| `.cls` | (class) document class file |
| `.cnf` | the configuration file `texmf.cnf` contains among other things the definitions of the search paths |
| `.dvi` | (device independent) output of a TeX or LaTeX run |
| `.def` | (definitions) run time modules of the LaTeX kernel which are loaded on demand ("`latex209.def`" for compatibility mode, "`slides.def`" for SliTeX); similar for some packages like `inputenc`, `fontenc`, and `graphics` |
| `.enc` | font encoding file |
| `.eps` | Encapsulated PostScript – graphics format |
| `.fd` | (font definition) TeX file with font definitions; contains tables which map the LaTeX specification of a font with encoding, family, series, and shape to the name of a `.fmt` file; the name of the `.fd` file consists of encoding and family |
| `.fmt` | (format) precompiled format file |
| `.glo` | (glossary) auxiliary file, contains the contents of the glossary |
| `.idx` | (index) auxiliary file, contains the contents of the index |
| `.ilg` | (index) log file, contains the output of a `makeindex` run |
| `.ind` | (index) auxiliary file, contains the entries after processing by `makeindex` |

| | |
|---|---|
| .jpg | Joint photographic experts group – graphics format |
| .ldf | (language definition) TeX file with the definition of a language and its dialects for the babel package |
| .loa | (list of algorithm) directory of algorithms |
| .lof | (list of figures) directory of figures |
| .log | log file of a TeX run |
| .lot | (list of tables) directory of tables |
| .ltx | (latex) document source file, alternative to the extension .tex (also for LaTeX kernel files) |
| .map | font mapping file |
| .mf | MetaFont file |
| .mp | MetaPost file – graphics format |
| .pdf | Portable document format – graphics format |
| .pfa | PostScript font – ASCII format |
| .pfb | PostScript font – binary format |
| .png | Portable network graphics – graphics format |
| .pool | pdftex.pool contains the internal strings of TeX |
| .sty | (style file) TeX file with the definitions of a LaTeX package |
| .tex | LaTeX source file |
| .tfm | (tex font metrics) font metric |
| .toc | (table of contents) contents |
| .vf | virtual font |

# Bibliography

[1] Paul W. Abrahams, Karl Berry, and Kathryn Hargreaves. TeX for the Impatient, 2003.
http://tug.org/ftp/tex/impatient/book.pdf.

[2] Claudio Beccari. "Typesetting mathematics for science and technology according to iso 31/xi". *TUGboat Journal*, 18(1):39–47, 1997.
http://www.tug.org/TUGboat/Articles/tb18-1/tb54becc.pdf.

[3] Thierry Bouche. "Diversity in math fonts". *TUGboat Journal*, 19(2):121–135, 1998.
http://www.tug.org/TUGboat/Articles/tb19-2/tb59bouc.pdf.

[4] David Cobac. Atelier documents mathématiques, 2004. http://crdp.ac-lille.fr/crdp2003/archives/latex/Ateliers/Atelier2/Presentation4.pdf.

[5] David Cobac. Ecrire des mathématiques avec LaTeX, 2004. http://crdp.ac-lille.fr/crdp2003/archives/latex/Ateliers/Atelier2/prepDocMaths.pdf.

[6] Michael Downes. Technical Notes on the amsmath package. American Mathematical Society, 1999. ftp://ftp.ams.org/pub/tex/doc/amsmath/technote.pdf.

[7] Michael Downes. Short Math Guide for LaTeX. American Mathematical Society, 2002.
http://www.ams.org/tex/short-math-guide.html.

[8] Victor Eijkhout. TeX by Topic, 1992. http://www.eijkhout.net/tbt/.

[9] J. Anthony Fitzgerald. Web Math Formulas Using TeX, 1997.
http://www.unb.ca/web/Sample/math/.

[10] George Grätzer. Math into LaTeX. Birkhäuser Boston, 3rd edition, 2000.

[11] George Grätzer. More Math into LaTeX. Springer, 4th edition, 2007. ISBN 978-0-387-32289-6.

[12] Johannes Küster. Designing Math Fonts, 2004.
http://www.typoma.com/publ/20040430-bachotex.pdf.

[13] Donald E. Knuth. The TeXbook. Addison Wesley Professional, 21st edition, 1986.

[14] Donald E. Knuth, Tracy Larrabee, and Paul M. Roberts. Mathematical Writing. Stanford University, Computer Science Department, 1987.
http://sunburn.stanford.edu/~knuth/papers/mathwriting.tex.gz.

[15] R. Kuhn, R. Scott, and L. Andreev. An Introduction to using LaTeX in the Harvard Mathematics Department. Harvard University, Department of Mathematics.
http://abel.math.harvard.edu/computing/latex/manual/texman.html.

[16] LaTeX3 Project Team. LaTeX $2_\varepsilon$ fontselection, 2000.
http://www.latex-project.org/guides/fntguide.pdf.

[17] Richard Lawrence. "Math=Typography?" *TUGboat Journal*, 24(2):165–168, 2003.
http://www.tug.org/TUGboat/Articles/tb24-2/tb77lawrence.pdf.

[18] Lars Madsen. "Avoid eqnarray". *The PracTeX Journal*, (4), 2006.
http://www.tug.org/pracjourn/2006-4/madsen/madsen.pdf.

[19] J. S. Milne. Guide to commutative diagram packages, 2005.
http://www.jmilne.org/not/CDGuide.pdf.

[20] Frank Mittelbach and Michel Goosens. The LaTeX Companion. Addison-Wesley, Boston, 2nd edition, 2004.

[21] Luca Padovani. "Mathml formatting with TeX rules and TeX fonts". *TUGboat Journal*, 24(1):53–61, 2003.
http://www.tug.org/TUGboat/Articles/tb24-1/padovani.pdf.

[22] Scott Pakin. The Comprehensive LaTeX Symbol List, 2009.
CTAN: /info/symbols/comprehensive/symbol-a4.pdf.

[23] Walter Schmidt. Mathematikschriften für LaTeX $2_\varepsilon$, 2008.
http://home.vr-web.de/was/mathfonts.html.

[24] Carole Siegfried and Herbert Voß. "Mathematik im Inline-modus". *Die TeXnische Komödie*, 3/04:25–32, 2004.

[25] B. N. Taylor. Guide for the Use of the International System of Units (SI) – 10.5.4 Multiplying numbers. National Institute of Standard and Technology, 2005.
http://physics.nist.gov/Pubs/SP811/sec10.html.

[26] Paul Taylor. Commutative Diagrams in TeX. Department of Computer Science, Queen Mary and Westfield College, 2000.    http://www.dcs.qmw.ac.uk/~pt/diagrams/.

[27] Herbert Voß. PSTricks – Grafik mit PostScript für TeX und LaTeX. LOB-media.de, Berlin, 5th edition, 2008.

[28] Herbert Voß. "Farbige Mathematik". *Die TeXnische Komödie*, 2/04:81–87, 2004.

[29] Herbert Voß. Tabellen mit LaTeX. DANTE – Lob.media, Heidelberg/Hamburg, 2008.

# Index of commands and concepts

To make it easier to use a command or concept, the entries are distinguished by their "type" and this is often indicated by one of the following "type words" at the beginning of an entry:

boolean, counter, document class, env., file, file extension, font, font encoding, key value, keyword, length, option, package, program, rigid length, or syntax.

The absence of an explicit "type word" means that the "type" is either a TEX or LATEX "command" or simply a "concept".

Use by, or in connection with, a particular package is indicated by adding the package name (in parentheses) to an entry.

An italic page number indicates that the command is demonstrated in a source code snippet or in an example on that page.

When there are several page numbers listed, **bold** face indicates a page containing important information about an entry.

# M

## O

# People

ALSO PUBLISHED BY UIT

# Typesetting tables with LaTeX

## Herbert Voss

This is the first-ever book dedicated to typesetting tables in LaTeX. With LaTeX you can create just about any kind of table, from simple to extremely complex. But while the table capabilities in LaTeX are powerful, they can be daunting at first sight or when you require a sophisticated layout. This book describes the additional LaTeX packages that are available to simplify your task, and gives ready-to-run examples of each, to get you working as quickly as possible, and present your data in the most effective way.

With this book you will learn:

- How to typeset tables, from basic to advanced.

- How to use advanced features, such as color and multi-page tables.

- How add-on LaTeX tables packages can simplify or enhance your work.

## Contents

1. Introduction to LaTeX's table-handling

2. LaTeX packages for tables

3. Using color in tables

4. Multi-page tables

5. Tips and tricks

6. Examples

## Praise for the German Edition

*"A concise reference book for those who may already have used LaTeX but aren't aware of the powerful capabilities provided by LaTeX's extra tables packages."*

ISBN: 9781906860257
240 pages

ALSO PUBLISHED BY UIT

# LaTeX quick reference

## Herbert Voss

This book lists all LaTeX macros and environments in a comprehensive reference format. (The packages **array** and **graphicx** are included even though they are not part of standard LaTeX, because they are so widely used.) The book also lists examples of fonts for both plain text and math, making it a convenient graphical resource.

This book will:

• Save you time by quickly giving you the detailed command syntax you require.

• Improve your LaTeX by providing a quick-reference to all the available command options.

• Show you how to choose suitable fonts, using the convenient samples of font output.

## Contents

1. The Standard Programs

2. Document Structure

3. Commands for Fine-Tuning your Typography

4. Command List

5. Lengths and Counters

6. Fonts

7. Packages

8. Bibliography

## Praise for the German Edition

*"An essential resource for LaTeX users"*

**ISBN: 9781906860219**
**160 pages**

ALSO PUBLISHED BY UIT

# PSTricks

## Graphics and PostScript for LaTeX

## Herbert Voss

A comprehensive guide to creating and including graphics in TeX and LaTeX documents. It is both a reference work and a tutorial guide.

PSTricks lets you produce very high-quality PostScript graphics, by programming rather than interactive drawing. For designers, data publishers, scientists and engineers, generating graphics from data or formulas instead of having to draw manually allows large data collections or complex graphics to be created consistently and reliably with the minimum of effort.

There are many special-purpose extensions, for visualizing data, and for drawing circuit diagrams, barcodes, graphs, trees, chemistry diagrams, etc.

Numerous examples with source code (freely downloadable) make it easy to create your own images and get you up to speed quickly.

## Contents

**1**. Introduction **2**. Getting Started **3**. The Coordinate System **4**. Lines and Polygons **5**. Circles, Ellipses and Curves **6**. Points **7**. Filling **8**. Arrows **9**. Labels **10**. Boxes **11**. Custom styles and objects **12**. Coordinates **13**. Overlays **14**. Basics **15**. Plotting of Functions and Data **16**. Nodes and Connections **17**. Trees **18**. Manipulating Text and Characters **19**. Filling and Tiling **20**. Coils, Springs and Zigzag Lines **21**. Exporting PSTricks Environments **22**. Color Gradients and Shadows **23**. Three-Dimensional Figures **24**. Creating Circuit Diagrams **25**. Geographic Projections **26**. Barcodes **27**. Bar Charts **28**. Gantt Charts **29**. Mathematical Functions **30**. Euclidean Geometry **31**. Additional Features **32**. Chemistry Diagrams **33**. UML Diagrams **34**. Additional PSTricks Packages **35**. Specials **36**. PSTricks in Presentations **37**. Examples

### Praise for the German Edition

*"A nice Christmas present – for me!"*

*"A detailed current description of PSTricks and the huge variety of PSTricks packages that are available, and written by an experienced LaTeX package developer."*

*"Searching through loads of different pieces of documentation is a thing of the past. This single compendium is a quick reference to everything I need."*

**ISBN: 9781906860134**
**900 pages**

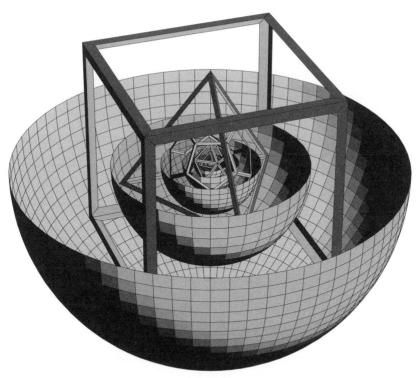

Example illustrations from PSTricks

ALSO PUBLISHED BY UIT

# Practical TCP/IP

Designing, using, and troubleshooting
TCP/IP networks on Linux and Windows

## Niall Mansfield

*Reprinted first edition*

## Key benefits

1. Explore, hands-on, how your network really works. Build small test networks in a few minutes, so you can try anything out without affecting your live network and servers.

2. Learn how to troubleshoot network problems, and how to use free packet sniffers to see what's happening.

3. Understand how the TCP/IP protocols map onto your day-to-day network operation – learn both theory and practice.

## What readers have said about this book

*"Before this book was released I was eagerly searching for a book that could be used for my Linux-based LAN-course. After the release of this book I stopped my searching immediately"* **Torben Gregersen, Engineering College of Aarhus**.

*"Accuracy is superb – written by someone obviously knowledgable in the subject, and able to communicate this knowledge extremely effectively."*

*"You won't find a better TCP/IP book!"*

*"An excellent book for taking your computer networking career from mediocre to top notch."*

*"Covers TCP/IP, and networking in general, tremendously."*

*"This book has been touted as the 21st-century upgrade to the classic TCP/IP Illustrated (by Richard W. Stevens). These are big boots to fill, but Practical TCP/IP does an impressive job. In over 800 pages of well-organized and well-illustrated text, there is no fat, but rather a lean and – yes – practical treatment of every major TCP/IP networking concept."*

*"It's an ideal book for beginners, probably the only one needed for the first and second semesters of a university networking course. ... (But it is not a book just for beginners. ...)"*

**ISBN: 9781906860363**
**880 pages**

ALSO PUBLISHED BY UIT

# The Exim SMTP Mail Server

## Official Guide for Release 4

### Philip Hazel

*Second edition*

Email is one of the most widely used applications, and Exim is one of the most widely used mail servers, handling mail for tens of millions of users daily.

Exim is free software. It's easy to configure. It's scalable, running on single-user desktop systems as well as on ISP servers handling millions of users. (It's the default server on many Linux systems, and it's available for countless versions of UNIX.)

Exim is fast, flexible, and reliable. It is designed not to lose messages even if your server machine crashes. It can be used as a secure Internet-facing front-end to other, proprietary, mail systems used internally in your organization.

Exim supports lookups from LDAP servers, SQL databases, and other data sources, letting you automate maintenance and configuration. It can work in conjunction with other tools for virus-checking and spam-blocking, to reject unwanted emails before they even enter your site.

This book will help you deploy Exim as your SMTP email server throughout your organization, and to configure, tune, and secure your Exim systems.

## Praise for the First Edition

*"The book is simply amazing. I find the format/style/whatever 100 times better than [other documentation]."*

*"If there's even a whiff of a chance of you having to come into contact with Exim or its runtime configuration, then I can do nothing else but strongly recommend this book. The detail's there in spades, it reads very well, and is a fine complement to the reference manual."*

*"The book exceeds my expectations."*

*"Well presented and easy to follow"*

*"An excellent book that is very well written"*

*"So well written I learn new things every time I open it"*

**ISBN: 9780954452971**
**xviii + 622 pages**

ALSO PUBLISHED BY UIT

# The Joy of X

## The architecture of the X window system

## Niall Mansfield

This is a reprint of the 1993 classic, describing the architecture of the X window system – the de facto standard windowing system for Linux, UNIX and many other operating systems. The book has three sections:

1. X in a nutshell – a quick overview.

2. How X works, in detail, and how the user sees it.

3. Using the system, system administration, performance and programming.

The book is written in a clear, uncomplicated style, with over 200 illustrations. For maximum accessibility, it has a flexibile, modular structure that makes it easy to skip to the sections that interest you. The book has been widely recommended as a course text.

Niall Mansfield founded the European X window system User Group. He also wrote *The X window system: a user's guide*, and the widely-acclaimed *Practical TCP/IP*.

### Praise for This Book

*"User interfaces come and go, but X remains the standard window system across a range of operating systems. Niall's book, The Joy of X, still offers an excellent look into how X works and how to make it work better for you.*

**Keith Packard, X.org project leader**

*"If you are new to the X Window System environment, we strongly suggest picking up a book such as The Joy of X"* **Eric Raymond, in the *Linux XFree86 HOWTO***

*"a great little book called The Joy of X by Niall Mansfield that taught me much of what I know."* **Jeff Duntemann's ContraPositive Diary**

*"My personal touchstone when looking for a broad introduction to all things X is The Joy of X . . . by Niall Mansfield"* **Peter Collinson**

ISBN: 9781906860004
xii + 372 pages

ALSO PUBLISHED BY UIT

# Alternative DNS Servers

## Choice and deployment, and optional SQL/LDAP back-ends

### Jan-Piet Mens

This book examines many of the best DNS servers available. It covers each server's benefits and disadvantages, as well as how to configure and deploy it, and integrate it into your network infrastructure. It describes the different scenarios where each server is particularly useful, so you can choose the most suitable server for your site. A unique feature of the book is that it explains how DNS data can be stored in LDAP directories and SQL databases, often required for integrating DNS into large-organization infrastructures.

Other important topics covered include: performance, security issues, integration with DHCP, DNSSEC, internationalization, and specialized DNS servers designed for some unusual purposes.

## Praise for This Book

*"The first book to describe NSD and Unbound in excellent detail."*
**NLnet Labs, authors of NSD and Unbound**

*"Finally - a clear, in-depth and accessible guide to using BIND-DLZ! A must read for anyone considering alternate DNS servers."*
**Rob Butler, BIND-DLZ project creator and author**

*"Takes the reader through the process of configuring the program from basics to advanced topics."*
**Simon Kelley, author of dnsmasq**

*"An informative accurate guide for anyone interested in learning more about DNS."*
**Sam Trenholme, MaraDNS author**

*"A valuable source of information for every PowerDNS administrator!"*
**Norbert Sendetzky, author of PowerDNS LDAP & OpenDBX back-ends**

*"Jan-Piet has done a great job describing PowerDNS."*
**Bert Hubert, principal author of PowerDNS**

ISBN: 9780954452995
xxxvi + 694 pages

ALSO PUBLISHED BY UIT

# OpenStreetMap

Using and enhancing the free map of the world

## Frederik Ramm and Jochen Topf, with Steve Chilton

*Second edition*

OpenStreetMap is a map of the whole world that can be used and edited freely by everyone. In a Wikipedia-like open community process, thousands of contributors world-wide survey the planet and upload their results to the OpenStreetMap database. Unlike some other mapping systems on the Web, the tools and the data are free and open. You can use them and modify them as you require; you can even download all the map data and run your own private map server if you need to.

This book introduces you to the OpenStreetMap community, its data model, and the software used in the project. It shows you how to use the constantly-growing OSM data set and maps in your own projects.

The book also explains in detail how you can contribute to the project, collecting and processing data for OpenStreetMap. If you want to become an OpenStreetMap "mapper" then this is the book for you.

**About the authors**: Frederik Ramm and Jochen Topf both joined the OpenStreetMap project in 2006, when they were freelance developers. Since then they have made their hobby their profession – by founding Geofabrik, a company that provides services relating to OpenStreetMap and open geodata.

### Praise for the First (German) Edition

*"A must-have for OSM newcomers. The basics are presented well and are easy to understand, and you do not need to be an IT specialist to contribute your first data to OSM after a short time."*

*"The book is very well written. It is obvious that the authors have a lot of knowledge and experience ..."*

*"A very good OSM introduction. Getting up to speed with OpenStreetMap is much easier if you have read this book."*

ISBN: 9781906860110
352 pages + 32 pages of color plates

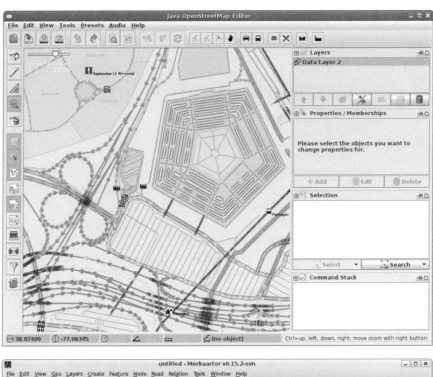

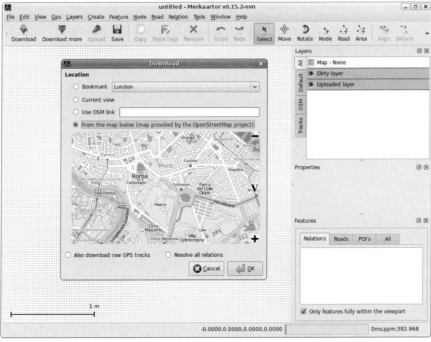

Example illustrations from OpenStreetMap

# More about this book

**Register your book**: receive updates, notifications about author appearances, and announcements about new editions. *www.uit.co.uk/register*

**News**: forthcoming titles, events, reviews, interviews, podcasts, etc. *www.uit.co.uk/news*

**Join our mailing lists**: get email newsletters on topics of interest. *www.uit.co.uk/subscribe*

**How to order**: get details of stockists and online bookstores. If you are a bookstore, find out about our distributors or contact us to discuss your particular requirements. *www.uit.co.uk/order*

**Send us a book proposal**: if you want to write – even if you have just the kernel of an idea at present – we'd love to hear from you. We pride ourselves on supporting our authors and making the process of book-writing as satisfying and as easy as possible. *www.uit.co.uk/for-authors*

UIT Cambridge Ltd.
PO Box 145
Cambridge
CB4 1GQ
England

Email: *inquiries@uit.co.uk*
Phone: **+44 1223 302 041**